ADVANCES IN CHEMICAL KINETICS AND DYNAMICS

Volume 1 • 1992

ADVANCES IN CHEMICAL KINETICS AND DYNAMICS

Editor: JOHN R. BARKER
Department of Atmospheric, Oceanic,
and Space Sciences
The University of Michigan

VOLUME 1 • 1992

JAI PRESS INC.

Greenwich, CT *London England*

55 Old Post Road No. 2
Greenwich, Connecticut 06836

JAI Press Ltd.
The Courtyard
28 High Street
Hampton Hill
Middlesex TW12 1PD
England

ISBN: 1-55938-161-2

Manufactured in the United States of America

CONTENTS

INTRODUCTION TO THE SERIES: AN EDITOR'S FOREWORD

The JAI series in chemistry has come of age over the past several years. Each of the volumes already published contains timely chapters by leading exponents in the field who have placed their own contributions in a perspective that provides insight into their long-term research goals. Each contribution focuses on the individual author's own work as well as the studies of others that address related problems. The series is intended to provide the reader with in-depth accounts of important principles as well as insight into the nuances and subtleties of a given area of chemistry. The wide coverage of material should be of interest to graduate students, postdoctoral fellows, industrial chemists, and those teaching specialized topics to graduate students. We hope that we will continue to provide you with a sense of stimulation and enjoyment of the various sub-disciplines of chemistry.

Department of Chemistry
Emory University
Atlanta, Georgia

Albert Padwa
Consulting Editor

PREFACE

Chemical kinetics and reaction dynamics are central to chemistry because they deal with the rates and mechanisms of chemical change. Traditionally, chemical kinetics has dealt with reacting systems on a macroscopic scale, while dynamics has focused on interactions at a molecular level. The dividing line between these areas has never been very clear, however, because kineticists have always used results from macroscopic systems to infer the dynamics that take place on the molecular level. The distinctions have become even more blurred during the past few decades with the development of state-resolved measurement techniques applied on ever shorter time scales.

Fundamental principles from kinetics and dynamics are applied regularly in areas as diverse as biochemistry, materials science, chemical synthesis, surface science, the earth sciences, and cosmochemistry. Publications appear in the primary literature at such a great rate that no one can stay abreast of all the important developments, although they may be of interest in many different applications. This series on chemical kinetics and dynamics will help to fill the need for summaries of recent developments by presenting timely reviews on recent advances in the understanding of fundamental principles, theoretical developments, and applications.

Some of the many flavors of kinetics and dynamics are presented in this first volume, which contains a diverse selection of topics. Some future volumes also will consist of assorted collections, while others will be devoted to single topics. It is my hope that this series will be stimulating and make the advances in chemical kinetics and dynamics more accessible to the scientific community.

Department of Atmospheric, Oceanic,
and Space Sciences
The University of Michigan
Ann Arbor, Michigan

John R. Barker
Editor

STATE-TO-STATE DYNAMICS: DO WE REALLY NEED ALL THOSE DATA?

James J. Valentini

I. INTRODUCTION

The investigation of the state-to-state dynamics of elementary chemical reactions and of the ''half-reactions'' that take place in photofragmentation is both experimentally challenging and intellectually stimulating. The aim is nothing less than the elucidation of the processes of chemical change at the most detailed level, and the elaboration or description of these processes in terms of the most fundamental con-

Advances in Chemical Kinetics and Dynamics, Volume 1, pages 1–45

ISBN: 1-55938-161-2

cepts, the first principles of quantum mechanics. The ideal of state-to-state dynamics is the measurement of the completely quantum-state-resolved differential cross section—that is, the scattering angle and speed distribution for the process

$$A(n_a) + B(n_b) \rightarrow C(n_c) + D(n_d), \tag{1}$$

at a single collision energy. Here the n_α specify all the quantum numbers of the particular reactant and product species. The specification of the n_α implies that the reactants are prepared in a single quantum state and the products are "catalogued" by their quantum numbers. Such an ideal is almost a Platonic form. Real experiments are only approximations to this ideal, but they continue to become ever better approximations.

The question raised by the title of the chapter—a question I hope to convince the reader should be answered in the affirmative—asks whether striving to achieve this ideal is as rewarding as it is challenging. State-to-state dynamics experiments produce an enormous body of experimental data, even when they fall far short of the ideal, fully state-, energy-, and angle-resolved measurement. It is not uncommon for a single such experiment to produce hundreds or even thousands of individual data. But what is the meaning of all this data? Is the whole more than the sum of the parts? Can the data be interpreted, manipulated, or otherwise used to provide information and insight about chemical reactions that would not have been obtained otherwise? The answers to these questions in my opinion are, respectively, very great, yes, and yes. Since so much of the effort in my research group is devoted to carrying out detailed state-to-state dynamics experiments the reader should fully expect such answers from me. However, such answers can be fully justified, and the purpose of this review is to do so. This chapter is devoted to renewing the conviction of the faithful and dispelling the skepticism of the disbelievers about the wisdom and value of state-to-state dynamics measurements. This I attempt to do by presenting and discussing specific examples of state-to-state dynamics experiments done in my own lab that provide important information and insight that could only have been gathered in a state-to-state experiment.

The experiments I will discuss involve examples of the prototype reaction,

$$A + BC \rightarrow AB + C, \tag{2}$$

and the prototype photodissociation

$$ABC + h\nu \rightarrow AB + C, \tag{3}$$

where A, B, and C are atoms. The specific examples are the abstraction reactions $H + HX \rightarrow H_2 + X$, where X is Cl, Br, or I, and the photodissociation $O_3 \rightarrow O_2 + O$. In all cases the experiments involve measuring the quantum state distribution of the products under single-collision conditions, at well-defined collision energies for the reactions or upon excitation with monochromatic light for the photodissociation. While the collision energy or photon energy is well defined, in only one case do the experiments involve any special initial internal state preparation of the reactants. However, for the reactants involved only a few initial states have appreciable Boltzmann population factors so this is not a significant deficiency. All the experiments are performed in homogeneous gas samples, that is they are so-called "bulb" experiments, not crossed molecular beam experiments, so differential state-to-state cross sections cannot be measured; only cross sections summed over scattering angle are determined. The experiments thus fall short of the as yet unrealized ideal state-to-state dynamics experiment described above. Despite these "shortcomings," the experiments yield a wealth of information and a depth of insight in understanding.

II. EXPERIMENTAL METHODOLOGIES

All the experiments discussed here used coherent anti-Stokes Raman scattering (CARS)[1] spectroscopy as the means to determine the quantum state distribution of the products of the chemical reaction or photodissociation. At first glance any Raman spectroscopy would seem to be a poor choice for a quantum-state-resolving detection method in chemical dynamics experiments, for all Raman-based spectroscopies suffer from relatively low absolute sensitivity. Because of the very small cross sections for the inelastic light scattering process that is the basis for all Raman spectroscopies, cross sections of the order of 10^{-30} cm^2 sr^{-1}, the fraction of incident photons that are converted to signal photons is very small except at high densities. At 10^{-30} cm^2 sr^{-1} cross section implies that a gas at 1 Torr pressure would have a spontaneous Raman scattering efficiency of only about 10^{-14}, and in an ordinary Raman experiment a 10 mJ laser pulse incident on this sample would produce a signal of fewer than 10^3 photons, of which fewer than 10 could be detected after spectral filtering via a monochromator to remove elastically scattered laser photons.[2] Of course, this low efficiency means a low optical depth for Raman scattering, and makes high density media accessible to probing by Raman techniques without attenuation of the laser beam or the scattered light. This is one reason that Raman techniques have found extensive application in combustion diagnostics.[1,2] The coherent light scattering that characterizes CARS and other coherent Raman techniques results in much higher scattering efficiencies and higher signals, but even then the sensitivity could best be described as modest. In CARS a general rule-of-thumb regarding sensitivity limits is that the minimum detectable density is $\approx 10^{13}$ cm^{-3} ($\approx 10^{-3}$ Torr at 298 K) and the minimum detectable relative concentration is $\approx$ 100 ppm. Like any other rule-of-thumb there are exceptions, and in fact we regularly detect species at absolute pressures and relative concentrations an order of magnitude lower than these, but still compared to laser photoionization and laser-induced fluorescence techniques these are very poor detectivities. Although this somewhat modest absolute sensitivity will restrict the range of state-to-state dynamics experiments that are feasible with CARS as a detection methodology, it is adequate—more than adequate—for many important experiments.

The absolute sensitivity of CARS limits its use to pressures quite a bit higher than those commonly associated with maintaining single-collision conditions. State-to-state dynamics experiments require single-collision conditions so that the initial state preparation is preserved and the product state distributions remain unrelaxed, and this is usually achieved by working at sample pressures in the millitorr range or below. However, the restriction imposed by the need for single-collision conditions in state-to-state dynamics experiments is only that the time scale of the experiment—from initial state preparation to product state determination—be such that only a single collision can take place, and this requirement can be satisfied at pressures high enough to give detectable signals in CARS experiments. For a typical thermal velocity the mean time between hard-sphere, elastic collisions at 1 Torr sample pressure is about 100 ns, while for the hyperthermal velocities encountered in some of our experiments this mean collision time may be about 10 ns. But it is really inelastic collisions, quantum-state-changing collisions, which must be avoided in state-to-state dynamics experiments, and the time scales for them are generally longer than those for elastic collisions. Thus, using nanosecond-duration pulsed lasers single-collision conditions can be maintained even at pressures around 1 Torr.

CARS has sensitivity sufficient for reaction product and photofragment detection at such a sample pressure if the reaction products or photofragments are at a partial pressure of about 10^{-3} Torr. This sounds high, but the sample volume probed by the CARS lasers is very small, typically approximating a cylinder of $\approx$ 0.01 cm diameter and $\approx$ 1 cm length, so the absolute number of molecules required is not really high, about 10^9 or so. Production of this number of photofragments is very easily achievable using laser photolysis. Note that in some of our CARS photofragmentation dynamics experiments we have studied photodissociation in systems having absorption cross sections of only 10^{-21} cm.[3,4] Production of 10^9 reaction products in the 10^{-4} cm^3 CARS probe volume is not trivial, but is feasible even given the restrictions imposed by the requirement for single-collision conditions, at least in those reactions in which one of the reactants is generated by laser photolysis of a stable precursor.

The relatively low absolute sensitivity of Raman and coherent Raman techniques will limit the ultimate range of applicability of the techniques in state-to-state dynamics experiments, but nothing else will. The real strength of Raman techniques lies in two attributes of the inelastic light scattering process: the universality of the phenomena and the "instantaneous" nature of the signal. All molecules have at least one Raman-active transition, and thus Raman spectroscopy is universal—all molecules can be spectroscopically detected by Raman spectroscopy. Equally important, the universal detection capability can be achieved over only a 4000 cm^{-1} spectral range, at any arbitrary excitation wavelength. The adjective "instantaneous" is somewhat awkward, but is used here to emphasize that the signal is produced only when the generating laser(s) are incident on the sample. The Raman signal appears as soon as the generating pulse appears at the sample and disappears as soon as the generating pulse is turned off. Together these two attributes, universality and instantaneous time response, make Raman spectroscopy a nearly ideal quantum-state-resolved detection technique for state-to-state dynamics experiments. If only the sensitivity were a few orders of magnitude better Raman scattering would indeed be ideal.

A schematic diagram depicting a typical state-to-state dynamics apparatus incorporating coherent anti-Stokes Raman scattering (CARS) as the detection methodology is shown in Figure 1. Two Nd:YAG-pumped dye laser systems are shown. One generates the tunable UV beam that effects photodissociation, both for studies of the

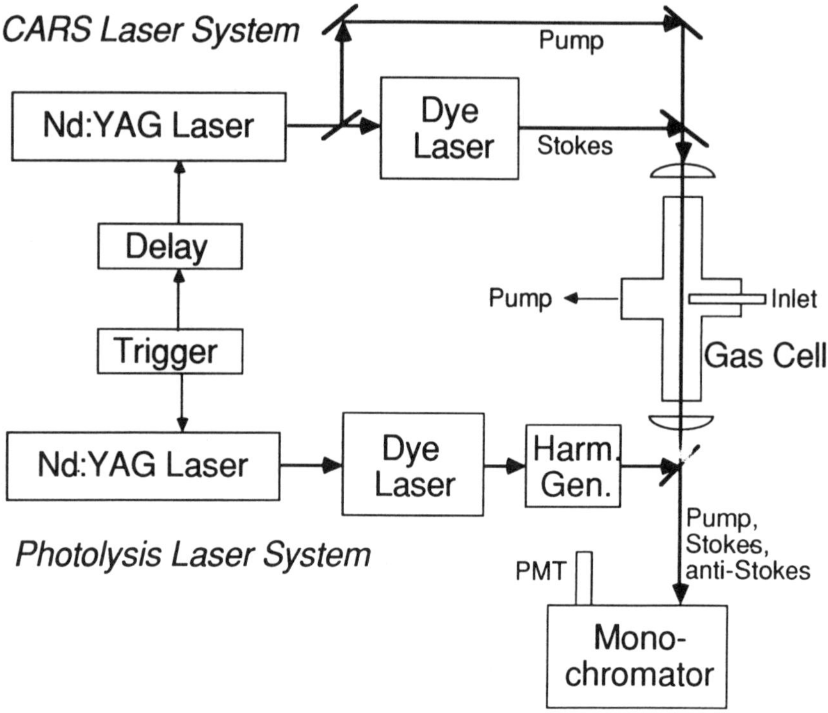

Figure 1. Simplified schematic showing the key elements of a CARS spectroscopy apparatus used for studies of state-to-state dynamics experiments.

dynamics of photofragmentation itself and to generate a reactive species for studies of state-to-state reaction dynamics. The second Nd:YAG-pumped dye laser system produces the two beams needed for the CARS product detection; one of these is the fixed-frequency pump beam, ω_P, and the other is the tunable Stokes beam, ω_S, which are combined and made collinear at a dichroic mirror. (In many experiments done in our lab a single Nd:YAG laser is used to generate both the photolysis and CARS probe beams.)

Although the choice of wavelength for ω_P is entirely arbitrary, the use of the Nd:YAG second harmonic at 532 nm is most convenient, and a portion of this output of the Nd:YAG laser usually serves as ω_P while the remainder of the 532 nm light is used to pump the tunable dye laser. A CARS signal is generated at the anti-Stokes frequency, $\omega_{AS} = 2\omega_p - \omega_S$, which is scanned by variation of ω_S. A signal identifying the product molecule in a particular quantum state is generated whenever the frequency difference $\omega_P - \omega_S$ is resonant with a Raman transition frequency, ω_R, for this state. The anti-Stokes signal beam can be separated from the input beams by a combination of dichroic filters (not shown) and a monochromator.

In the diagram the photolysis laser beam is shown counter-propagating relative to the CARS probe beams. However, we more frequently have the photolysis and probe beams *co-propagating,* and in fact in all the experiments discussed in this review the co-propagating geometry is the one that actually has been used. Whether co- or counter-propagating the UV and visible beams are combined/separated at dichroic mirrors, and we align the photolysis beam to overlap the CARS beams every bit as carefully as we overlap the two CARS beams themselves. This ensures that the volume in which the photofragments or reaction products are being produced, defined by the photolysis beam, is the same as the volume in which we are looking for these photofragments or reaction products. It may seem unnecessary to mention this, since such optimization of photolysis–probe overlap is important in any pump-and-probe optical experiment. However, the importance of good photolysis–probe overlap is more important in CARS than in most other pump-and-probe experiments. The modest sensitivity of CARS makes good overlap critical, as does the fact that the volume elements being overlapped are so small, typically $\approx$ 0.01 cm in diameter by $\approx$ 1 cm long.

The gas sample in all the experiments that will be discussed here is a flowing gas at a pressure of a fraction of a Torr to a few Torr at ambient temperature. However, we have also carried out experiments in which the sample is a free-jet expansion. No matter what the form of the sample, it must be delivered such that there is a fresh sample in the photolysis-probe volume on every laser pulse. This guarantees that there are no photofragments or reaction products from a previous laser pulse still in the probe volume.

For our photodissociation studies the delay between the photolysis and CARS laser pulses is adjusted such that the probability of the photofragments experiencing a hard-sphere, elastic collision is much less than one, approximately 0.1. When we are investigating reaction dynamics the delay typically is set to be slightly less than the mean time between hard-sphere gas kinetic collisions involving the two reactants. Such a delay does allow some probability, typically approximately 0.1, that a product we detect may be formed in a reactive collision that was preceded by an elastic collision of the photolyticly generated reactant with some other species. Because of this, we regularly measure product state distributions at various delay times, extrapolating to zero delay to obtain the true single-collision state distributions. In doing so we have found that measurements made under conditions such that the delay time is equal to or less than the mean time between elastic hard-sphere collisions of the reactants give state distributions that are indistinguishable (within the typical $\pm$ 10% measurement uncertainty) from those extrapolated to zero delay.[5]

III. SYMMETRY BREAKING IN TRIATOMIC PHOTODISSOCIATION

A major goal of this review is to demonstrate, or perhaps I should say reaffirm, how valuable state-to-state dynamics experiments are because the quantum states in which the photofragments and reaction products are formed tell so much about how the dissociation or reaction takes place. However, I am going to initiate this task by discussing a case for which it is the quantum states in which the photofragments are *not* formed that tell the important story. This case involves the photodissociation of ozone by visible light of wavelength between 500 and 700 nm,

$$O_3 + h\nu \rightarrow O_2(^3\Sigma_g) + O(^3P). \tag{4}$$

Absorption in this region is known as the Chappuis band; the absorption spectrum is shown in Figure 2.

As is the case with all the experiments that are discussed in this review, we chose to study this system because we expected it to be an appropriate prototype for elucidating generally important dynamical forces, here specifically the forces operative in photodissociation. As the reader will subsequently see, in this particular case we were right for the wrong reasons. Our expectations of prototype status for the ozone visible dissociation derived from several attributes of the system. Because the masses are all equal the potential energy surface's control of the dynamics cannot be overwhelmed by extreme kinematic effects associated with highly disparate masses, which are encountered in most other systems whose photodissociation dynamics have been studied. Similarly, the equality of the two bond strengths, rigorously imposed by the symmetry of the molecule, makes the dissociation truly polyatomic in nature and precludes the dissociation dynamics being simply "pseudo-diatomic" in character, as often happens when a polyatomic molecule with highly disparate bond strengths is dissociated.[7–10]

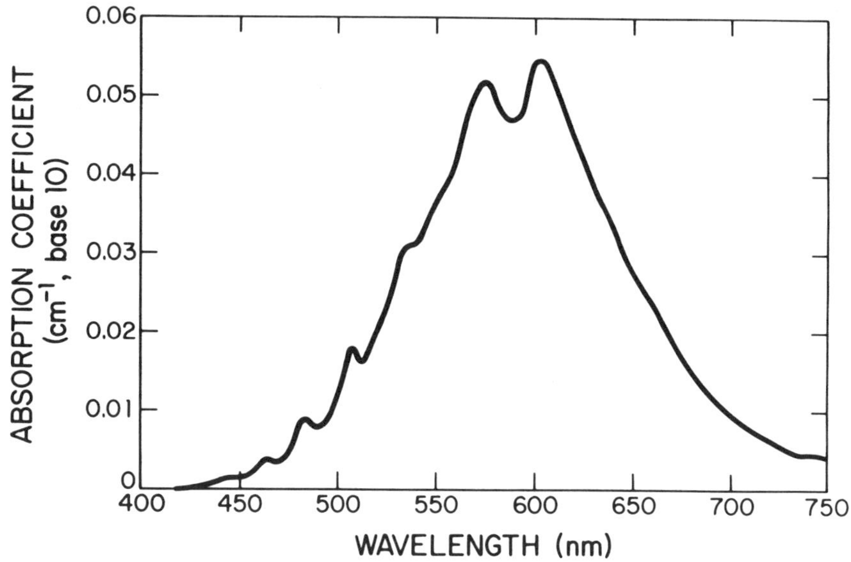

Figure 2. Absorption spectrum of ozone in the visible region of the spectrum. Adapted from Ref. 6.

While these features of the O_3 system ensure that the dissociation dynamics cannot be simplified to the level of the trivial, the system does possess one simplifying aspect that is very desirable: there is only one excited potential energy surface important in the dissociation. This means that electronic non-adiabatic effects, which are very interesting in their own right but complicating, are not superimposed on the dissociation dynamics. Although the low-lying electronic states of O_3 have not been well characterized, it is known that the only excited state that is accessible by a dipole-allowed transition from the ${}^1A_1$ ground state with 500 to 700 nm photons in the ${}^1B_1$ state, and the only other singlet state at low energy is the ${}^1A_2$ state.[11-13] Note also that only ground state oxygen atoms and molecules are formed in the dissociation.[3,14,15]

So, we expected the dissociation of O_3 by visible light to be truly polyatomic in character, and to have dynamics controlled by the shape of the dissociative excited state potential energy surface, and we expected the state-to-state dynamics to be relatively easily understood because the dissociation should involve only this single dissociative excited state. With these expectations in mind we set up an experiment to use CARS to determine the $O_2({}^3\Sigma_g)$ rotational and vibrational state distributions as a function of dissociation wavelength, using wavelengths between 560 nm and 638 nm. This required a slight variation on the experimental apparatus shown schematically in Figure 1. Both the dissociation beam and the CARS beams would be in the same wavelength domain; the CARS pump beam at 532 nm requires a Stokes wavelength of around 578 nm, both in the Chappuis band of O_3. This presents two problems. One occurs in combining the photolysis and CARS beams. The other is that we must somehow avoid, or at least minimize, photolysis by the CARS probe.

Combining the visible photolysis beam with the CARS probe beams is accomplished by using a polarizing beamsplitter and setting the electric field polarization vector of the photolysis beam to be orthogonal to the polarization direction of the probe beams. We avoid the complication of having the CARS probe beams also generate photofragments by having the dissociation pulse precede the CARS probe pulse by 4 ns and by using a pulse energy for the photolysis beam that is high enough to saturate the dissociation. Because of the small absorption coefficient for absorption in the Chappuis band (see Figure 1), which corresponds to an absorption cross section of (2–5) $\times$ 10^{-21} cm^2 at the wavelengths we studied, saturation of the absorption might seem difficult. However, given the small cross sectional area of the CARS probed volume, approximately 4 $\times$ 10^{-5} cm^2 (see section II), this only requires pulse energies of about 5 mJ. We used photodissociation pulse energies of 20 mJ in the experiments.

The $O_2(^3\Sigma_g)$ photofragment rotational and vibrational state distributions that we obtained are highly revealing and for the most part are readily interpretable in terms of the influence of the potential energy surface.[4] The vibrational state distributions, shown in Figure 3 for all nine wavelengths at which we effected photolysis, are very strongly peaked at $v = 0$ and are independent of wavelength. The experimental uncertainties typically are larger than the differences among the distributions. This implies that the dissociative potential energy surface does not channel much energy into the diatomic fragment vibration. Rather, the vibrational state distribution is established by Franck–Condon-like overlaps between the vibrational wavefunction of O_3 in its ground electronic state and the vibrational wavefunction of the $O_2(^3\Sigma_g)$ fragment. The bond lengths in O_3 and O_2 are nearly identical, for O_3 1.27 Å in the $^1A_1$ ground state[16] and probably only 0.07 Å longer in the $^1B_1$ excited state[11] and for O_2 1.21 Å.[17] Thus, Franck–Condon factors should favor transition from O_3 to O_2 + O with little change in vibrational quantum number and give a strong propensity for $v = 0$ O_2 fragment formation from the dissociation of ground vibrational state

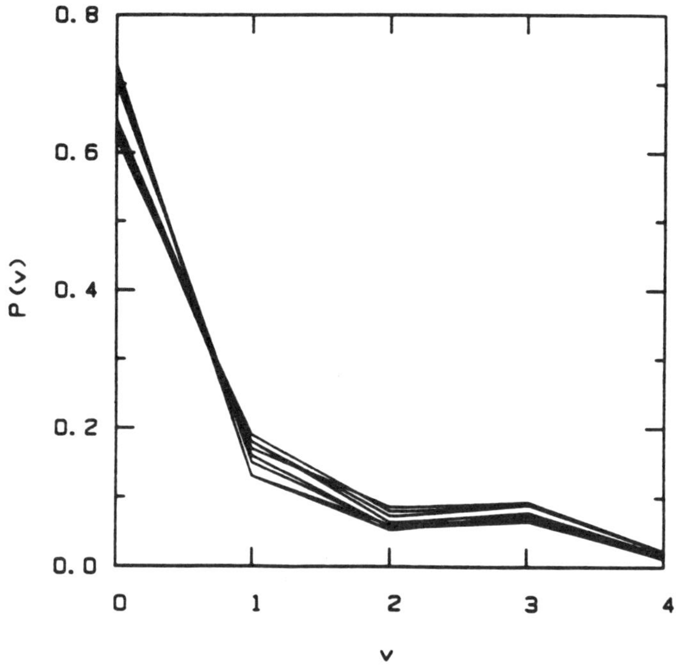

Figure 3. $O_2(^3\Sigma_g)$ photofragment vibrational state distributions measured for photodissociation of O_3 at nine visible photolysis wavelengths, 560, 572, 591, 596, 599, 606, 609, 625, and 638 nm. The distributions are normalized such that $\Sigma P(v) = 1$. From Ref. 4.

O_3. Note that the energy available to the fragments at the shortest photolysis wavelength, 560 nm, is sufficient to generate $O_2(^3\Sigma_g)$ in $v = 6$, while at the longest wavelength, 638 nm, only $v = 4$ is accessible, and the available energy changes by 30% between 560 and 638 nm, yet the vibrational distributions are completely insensitive to this.

The $O_2(^3\Sigma_g)$ rotational state distributions, one of which is shown in Figure 4, seem simple. The distributions are peaked at high rotational angular momentum but have narrow spread in J. (Note that the most probable J for $O_2(^3\Sigma_g)$ at 300 K is 7.) We interpret this to imply rotationally impulsive dissociation dynamics, in contrast to the vibrationally adiabatic nature of the dissociation dynamics implied by the "cold" and dissociation-wavelength–independent fragment vibrational state distributions. For non-rotating, non-vibrating O_3 rotationally impulsive, vibrationally adiabatic dissociation dynamics imply a single

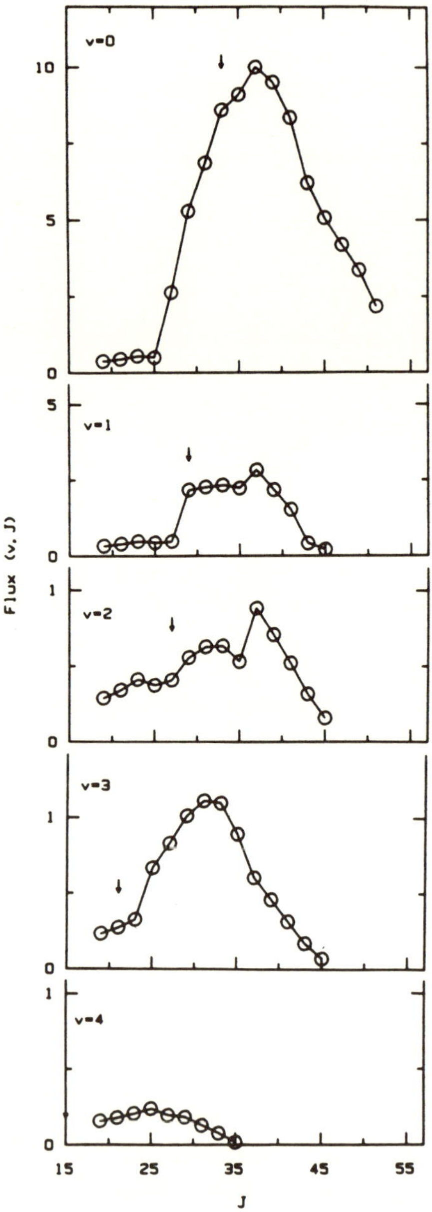

Figure 4. $O_2(^3\Sigma_g)$ photofragment rotational state distribution for photodissociation of O_3 at 625 nm. The arrows indicate the expected rotational states for a fragment produced by rotationally impulsive, vibrationally adiabatic dissociation of rotationless and vibrationless O_3; see the text for details. From Ref. 4.

fragment rotational state for each fragment vibrational state.[18,19] The arrows in Figure 4 indicate these expected rotational states. The rotational energy imparted by the dissociation can be high because the bond angle in O_3 is small, 117° in the ground electronic state[16,20] and probably only 1–2° larger in the excited state,[11] and thus an impulsive force can impart significant angular momentum. The arrows generally fall within a few *h* of the actual observed most probable *J*, suggesting that this rotationally impulsive description of the dissociation is correct. However, rotationless O_3 does not describe our 300 K O_3 sample, and even though 93% of the O_3 molecules in our sample are in the ground vibrational state, they are not vibrationless. Given this, we thought that it would be appropriate to calculate what rotational state distribution would result from rotationally impulsive, vibrationally adiabatic dissociation dynamics of 300 K thermal ozone.

It is relatively easy to do this, especially if one treats the O_3 rotations classically, retaining a quantum description only for the vibrations of O_3, and we have carried out such semi-classical calculations to determine the expected $O_2(^3\Sigma_g)$ rotational state distributions.[19] The contributions to the O_2 fragment rotational motion from the O_3 motions are of two types: one from the rotation of the O_3 about each of its principal axes and a second from the bending vibration of the O_3. The effects of these contributions are shown separately in Figures 5 and 6 for just one vibrational state at one photodissociation wavelength that we studied experimentally, but the results are very similar for all fragment vibrational states at all wavelengths.

The effect of the bending zero-point vibration on the O_2 fragment rotational angular momentum is shown in Figure 5. There is a simple broadening, but no shift in the peak *J*. In contrast, the O_3 parent rotation yields a double-peaked $O_2(^3\Sigma_g)$ rotational state distribution. The bimodality is a consequence of the restriction that the angular momentum imparted by the rotationally impulsive dissociation must be perpendicular to the ozone molecular plane. Vector addition of all the angular momenta contributing to the overall O_2 fragment rotational angular momentum gives:

$$J_{O_2} = [(J_{DISS} \pm J_{O_2,C})^2 + (J_{O_2,A})^2 + (J_{O_2,B})^2]^{1/2}, \qquad (5)$$

where J_{DISS} is the angular momentum imparted by the dissociative energy release, and the $J_{O_2,\alpha}$ are the contributions from J_α, the O_3 rotational angular momenta about the *A*, *B*, and *C* axes. The *C* axis is perpendicular to the plane of the O_3 and so $J_{O_2,C}$ must be strictly parallel or anti-parallel to J_{DISS}. The "+" combination in the first term on the

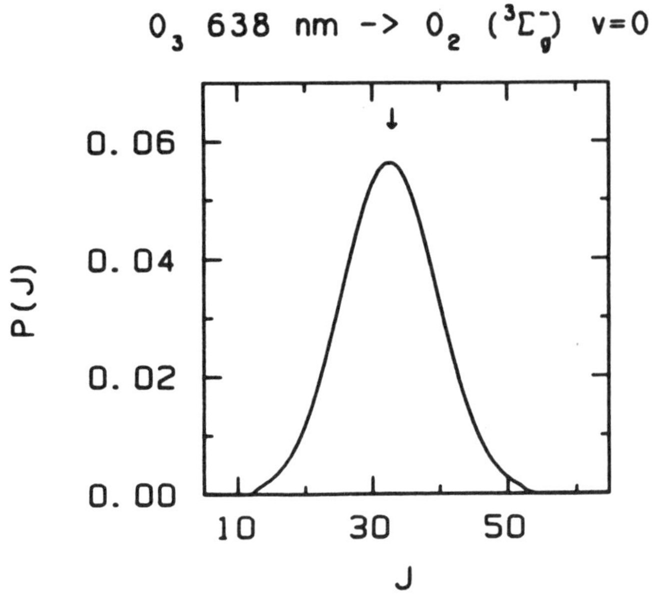

Figure 5. Calculated rotational state distribution for the $O_2(^3\Sigma_g)$ photofragment from the 638 nm photodissociation of non-rotating O_3 with zero-point vibration only, assuming rotationally impulsive, vibrationally adiabatic dissociation dynamics. The arrow indicates the unique $O_2(^3\Sigma_g)$ rotational state expected for vibrationless O_3 photodissociation. From Ref. 19.

right side of Equation 5 results from the parallel combination of J_{DISS} and $J_{O_2,C}$ and gives the high-J peak in Figure 6, while the "−" combination is associated with the anti-parallel orientation of these two angular momentum vectors.

We have plotted separately the O_3 vibrational and O_3 rotational contributions to the $O_2(^3\Sigma_g)$ photofragment rotational angular momentum to illustrate the qualitative difference in the nature of the resultant fragment rotational state distributions. However, our experimental measurements cannot make such a separation and the rotational and vibrational contributions must be combined to yield the actual $O_2(^3\Sigma_g)$ rotational state distributions that our calculations predict should be observed. In Figure 7 is shown one such predicted rotational state distribution, and we see that the bimodality of the distribution for rotating but not vibrating O_3 (Figure 6) is gone as a result of the addition of the contributions from the O_3 vibration (Figure 5). Figure 8

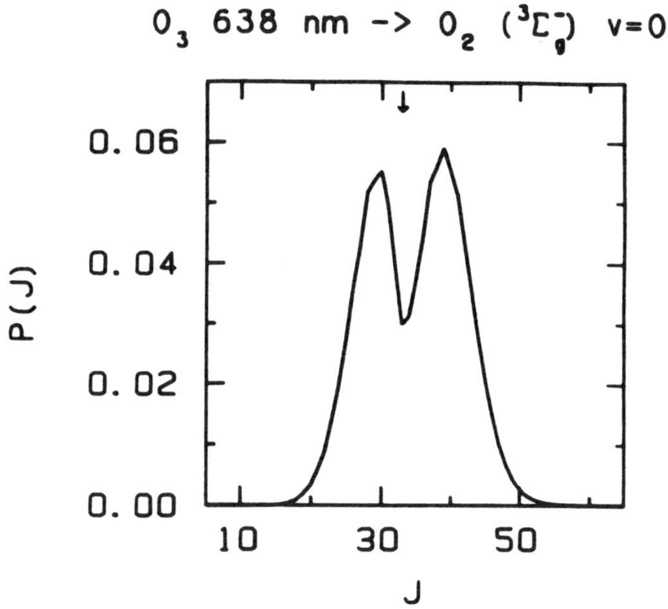

Figure 6. Calculated rotational state distribution for the $O_2(^3\Sigma_g)$ photofragment from the 638 nm photodissociation of vibrationless O_3 with a 300 K rotational state distribution, assuming rotationally impulsive, vibrationally adiabatic dissociation dynamics. The arrow indicates the unique $O_2(^3\Sigma_g)$ rotational state expected for photodissociation of non-rotating, non-vibrating O_3. From Ref. 19.

compares this predicted rotational state distribution with that actually observed, and we see that the agreement is not very good. The calculated distribution is substantially narrower than the measured one (FWHM of 16 as opposed to FWHM of 21), and peaked at lower J ($J = 33$ as compared to $J = 39$). While the discrepancy in the value of the most probable value of J is not difficult to rationalize, the too-narrow width of the calculated $O_2(^3\Sigma_g)$ rotational distribution is not. The same kind of discrepancy between the experimentally observed and calculated rotational state distributions is seen for all other vibrational states of fragments produced at all other photodissociation wavelengths that we studied.

Our calculations assume that in the absence of O_3 vibration and rotation there would be just a single O_2 fragment rotational state observed, so we cannot have underestimated the width of the "intrinsic" distribution. As a result of the semi-classical nature of the calcula-

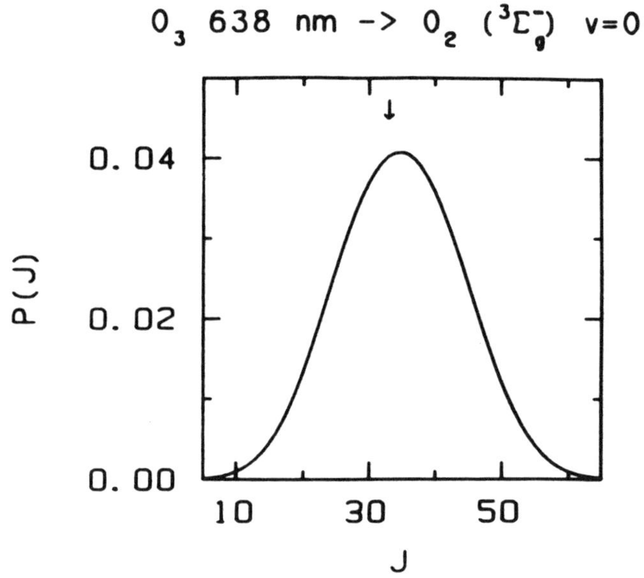

Figure 7. Calculated rotational state distribution for the $O_2(^3\Sigma_g)$ photofragment from the 638 nm photodissociation of O_3 with a 300 K rotational state distribution and zero-point vibrational energy, assuming rotationally impulsive, vibrationally adiabatic dissociation dynamics. The arrow indicates the unique $O_2(^3\Sigma_g)$ rotational state expected for photodissociation of non-rotating, non-vibrating O_3. From Ref. 19.

tion, which treats the O_3 rotations classically, one would expect it to underestimate the width of the actual O_2 rotational state distribution, and it is hard to see how it could possibly overestimate the width. This suggests that the discrepancy between the calculated, or expected, distribution and the one actually measured is of physical significance, and indicates that the angular momentum disposal in the ozone dissociation does not follow our expectations. To be more specific, the discrepancy indicates that some $O_2(^3\Sigma_g)$ fragment rotational states are not produced, even though they are allowed by energy and angular momentum conservation. This means that some combinations of J_{DISS} with the O_3 bending vibrational angular momenta and the O_3 rotational angular momenta, J_A, J_B, and J_C, are not actually sampled in the dissociation. There must be a dynamical restriction on the angular momentum disposal.

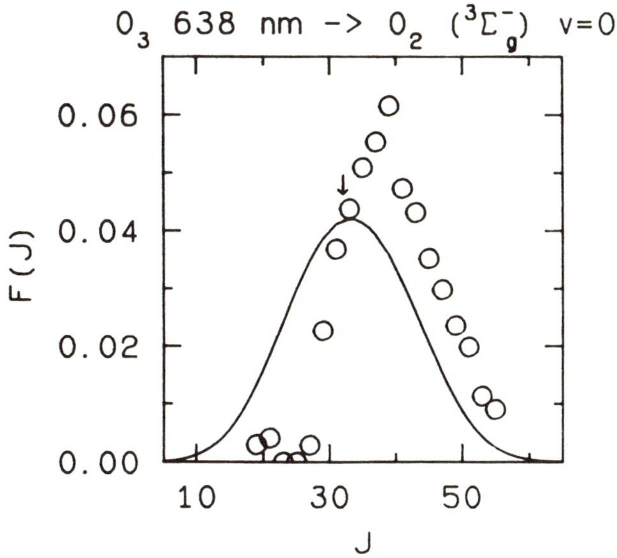

Figure 8. Comparison of experimentally measured (open circles) and expected (solid line) rotational state distributions for the $O_2(^3\Sigma_g)$ photofragment from the 638 nm photodissociation of O_3. The arrow indicates the unique $O_2(^3\Sigma_g)$ rotational state expected for photodissociation of non-rotating, non-vibrating O_3. The distribution shown by the solid line is the same as that in Figure 7, calculated for rotationally impulsive, vibrationally adiabatic dissociation dynamics of O_3 having a 300 K rotational state distribution and zero-point vibrational energy. The measured and expected distributions have both been normalized to have integrated areas of unity. From Ref. 19.

The results presented in Figure 9 show a comparison similar to that in Figure 8 between the calculations and experimental measurements, except that we have introduced a dynamical bias that restricts the combinations of angular momentum vectors that are allowed in the calculation. Specifically, we restrict J_{DISS}, the angular momentum associated with the dissociative energy release, and J_C, the rotational angular momentum due to the in-plane rotation of the O_3, to be parallel, excluding the otherwise-allowed anti-parallel orientation. Since J_{DISS} is perpendicular to the molecular plane by symmetry and J_C is perpendicular by definition only the parallel and anti-parallel orientations of these vectors are possible. In terms of Equation 5, the

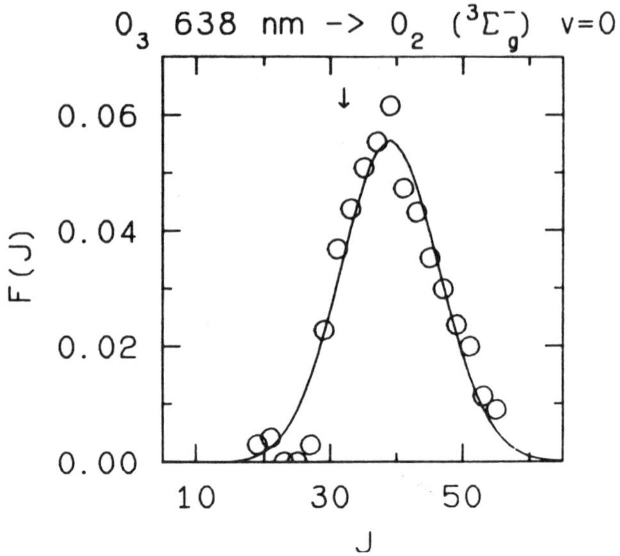

Figure 9. Comparison of experimentally measured (open circles) and expected (solid line) rotational state distributions for the $O_2(^3\Sigma_g)$ photofragment from the 638 nm photodissociation of O_3. The arrow indicates the unique $O_2(^3\Sigma_g)$ rotational state expected for photodissociation of non-rotating, non-vibrating O_3. The distribution shown by the solid line is calculated for rotationally impulsive, vibrationally adiabatic dissociation dynamics of O_3 having a 300 K rotational state distribution and zero-point vibrational energy, with the additional restriction that J_{DISS} must be parallel to J_C. See text for details. The measured and expected distributions have both been normalized to have integrated areas of unity. From Ref. 19.

dynamical restriction we impose means taking only the "+" combination in the first term on the right. This dynamical bias eliminates exactly one-half of the possible final states, and as Figure 9 shows makes the agreement between the calculations and the experimental measurements essentially perfect. This same dynamical bias makes the calculations and measurements of the rotational state distributions for all other $O_2(^3\Sigma_g)$ fragment vibrational states at all the other photodissociation wavelengths we studied agree as well.

What is the physical significance of this dynamical bias? Well, the direction of J_{DISS} is determined by which of the two nominally equivalent O–O bonds breaks, that is J_{DISS} is "up" for breaking of one bond and "down" for breaking of the other bond. We also know, of course, that whether the vector J_C is "up" or "down" is determined by the sense (direction) of the in-plane rotation of the ozone molecule. Therefore, restricting J_{DISS} and J_C to be parallel means that the identity of the O–O bond that breaks is correlated with the sense (direction) of the in-plane rotation of the O_3 molecule prior to dissociation. This implies that the sense of the in-plane rotation distinguishes the two otherwise indistinguishable O–O bonds in ozone, breaking the C_{2v} symmetry of the molecule! How can such a symmetry breaking occur? We believe that it actually results from the effect of the O_2 + O *orbital* angular momentum on the dissociative potential energy surface, and the correlation between the magnitude of the orbital angular momentum and the direction of the in-plane rotation of the ozone molecule.

The O_2 + O orbital angular momentum is correlated with the O_2 rotational angular momentum, since the vector sum of these two angular momenta is equal to the total (rigorously conserved) angular momentum. The orbital angular momentum is the source of the centrifugal potential that adds to the electronic potential to give the effective potential energy surface. The electronic potential energy surface must have C_{2v} symmetry and thus must be the same for breaking of either one of the two O–O bonds. However, the effective potential that results from inclusion of the O_2 + O orbital angular momentum is asymmetric: the potential is more repulsive for dissociation of one O–O bond than the other. This is illustrated in Figure 10. The O_2 + O orbital angular momentum that makes the centrifugal potential is given by a vector sum of exactly the same form as Equation 5, namely,

$$\iota_{O_2+O} = [(\iota_{DISS} \pm \iota_{O_2+O.C})^2 + (\iota_{O_2+O.A})^2 + (\iota_{O_2+O.B})^2]^{1/2}. \tag{6}$$

The contribution of the 300 K thermal distribution of O_3 rotational angular momenta to the O_2 + O orbital angular momentum gives rise to a distribution in ι_{O_2} with the same double peak as for J_{O_2} in Figure 6. The "±" in the first term on the right side of Equation 6 is of course the source of the bimodality. However, when the "+" appears in Equation 5 the "−" must appear in Equation 6, since ι_{DISS} and J_{DISS} are equal in magnitude, but oppositely directed. Therefore, since our $O_2(^3\Sigma_g)$ fragment rotational state distributions display a dynamical bias that restricts J_{DISS} and J_C to be parallel, this means that ι_{DISS} and J_C must be anti-parallel.

The dynamical bias forces J_C to be anti-parallel to ι_{DISS} because this selects the smaller of two possible ι_{O_2} + O, not because this selects the

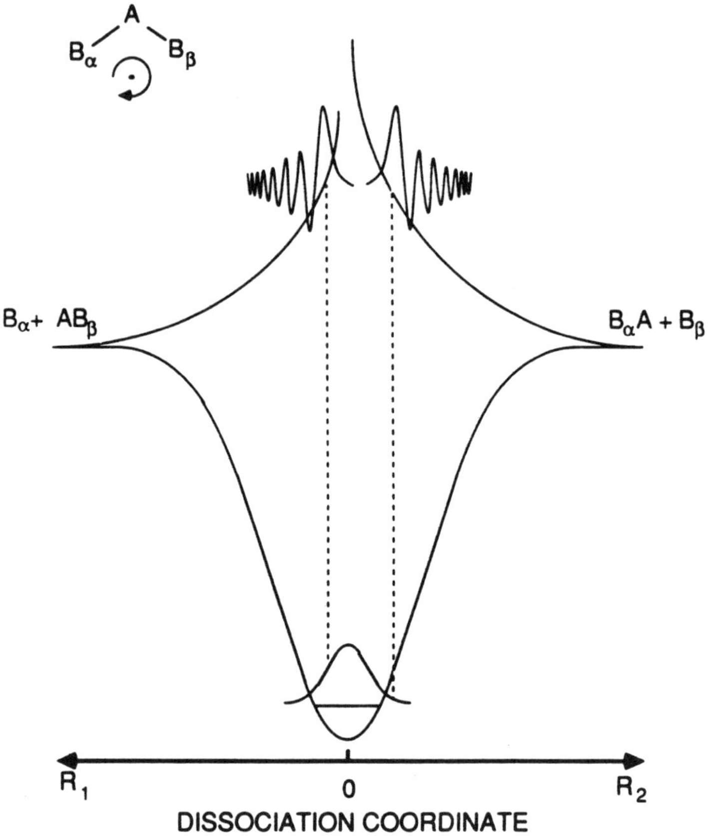

Figure 10. Ground and excited state potential energy surfaces for dissociation of symmetric AB_2 molecules, shown with some representative nuclear eigenfunctions. Note the asymmetry in the repulsive excited surfaces due to the asymmetry in the centrifugal potential, and the consequent effect on the Franck–Condon overlaps. For AB_2 in-plane rotation in the opposite direction the asymmetry is inverted. From Ref. 19.

larger of two possible J_{O_2}. The dynamical bias works this way because, as Figure 10 shows, the dissociation channel (bond-breaking) that has the smaller ι_{O_2} + O has the better Franck–Condon overlap with the ground state. The relative magnitudes of the Franck–Condon overlaps for the two channels determines the probability of selecting them—that is, the probability of breaking each of the two nominally equivalent bonds. The direction of the in-plane rotation of the ozone thus breaks the C_{2v} symmetry of the molecule and determines which bond will break, because the direction of rotation determines the magnitude of the fragment orbital angular momentum, which in turn determines the Franck–Condon factors. We have calculated the Franck–Condon overlaps for the two dissociation channels in ozone and find the one with smaller orbital angular momentum gives about an order of magnitude greater Franck–Condon factor than the other.[19] Thus, the dynamical bias is really not a perfectly selective one, but is a very strong propensity. We note that our experimental measurements of the $O_2(^3\Sigma_g)$ fragment rotational state distributions cannot actually distinguish a propensity this strong from a perfect selectivity.

Figure 10 is labeled in terms of the dissociation of a general C_{2v} molecule, AB_2, rather than in terms of the specific dissociation of O_3. Although our discussion has focussed exclusively on O_3 photodissociation, none of the arguments presented are restricted to O_3. In fact, the dynamical bias we observe should be even stronger in other C_{2v} triatoms. For example, our Franck–Condon overlap calculations show that for H_2O photodissociation through the first continuum the Franck–Condon overlap for ι_{DISS} anti-parallel to J_C is more than one thousand times greater than the ι_{DISS} parallel to J_C! This dramatic effect occurs because the extreme mass difference between O and H directs almost all the H_2O angular momentum and the angular momentum imparted by the dissociative energy release into H + OH orbital angular momentum. Unfortunately, this same extreme mass difference between O and H precludes this dissociation propensity from being apparent in the OH fragment rotational angular momentum distributions. Nonetheless, we believe that the rotational motion of any AB_2 molecule can and probably often does break the C_{2v} symmetry in photodissociation of these molecules and does determine the identity of the bond that will break.

These state-to-state dynamics experiments on the photodissociation of ozone show how a dynamical process controls the loss of a symmetry element. This is important because collisions and half-collisions often involve the development and loss of symmetries as atoms come together and then separate, as bonds are made and then broken. However, if the bond connecting two identical particles is not broken in a collision certain symmetries will be preserved in the collision, and the conservation of these symmetries can have equally interesting and important dynamical consequences. To illustrate how the results of state-to-state dynamics experiments can illuminate the dynamical constraints imposed by the conservation of a symmetry I turn now to a discussion of another of our experiments—one that coincidentally also involves the photodissociation of ozone.

IV. NUCLEAR SYMMETRY RESTRICTIONS IN NON-ADIABATIC COLLISIONS

The dissociation of ozone by ultra-violet light is a very different process from the photodissociation by light of visible wavelengths. First, the UV photodissociation is of more than simple academic scientific interest, since the interaction of ozone with light in the UV plays a major role in the atmosphere of the earth. The UV photochemistry of O_3 determines in large measure the physical structure and chemical composition of the stratosphere and the UV absorption that is the basis of that photochemistry shields the earth from harmful radiation below 300 nm.[21-24] That the UV absorption of O_3 is of such significance while the visible absorption is of essentially no practical import is a consequence of the second major difference between the absorption of visible light and the absorption of UV light by ozone: the first is a remarkably feeble process, while the latter involves very strong absorption, more than 2000 times stronger. However it is yet another major difference that is the important one for the discussion in this section. The UV photodissociation of ozone is not the simple, single–electronic-surface process that the visible dissociation is. In fact, the most significant characteristic of the UV photolysis for this discussion is a departure from such simplicity, the presence of an electronic curve crossing in the dissociation.

This curve crossing is depicted in Figure 11, which shows a cut through the potential surfaces important for the uv photolysis. These are ab initio surfaces, calculated by Hay et al.[25] The plots in Figure 11 give a cut through the surfaces at constant bond angle with one O–O bond length fixed, a cut very close to being right along the dissociation coordinate. The curve crossing, which is technically an avoided crossing because all the surfaces involved are A^1 symmetry in the C_S point group describing O_2 + O, occurs at quite large O–O separation, around 1.7 Å. The presence of this curve crossing, specifically the interaction of the states labeled "B" and "R" in the figure, is the reason that even though the UV absorption terminates on the dipole-allowed $^1B_2$ surface the photodissociation produces both $O_2(^1\Delta_g)$ + $O(^1D)$ and $O_2(^3\Sigma_g)$ + $O(^3P)$ fragments, the former being the dominant ones.

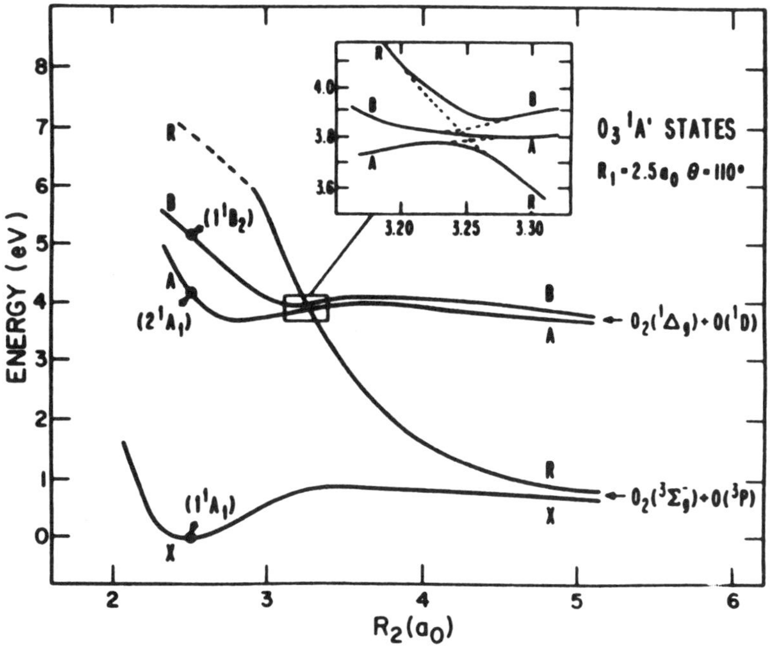

Figure 11. A cut through the low-lying singlet potential energy surfaces of O_3, from Ref. 25. The energy is plotted as a function of one O-O bond length, holding the other bond length and the bond angle constant.

I will continue to refer to the interaction among the "R," and "B," and "A" surfaces in Figure 11 as a curve crossing rather than an avoided crossing, because the dynamical phenomena that I will discuss are more easily framed in an electronically diabatic representation rather than in an electronically adiabatic basis. Molecules that cross from the "B" to the "R" curve undergo an electronic transition, and like any other electronic transition there are certain selection rules that must be followed if the crossing is to take place. These selection rules will control the dynamics at the crossing. This seems obvious, and it should have been obvious to us when we undertook a state-to-state dynamics study[26,27] of the UV photodissociation of ozone, but it took an anomalous observation in our actual experiments to get us thinking about the curve crossing selection rules. In the process we stumbled upon one of the most significant findings ever to come out of our lab.

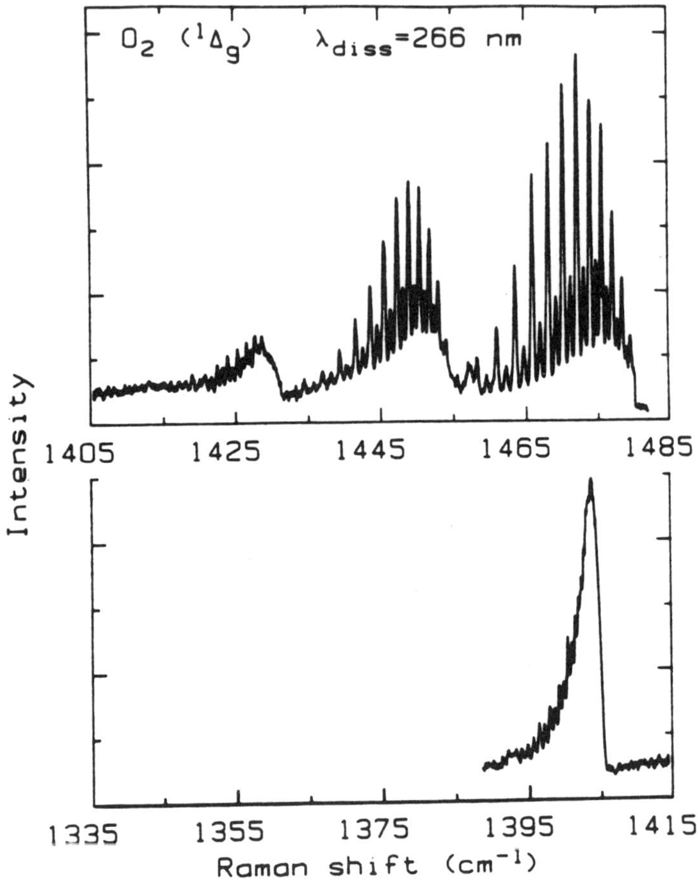

Figure 12. Vibrational Q-branch CARS spectrum of the $O_2(^1\Delta_g)$ photofragment from O_3 photolysis at 266 nm. Four vibrational bands are shown: $v = 0$ at 1473 cm^{-1}, $v = 1$ at 1450 cm^{-1}, $v = 2$ at 1428 cm^{-1}, and $v = 3$ at 1403 cm^{-1}. From Ref. 27.

The anomalous observation that initially puzzled and later delighted us is illustrated in Figure 12, which presents a CARS spectrum of the $O_2(^1\Delta_g)$ photofragment produced upon the 266 nm dissociation of O_3. The spectrum is a lovely one, but it is missing something. Specifically, it is missing most of the intensity that should be in the Raman transitions from states having odd values of the rotational quantum number. These are spectra of $^{16}O^{16}O$, a homonuclear diatomic molecule having nuclei with nuclear spin zero, and as such half the rotational states should indeed be missing, just as they are in the spectra of $O_2(^3\Sigma_g)$ discussed above. However, due to lambda-doubling[17] there are two states for every value of the rotational quantum number, and while exactly half these are missing, there is one allowed state for every value of J. Therefore, the vibrational Q-branch CARS spectra of Figure 12 should show a smooth band profile, with no distinction between transitions with even as opposed to odd rotational quantum numbers, since the nuclear spin degeneracy of all states is unity. The spectrum actually observed has two band profiles for each vibrational state, one of high intensity composed of all the even J states and another of much lower intensity composed of all the odd J states.

The distinction between states with even and odd rotational quantum numbers must be the result of some dynamical propensity in the photofragmentation. That the even-J propensity does indeed have a dynamical basis is supported by the fact that the propensity of even J states is vibrational state dependent. Although the vibrational state dependence is not easily discerned in the $O_2(^1\Delta_g)$ CARS spectra of Figure 12, it is obvious in the quantum state populations derived by an analysis of this spectrum, shown in Figure 13. But what sort of dynamical bias is at work here? Do the dynamics favor the production of even J or do they disfavor the production of odd J? Is the even-J selectivity an intrinsic property of the fragmentation of O_3 into O_2 + O, for example a symmetry effect?

We believe that the dynamical bias is indeed an intrinsic one and that it is symmetry based. However, it appears that the bias is an intrinsic property not of the O_2 + O photofragmentation process, but of the curve crossing that follows it, and that the basis of the effect is a nuclear exchange symmetry restriction. Further, the distinction between even and odd J is not the result of a propensity to form $O_2(^1\Delta_g)$ even J, but rather is the result of a selective depletion of $O_2(^1\Delta_g)$ odd J by the curve crossing. To see how this comes about I refer again to the potential energy curves shown in Figure 11.

The fragments start out on the $^1B_2$ surface (also labeled "B" in the figure), on which they would uniformly and totally evolve into $O_2(^1\Delta_g)$ + $O(^1D)$ if it were not for the curve crossing between this surface and the very repulsive surface labeled "R," which correlates with $O_2(^3\Sigma_g)$ + $O(^3P)$ fragments. Describing the curve crossing in a diabatic representation, the crossing is induced by the off-diagonal matrix element $|<\Phi_R|H_{el}|\Phi_B>|^2$, where the subscripts "R" and "B" identify the wave functions associated with the surfaces labeled "R" and "B" in

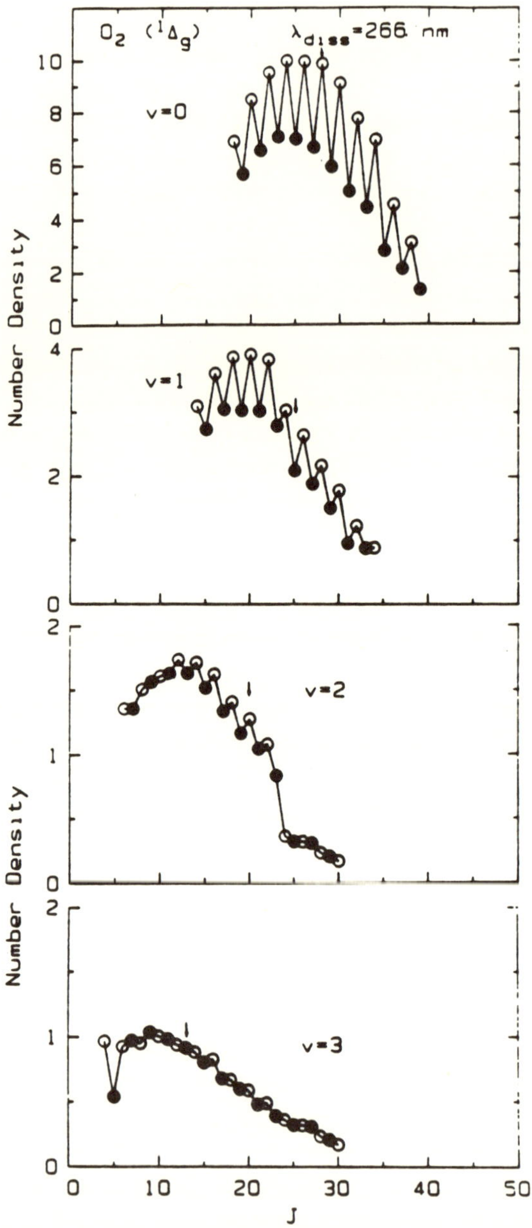

Figure 13. Quantum state population distribution for the $O_2(^1\Delta_g)$ photofragment from 266-nm photolysis of O_3, extracted by analysis of the CARS spectrum shown in Figure 12. The closed circles indicate the odd-*J* populations, while the open circles show the even-*J* populations. From Ref. 27.

Figure 11. In the diabatic representation the Born–Oppenheimer approximation remains valid, so this off-diagonal matrix element can be written as the product of an electronic part and several nuclear parts,

$$< \Phi_R \mid H_{el} \mid \Phi_B > \; = \; < \Phi_{r,el} \mid H_{el} \mid \Phi_{B,el} > < \Phi_{R,vib} \mid \Phi_{B,vib} > \\ X < \Phi_{R,rot} \mid \Phi_{B,rot} > < \Phi_{R,trans} \mid \Phi_{B,}trans >. \quad (7)$$

An atom-diatom basis is appropriate for the nuclear coordinates, since the curve crossing occurs at large O_2-O separation, for which the O_2-O interaction energy on the "B" surface is negligible. Thus, the vibrational and rotational matrix elements in Equation 7 are those for O_2. While the exact magnitudes of the matrix elements will determine the curve crossing probability, selection rules associated with the nuclear overlap integrals can turn the crossing on or off, determining whether the probability is zero or non-zero. An important selection rule associated with the rotational overlap integral seems to be the one that gives rise to the apparent propensity of even-J states in the $O_2(^1\Delta_g)$ fragment.

The rotational overlap integral, $< \Phi_{R,rot} \mid \Phi_{B,rot} >$, describes the overlap between the rotational wavefunction of $O_2(^1\Delta_g)$ on the "B" surface and the rotational wavefunction of $O_2(^3\Sigma_g)$ on the "R" surface. This integral is zero unless the rotational wave functions are of identical parity. The rotational wave functions are of even parity if the rotational angular momentum J is even, and odd parity if J is odd, so the rotational wave function parity introduces the selection rule $\Delta J = 0, \pm 2, \pm 4, \pm 6, \ldots$. This selection rule is important in determining the $O_2(^1\Delta_g)$ rotational state distribution because of the absence of even-J rotational states of the homonuclear $^{16}O_2(^3\Sigma_g)$. $^{16}O_2(^3\Sigma_g)$ has only odd-J rotational states because only these states give an overall wave function that has the required even symmetry with respect to exchange of the identical spin-zero ^{16}O nuclei.[17] The combination of the absence of even-J rotational states for $^{16}O_2(^3\Sigma_g)$ and the ΔJ = even selection rule imposed by the parity constraints of $<\Phi_{R,rot} \mid \Phi_{B,rot}>$ implies that the "B" to "R" curve crossing—the crossing from $O_2(^1\Delta_g)$ + $O(^1D)$ to $O_2(^3\Sigma_g)$ + $O(^3P)$—is allowed only for $O_2(^1\Delta_g)$ in odd-J rotational states. Thus, the effect of the curve crossing on the $O_2(^1\Delta_g)$ rotational state distributions is to deplete selectively the odd-J states, leaving the even-J states unaffected.

If this analysis is correct the difference in the even-J and odd-J rotational populations of $O_2(^1\Delta_g)$ should be consistent with the known quantum yield for $O_2(^3\Sigma_g)$ + $O(^3P)$ in the O_3 photodissociation. It is. For example, for photodissociation at 266 nm our $O_2(^1\Delta_g)$ rotational state distributions imply a $O_2(^3\Sigma_g)$ + $O(^3P)$ quantum yield of 0.15 ±

0.03, while a previous direct measurement[28] gives 0.12 ± 0.02. However, this agreement does not provide the definitive demonstration of the correctness of our analysis. That comes from experiments in which we photodissociate ^{18}O-enriched O_3 and compare the rotational state distributions for $^{16}O^{16}O$, $^{16}O^{18}O$, and $^{18}O^{18}O$.

Unlike $^{16}O^{16}O(^3\Sigma_g)$, the heteronuclear $^{16}O^{18}O(^3\Sigma_g)$ has both even- and odd-*J* rotational states, so the curve crossing from $O_2(^1\Delta_g)$ + O(^{1}D) to $O_2(^3\Sigma_g)$ + O(^{3}P) should deplete both the even- and odd-*J* rotational states of $^{16}O^{18}O(^1\Delta_g)$ equally, and there should be no favoring of even-*J* states in the $^{16}O^{18}O(^1\Delta_g)$ photofragment rotational state distribution. The homonuclear $^{18}O^{18}O(^1\Delta_g)$ should have the same even-*J* propensity as the homonuclear $^{16}O^{16}O(^1\Delta_g)$, since ^{18}O also has zero nuclear spin. This is exactly the behavior we observe when we record photofragment spectra of $^{16}O^{16}O(^1\Delta_g)$, $^{16}O^{18}O(^1\Delta_g)$, and $^{18}O^{18}O(^1\Delta_g)$ from the 266-nm photolysis of O_3 containing equal ^{16}O and ^{18}O abundances. Part of such a CARS $O_2(^1\Delta_g)$ spectrum is presented in Figure 14, and it shows the even-*J* propensity in the bands due to $^{16}O^{16}O$ and $^{18}O^{18}O$, but not in the bands due to $^{16}O^{18}O$. But there is even more compelling evidence that can be gathered from this experiment.

Because we know the ^{16}O and ^{18}O abundances (1:1) in the ozone sample we know what the relative abundances of $^{16}O^{16}O(^1\Delta_g)$, $^{16}O^{18}O$ $(^1\Delta_g)$, and $^{18}O^{18}O(^1\Delta_g)$ should be in the photofragments. With this knowledge we can tell whether the lower probability of forming $^{16}O^{16}O(^1\Delta_g)$ with odd rotational quantum number *J* really reflects the selective depletion that we expect. If it does, all the rotational states of $^{16}O^{18}O(^1\Delta_g)$ should behave like the odd-*J* rotational states of $^{16}O^{16}O(^1\Delta_g)$ and ^{18}O $^{18}O(^1\Delta_g)$, since all these states are depleted by the curve crossing. Therefore, after correcting for the different statistical probabilities of forming $^{16}O^{18}O(^1\Delta_g)$, $^{16}O^{16}O(^1\Delta_g)$, and $^{18}O^{18}O(^1\Delta_g)$ the rotational state population distribution of $^{16}O^{18}O(^1\Delta_g)$—both even-*J* and odd-*J*—in a given vibrational state should be the same as the odd-*J* rotational state population distributions of $^{16}O^{16}O(^1\Delta_g)$ and $^{18}O^{18}O(^1\Delta_g)$ in that same vibrational state. The rotational state population distributions presented in Figure 15 show that this is precisely what we find. After correcting for the two-times greater statistical probability of forming $^{16}O^{18}O(^1\Delta_g)$ we see that the $^{16}O^{18}O(^1\Delta_g)$ rotational state populations in $v = 0$ are the same as the odd-*J* rotational state populations for $^{16}O^{16}O$ $(^1\Delta_g)$ in v = 0. This is very strong evidence for attributing the even-*J* propensity that we observe in the $^{16}O^{16}O(^1\Delta_g)$ photofragments to the selective depletion of odd-*J* states by the curve crossing in the photodissociation exit channel.

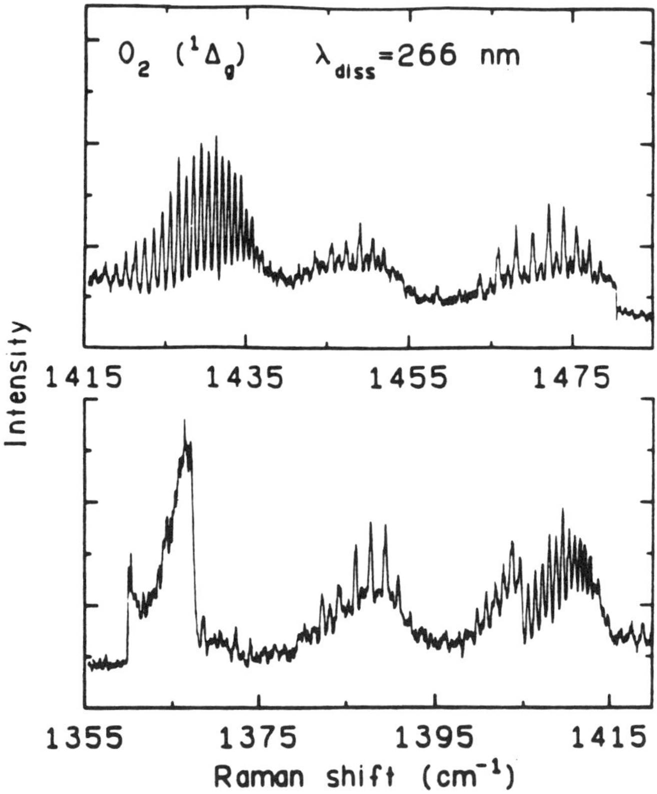

Figure 14. A portion of the vibrational Q-branch CARS spectrum of $O_2(^1\Delta_g)$ from the 266-nm photolysis of O_3 containing equal abundances of ^{16}O and ^{18}O. The spectrum between 1437 cm^{-1} and 1477 cm^{-1} is assignable to the $^{16}O^{16}O$ isotopomer, while the bands between 1405 cm^{-1} and 1437 cm^{-1} are from $^{16}O^{18}O$, and the spectrum between 1357 cm^{-1} and 1395 cm^{-1} is due to $^{18}O^{18}O$. From Ref. 27.

This observation of the different behavior of the heteronuclear ^{16}O $^{18}O(^1\Delta_g)$ as compared with the homonuclear $^{16}O^{16}O(^1\Delta_g)$ and $^{18}O^{18}O(^1\Delta_g)$, has a very important implication. It indicates that in a sample having the O isotopes present in their natural abundances the curve crossing should lead to isotope fractionation between the $O_2(^1\Delta_g)$ and $O_2(^3\Sigma_g)$ photoproducts, since only *half* the homonuclear $^{16}O^{16}O(^1\Delta_g)$ photoproducts are subject to depletion by the curve crossing to $O_2(^3\Sigma_g)$ while *all* the heteronuclear $^{16}O^{18}O(^1\Delta_g)$ photoproducts suffer depletion as a

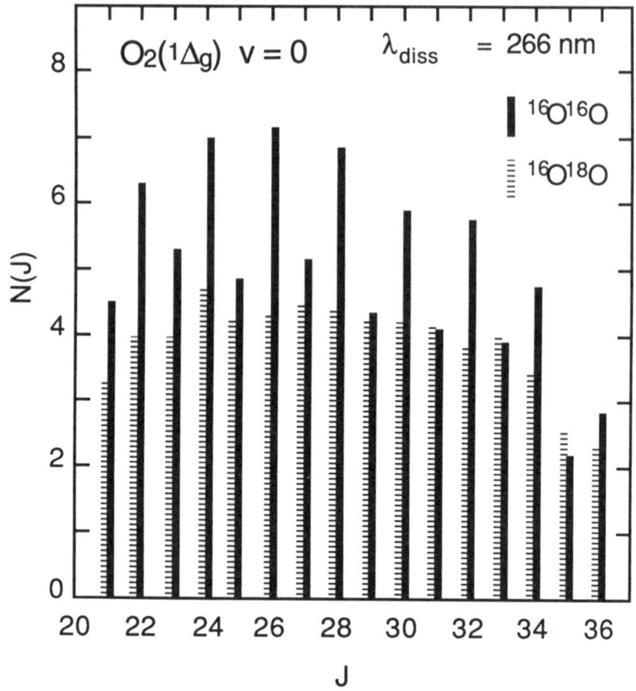

Figure 15. Quantum state population distributions for the $^{16}O^{16}O$ and $^{16}O^{18}O$ photofragments produced in the 266-nm dissociation of an O_3 sample containing equal abundances of ^{16}O and ^{18}O. The $^{16}O^{18}O$ populations have been divided by two to correct for the two-times greater statistical abundance of this species relative to $^{16}O^{16}O$. From Ref. 27.

result of it. (The 0.2% natural abundance of ^{18}O implies that the probability of having $O_2(^2\Delta_g)$ photofragment with two ^{18}O nuclei is so small that we need not consider the $^{18}O^{18}O(^1\Delta_g)$ species.) For $^{16}O^{16}O$ the $^3\Sigma_g$ quantum yield is 0.15, as already noted above, while for $^{16}O^{18}O$ the implied $^3\Sigma_g$ quantum yield is 0.30, so the $O_2(^3\Sigma_g)$ should be enriched in ^{18}O by a factor of two! This enrichment is not only large, it also should be mass independent, since the $O_2(^3\Sigma_g)$ will be enriched by the same factor of two in the ^{17}O isotope, through depletion of $^{16}O^{17}O(^1\Delta_g)$ by the curve crossing. We cannot verify this by direct experimental measurement, because the $O_2(^3\Sigma_g)$ photofragments are spread out over so many quantum states that the population in any one state is too low to be detectable, even for the $^{16}O^{16}O$ isotopomer. However, the indirect evidence for this isotope fractionation from our $O_2(^1\Delta_g)$ photofragment spectra is very strong indeed.

This mass-independent isotope fractionation is significant, because traditionally it has been believed that purely chemical mechanisms cannot effect mass-independent fractionation.[29] For example, in cosmochemistry mass-independent isotope fractionation is interpreted as a sign of non-chemical, that is nuclear phenomena, and is used extensively, to formulate hypotheses about the formation of the solar system.[30-32] Recently, however, Thiemens and co-workers[33-35] have shown that production or consumption of oxygen-containing species by electric discharge, thermal decomposition, and photochemical means can lead to mass-independent fractionation. The mass-independent fractionation implied by our experiments on ozone photodissociation should not be limited either to photodissociation of ozone or to oxygen isotopes. In fact, I have outlined a general mechanism for mass-independent fractionation in electronically non-adiabatic collisions that involve a curve-crossing between a non-Σ electronic state and a Σ electronic state.[36] The photodissociation of ozone is but one example that involves such an electronically non-adiabatic process. This mechanism clearly shows how purely chemical processes do effect mass-independent isotoptic separation, and suggests that chemical processes do effect mass-independent isotope separation.

V. WHEN EXPERIMENTS ALONE ARE NOT ENOUGH

When a single dynamical feature or constraint dominates some part of the state-to-state dynamics of a reaction or photodissociation we expect to be able to identify and characterize that dynamical feature with little if any ambiguity. Such is the case for the state-to-state measurements described in the preceding two sections. The rotational state dependence on the $O_2(^1\Delta_g)$ + $O(^1D)$ / $O_2(^3\Sigma_g)$ + $O(^3P)$ curve crossing in the UV photolysis of O_3 is controlled by nuclear exchange symmetry restrictions. The rotational state distribution of the $O_2(^3\Sigma_g)$ fragment in the visible photolysis of O_3 is dominated by the effect of the parent rotation on the C_{2v} symmetry breaking.

However, the state-to-state dynamics of most reactions and photodissociations are not dominated by a single dynamical constraint.

Rather, there are typically several factors at work that in concert determine the quantum state distribution of the products given a specific quantum state of the reactants. This presents one of the biggest challenges in the interpretation of the results of a state-to-state experiment. In general we do not expect the "deconvolution" of these different dynamical influences to be easy using the results of state-to-state experiments alone. In such cases computational experiments—dynamical simulations of the collision dynamics—done together with the laboratory experiments can provide the extra insight needed to interpret effectively the results of the state-to-state measurements. It is the purpose of this section to describe such a combination of computational and laboratory experiments that we have recently carried out[37–39] in order to show the effectiveness of such an approach.

The object of this combined experimental/computational study is the homologous series of hydrogen atom abstraction reactions

$$H + HX \rightarrow H_2 + X, \tag{8}$$

where X is Cl, Br, or I. Due to the large difference between the mass of the hydrogen atom and the masses of any of the halogen atoms the reduced mass of the reactants, the reduced mass of the products, and the skew angle[40] are effectively independent of the identity of X. The reactant reduced masses are 0.97, 0.99, and 0.99 for X = Cl, Br, and I respectively, while for the $H_2 + X$ products we have 1.89, 1.95, and 1.97, respectively. The skew angles—45.8°, 45.4°, and 45.2°, respectively are all very close to the limiting 45° characteristic of the prototypical light-light-heavy mass combination that they approximate. In contrast, the energetics of these three reactions are quite different, as a result of the significant differences in HX bond strength. All three reactions are exoergic, but by greatly different amounts, 0.05 eV, 0.72 eV, and 1.42 eV for H + HCl, H + HBr, and H + HI, respectively. These widely differing exoergicities will be associated with widely differing reaction barrier heights. The DIM-3C potential energy surfaces of Baer and Last[41] have barrier heights of 0.22 eV, 0.08 eV, and 0.03 eV for X = Cl, Br, and I, respectively, for the collinear minimum energy path.

The state-to-state dynamics laboratory experiments done for these three reactions are measurements of the rotational and vibrational state distributions and reaction partial cross sections at a collision energy of 1.6 eV, starting with the reactant diatom having a 300K thermal distribution of rotational states.[37,38] Some of these results are shown in

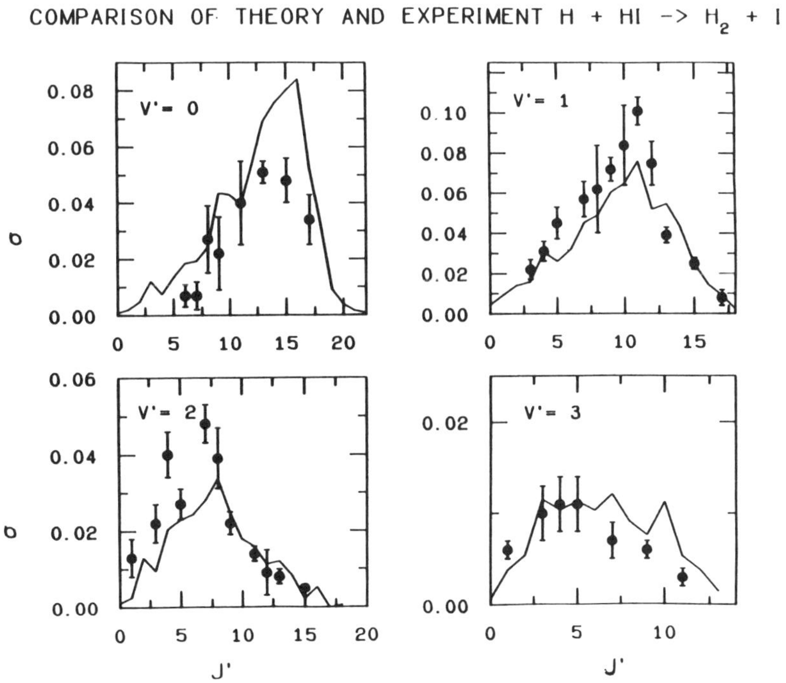

Figure 16. Comparison of quantum state population distributions for the H + HI → H_2 + I reaction at 1.6 eV collision energy. The points with error bars are the experimental measurements, while the solid lines give the distribution derived from QCT calculations. The populations are presented in terms of the relative cross section into H_2 with v',J'. From Ref. 39.

Figures 16 to 18. The state distributions reveal highly specific product quantum state production in the reactions. Products appear in only a fraction of the energetically allowed quantum states, and the distributions for each reaction are distinct and characteristic. However, despite the detail of the measurements their interpretation is not simple. There are many "Why?" questions about the results that we can pose but not answer based on the measurements alone.

To answer these questions we really need to do more detailed experiments, such as selecting reaction geometries and impact parameters, that are possible in computational experiments but not in laboratory experiments. That is, we need to do computational experiments that simulate the reaction dynamics. Our simulations involve classical

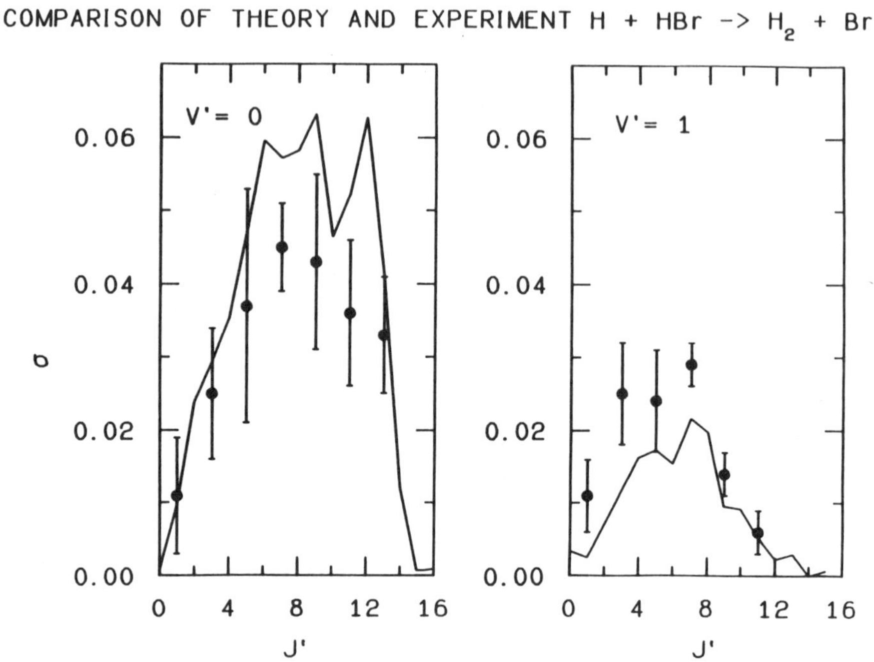

Figure 17. Comparison of quantum state population distributions for the H + HBr → H_2 + Br reaction at 1.6 eV collision energy. The points with error bars are the experimental measurements, while the solid lines give the distribution derived from QCT calculations. The populations are presented in terms of the relative cross section into H_2 state, v',J'. From Ref. 39.

trajectory calculations, more correctly quasi-classical trajectory calculations since the dynamics are treated classically, but the initial states are selected from a quantum-mechanically dictated distribution over reactant internal motions, and the final states are binned to "quantize" the product rotation and vibration. To do the calculations we need potential energy surfaces, and for this we have used approximate, but realistic LEPS and DIM-3C potentials for the H + HX → H_2 + X reactions.[39]

Simulations often illuminate general trends in reaction dynamics, telling us what the dynamics would be like if the potential energy surface used in the computations were the "real" one. But here we want to use the simulations to tell us why the real H + HX → H_2 + X

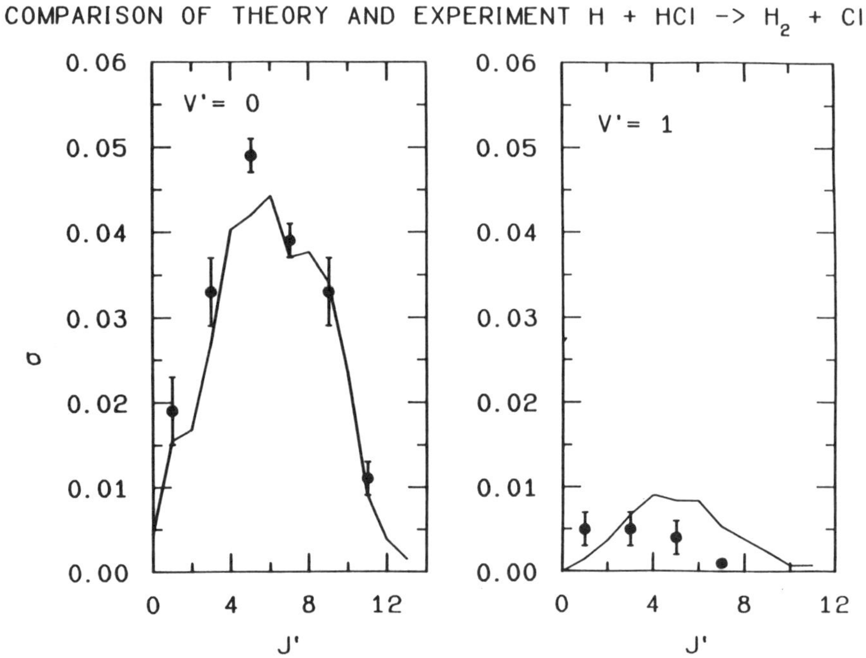

Figure 18. Comparison of quantum state population distributions for the H + HCl → H_2 + Cl reaction at 1.6 eV collision energy. The points with error bars are the experimental measurements, while the solid lines give the distribution derived from QCT calculations. The populations are presented in terms of the relative cross section into H_2 state v', J'. From Ref. 39.

reactions behave the way they actually do. The simulations are of value in this regard only if the behavior they predict is exactly the behavior that the real H + HX → H_2 + X systems display. This is how the integration of laboratory and computational experiments develops. Since our state-to-state dynamics experiments yield very detailed information on how the real system dynamically behaves these laboratory experiments can be used to establish the accuracy of the computational experiments. Then the simulations can be used to provide insight about the dynamical behavior revealed by the laboratory measurements. From the very good agreement of the QCT calculated and experimentally measured product quantum state distributions shown in Figures 16 to 18 it seems that the simulations are providing a faithful

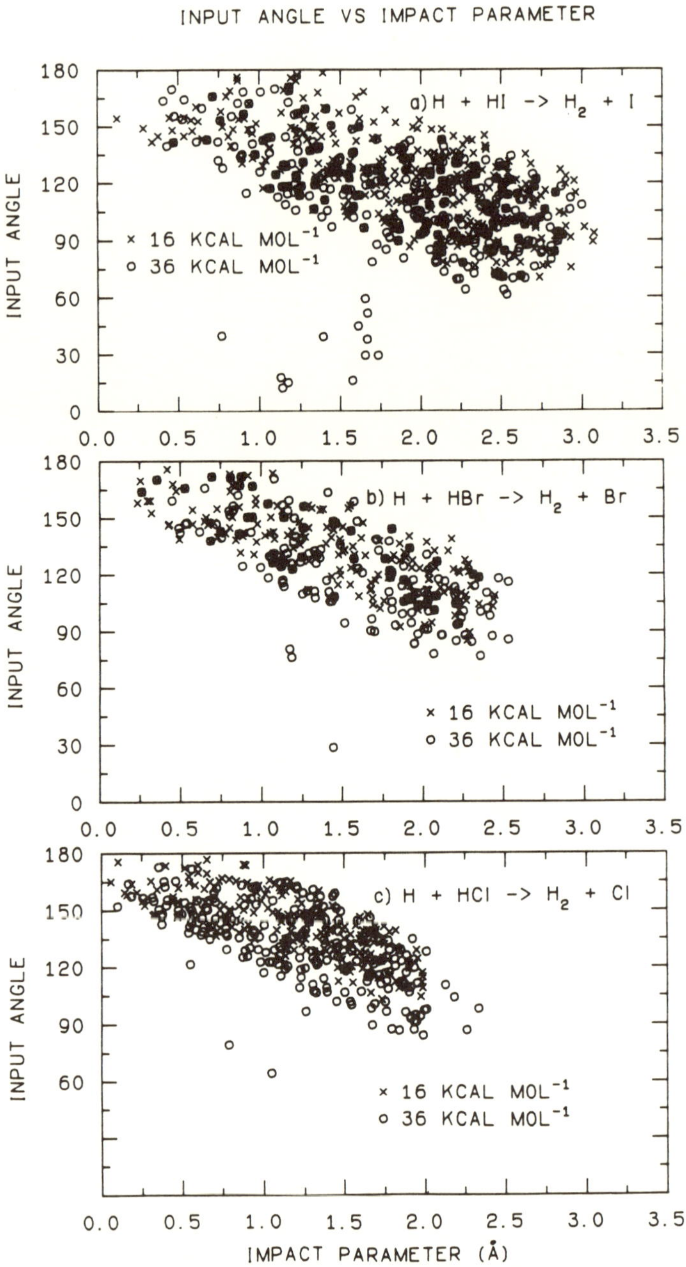

Figure 19. Scatter plot of input angle (H-H-X angle) versus impact parameter for reactive trajectories H + HX $\rightarrow$ H_2 + X at 1.6 eV relative energy, showing the strong correlation between the angle and the impact parameter for trajectories that successfully react. From Ref. 39.

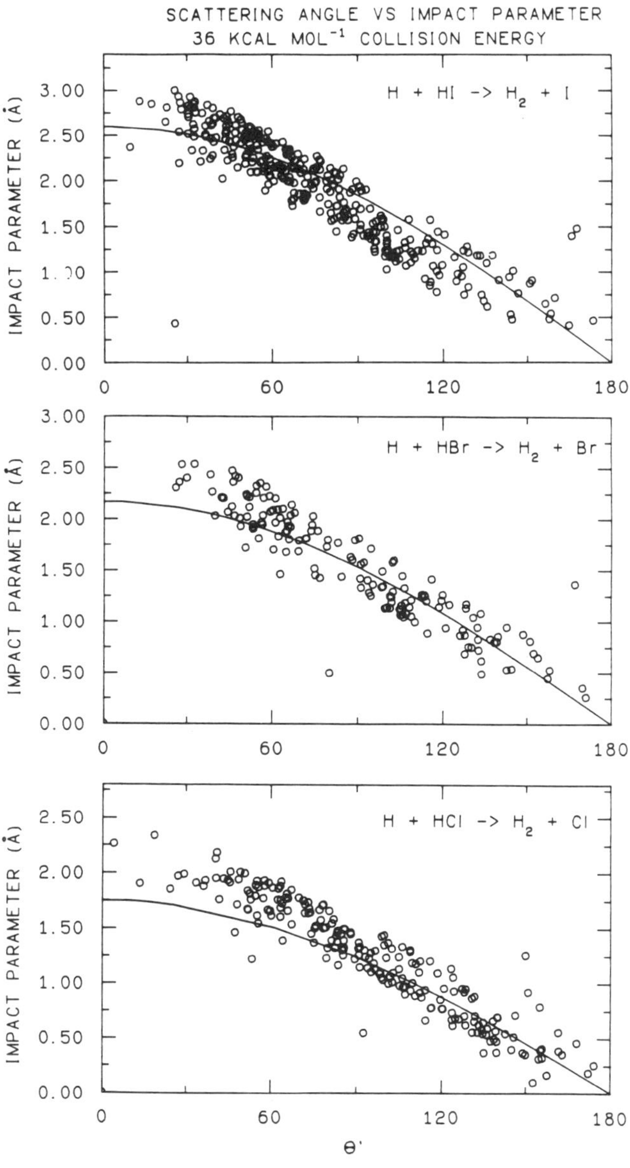

Figure 20. Scatter plot of impact parameter versus H_2 product scattering angle (180° corresponds to exactly backward scattering) for reactive trajectories H + H$X \rightarrow H_2 + X$ at 1.6 eV relative energy. The solid line gives the scattering angle function expected for hard-sphere scattering with a sphere of radius equal to one-half the H-H distance plus the H-X distance, both evaluated at the saddle point of the potential energy surface used in the trajectory calculations. From Ref. 39.

description of the H + $HX \rightarrow H_2 + X$ reaction dynamics, at least in so far as we can experimentally determine.

What kind of dynamics do the QCT simulations reveal? If forced to condense the results of 300,000 trajectories into a single sentence, I would say that the dynamics of these reactions can be described by the interaction of the attacking H atom with the H-atom-end of the *HX* to form an H_2 molecule, an H + H $\rightarrow H_2$ collision as it were, which is followed by an $H_2 + X$ collision in which this nascent H_2 molecule scatters nearly elastically off the large and heavy (relative to H_2) halogen atom. To elaborate that behavior I show in Figure 19 a scatter plot of the H-H-*X* angle versus the H + *HX* impact parameter for those trajectories that lead to reaction. There is a strong correlation between the impact parameter, *b*, and the H-H-*X* angle, Θ: reaction at small impact parameters occurs only for near-linear H-H-*X* (Θ = 180°), and at large impact parameters only for nearly L-shaped geometries (Θ = 90°). The shape of the correlation and the scatter in the correlation between *b* and Θ quantitatively imply that reaction is possible if and only if the two hydrogen atoms come within a distance that is roughly equal to the H-H bond distance at the saddle point, which is 0.95 Å for HHC1, 1.13 Å for HHBr, and 1.99 Å for HHI. The value of the H-H-*X* angle simply determines what range of values of the H + *HX* impact parameter satisfies this requirement.

Our description of the reaction as involving the nearly elastic scattering of this nascent H_2 product off the halogen atom is derived from another correlation. The data of Figure 20 show that the scattering angle for a given impact parameter is very nearly that expected classically for the scattering by a hard sphere with radius given by one-half of the H-H bond distance plus the H-*X* bond distance (using values of the bond distances at the saddle point of the potential energy surface used in the calculations). The points in the figure represent the outcome of reactive trajectories, and the solid lines give the result expected for the scattering by this sphere.

This hard-sphere-like deflection function does not imply an isotropic distribution of the H_2 product, because the probability of reaction is not independent of *b*. In fact, the opacity function, the probability of reaction as a function of *b*, is a very strong function of the impact parameter, as Figure 21 shows. The solid curves in Figure 21 represent the purely geometric opacity function—the reaction probability expected if reaction at a given value of *b* is determined only by the probability that the two hydrogen atoms will come within the critical H-H distance (see above) in a collision at this value of *b*. We call this

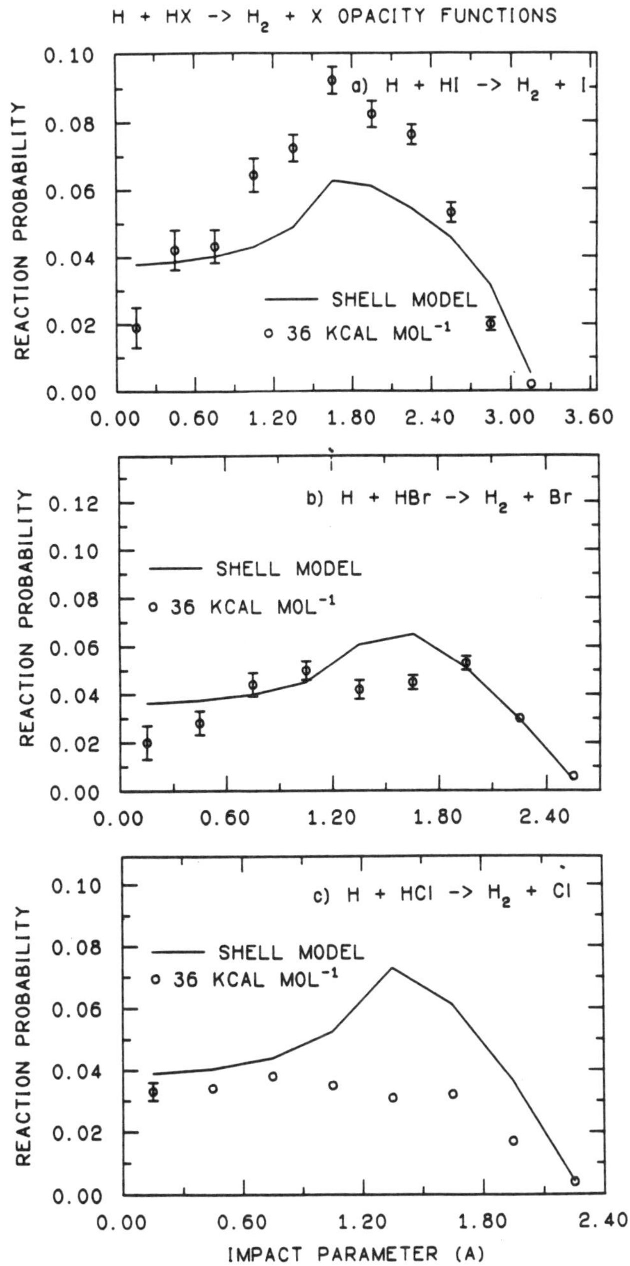

Figure 21. Opacity function for the H + H*X* → H_2 + *X* reactions at 1.6 eV relative energy. The points with error bars are the results of the QCT calculations, while the solid lines give the opacity function expected if only geometric factors influenced the reaction probability. See text for details. From Ref. 39.

the shell model, because the geometric probability distribution of the hydrogen atom around the halogen atom in H*X* is a uniform shell of inner radius equal to the H-*X* saddle point bond distance and thickness equal to the H-H bond distance at the saddle point. Deviations from the shell model reflect dynamical biases for reaction over and above the geometric bias that favors reaction at larger impact parameter. These will not be discussed here. The interested reader is referred to our paper describing in full these QCT calculations.[39]

These representations of the dynamics are somewhat awkward to state because the impact parameter in its conventional sense—the distance between the relative velocity vector and a line parallel to it passing through the center-of-mass of the H*X*—is not really a very convenient parameter to categorize those collisions that lead to reaction to form $H_2 + X$. Except for the scattering angle we find that an unconventional definition of the impact parameter is more appropriate—the distance between the relative velocity vector and a line parallel to it passing through the H atom on the H*X*. We will call this impact parameter b_{HH} to distinguish it from the conventional one. Given the discussion so far this impact parameter seems entirely natural, and it is, since it is the H-H distance that really determines reaction probability. Figure 22 shows just how important the characterization of the collisions in terms of the H-H impact parameter is.

This figure reveals a correlation between the H_2 product rotational angular momentum observed for a reactive trajectory and the H-H orbital angular momentum implied ($l_{HH} = \mu_{HH} \times v \times b_{HH}$) by the value of b_{HH} for that reactive trajectory. This remarkable correlation is quite good for H + HI and H + HBr, although a bit less so for H + HCl. The specific quantitative correlation we observe is even more remarkable: we find that $l_{HH} \approx J'$—that is, the angular momentum of the H-H relative motion is almost conserved. Initially this angular momentum is H-H orbital angular momentum (l_{HH}) while asymptotically in the product channel it is H_2 rotation (J'). This means that H-H motion is roughly decoupled from the other angular momenta in the collision. This is just another reason why we describe the reaction as a reactive encounter between two H atoms, with the *X* almost uninvolved, followed by a nearly elastic collision of the nascent H_2 with the *X*. The latter is unavoidable because of the large size of the halogen atoms.

There are many other details of the dynamics of the $H + HX \rightarrow H_2 + X$ reaction that are revealed by the QCT simulations. However, I will refrain from discussing them, for my purpose in presenting these results is not to tell a complete story of the dynamics of this reaction.

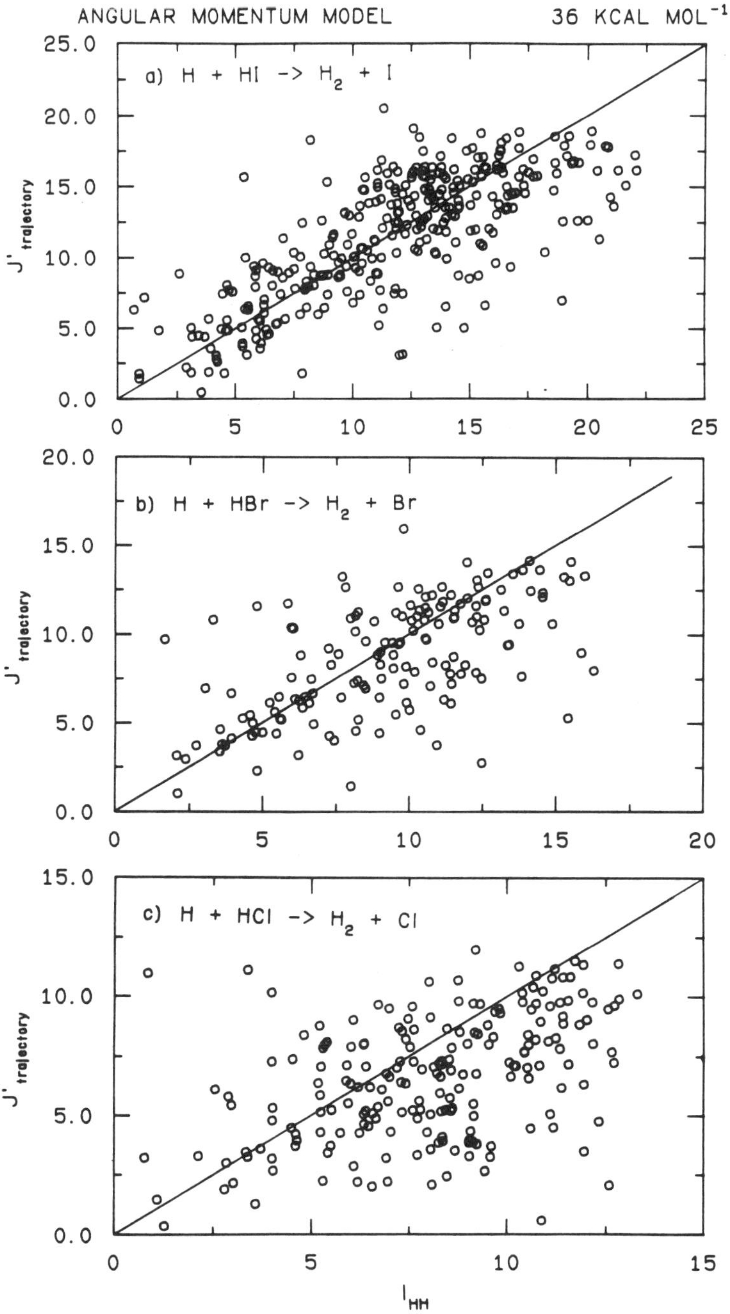

Figure 22. Scatter plot of H_2 product rotational angular momentum versus H-H orbital angular momentum for reactive trajectories H + H*X* → H_2 + *X* at 1.6 eV relative energy. The line represents $J'_{H_2} = 1_{HH}$. See text for details. From Ref. 39.

The important point to be made has already been made: that even without some unusually strong and dominant feature in the measured state-to-state cross sections such measurements can be used to extract very detailed insight about the intimate dynamics of the reaction. Perhaps the degree of certainty surrounding conclusions drawn with the aid of simulations is not perceived to be as high as that for conclusions derived from an interpretation of experimental results alone. However, the experimentally measured product state distributions provide a very detailed test of the QCT simulations. When the calculations give distributions that are in very good agreement with the measurements, as they do for the H + $HX \rightarrow H_2 + X$ reactions, we have every reason to believe that they describe the real system with high fidelity.

VI. SUMMARY

At this point I hope that the reader is saying to himself, "Yes, we really do need these state-to-state dynamics experiment!" For most of the reaction and photodissociation systems that I have selected for discussion the information gathered from the state-to-state experiments probably could not have been obtained from any other type of experiment. Yet the insight gained from these experiments has importance that goes far beyond the domain of state-to-state processes. The symmetry breaking in dissociating molecules, the nuclear symmetry restrictions in non-adiabatic collisions, the dynamical behavior of the H + $HX \rightarrow H_2 + X$ reactions all have broad implications.

The mass-independent isotopic fractionation that occurs in the photodissociation of ozone is not restricted to either photodissociation or oxygen isotopes. In fact, it is generally applicable to many electronically non-adiabatic processes and many different elements. The isotopic fractionation depends on the exchange symmetry between identical nuclei in a molecule, and the special restrictions that exchange symmetry imposes on the overall wavefunction when the nuclei are of zero nuclear spin. Many atoms have common isotopes with zero nuclear spin—all those having both even atomic number and even atomic weight—and include O, S, Mg, and Si. Whenever a molecule containing two or more identical O, S, Mg, Si, and so on, atoms is involved in a curve crossing from a non-Σ electronic state to a Σ electronic state isotopic fractionation will inevitably occur by the same mechanism that we discovered in the uv photodissociation of ozone.

The curve crossing rate will be twice as great when one of the O, S, Mg, Si, and so on, atoms in the molecule is an isotope other than the common one, and this will lead to enrichment of the Σ state in the rare isotopes and enrichment of the non-Σ state in the common isotope. Any chemistry—reaction, energy transfer, photodissociation—that distinguishes between the two electronic states can make the isotopic fractionation permanent.

Many such processes are known to occur in nature, and many more that are not yet known must also surely occur. Whether known or unknown these processes are constantly effecting isotope fractionation without any dependence on the mass of isotopes. This is a phenomenon whose importance transcends state-to-state dynamics, but could only be discovered and understood from state-to-state dynamics experiments.

We could similarly describe the symmetry breaking induced by the rotation of the molecule prior to dissociation that we observe in the photodissociation of O_3. Symmetry and the breaking of symmetry are fundamental to all molecular processes. The specific observation we made—that molecular rotation can break the symmetry of AB_2 triatomic molecules—has implications for all bond–breaking processes. Our observations point to the fundamental influence of motions of the parent molecule that do not explicitly break the symmetry in the parent molecule, but nonetheless break the overall molecular symmetry when projected on the fragments of the parent molecule. This is a subtle but important point, and again one that probably would not have become evident had state-to-state dynamics measurements not been made.

The implications of the state-to-state dynamics characterization of H + $HX \rightarrow H_2 + X$ reactions are not as dramatic nor as far reaching, but nonetheless important. They show that the elementary constructs of reaction dynamics, the conventional impact parameter, steric effects, and energetic barriers, lose importance as interpretive concepts when the energy is well above that minimally necessary to achieve reaction. While many reactions are dominated by collisions with just enough energy to get over the energetic barriers, many more are not, and even the simplest aspects of the latter are not easily interpreted with the more conventional constructs. For example, the cross sections of the H + $HX \rightarrow H_2 + X$ reactions do not exceed about 3 Å^2 even when the collision energy is great enough to surmount the reaction barrier for any H-H-X orientation. This reaction cross section is more than an order of magnitude smaller than the collision cross section, and as a consequence the thermal rate constant will never become equal to the gas kinetic rate constant no matter how high the temperature.

ACKNOWLEDGMENTS

The research that made possible this review was carried out by enthusiastic graduate students and postdocs who made seemingly impossible experiments almost routine. Without them there would have been nothing for me to write about, and I thank them all: Pam Aker, Dan Gerrity, Geoff Germann, Harold Levene, Jong-Chen Nieh, and Dave Phillips. If their successors are as successful I will have many more chapters to write. The National Science Foundation, the U. S. Department of Energy, and the American Chemical Society Petroleum Research Fund have provided major funding for the research described here, and their support I also gratefully acknowledge.

REFERENCES

1. Valentini, J. J., "Coherent Anti-Stokes Raman Spectroscopy." In *Spectrometric Techniques,* Volume 4; Vanasse, G. A., Ed.; Academic Press: New York, 1985.
2. J. J. Valentini, "Laser Raman Techniques." In *Applications of Laser Spectroscopy;* Radziemski, L. J.; Solarz, R. W.; and Paisner, J. A., Eds.; Marcel Dekker: New York, 1986.
3. Moore, D. S.; Bomse, D. S.; Valentini, J. J. *J. Chem. Phys.* **1983,** *79,* 1745.
4. Levene, H. B.; Nieh, J. -C.; Valentini, J. J. *J. Chem. Phys.* **1987,** *87,* 2583.
5. Gerrity, D. P.; Valentini, J. J. *J. Chem. Phys.* **1984,** *81,* 1298.
6. Inn, E. C. Y.; Tanaka, Y. *J. Opt. Soc. Am.* **1953,** *43,* 870.
7. Bersohn, R. *J. Phys. Chem.* **1984,** *88,* 5145.
8. Simons, J. P. *J. Phys. Chem.* **1984,** *88,* 1287.
9. Green, C. H.; Zare, R. N. *Annu. Rev. Phys. Chem.* **1982,** *33,* 119.
10. Shapiro, M.; Bersohn, R. *Annu. Rev. Phys. Chem.* **1982,** *33,* 409.
11. Hay, P. J.; Dunning, Jr., T. H. *J. Chem. Phys.* **1977,** *67,* 2290.
12. Thunemann, K. -H.; Peyerimhoff, S. D.; Buenker, R. J. *J. Mol. Spectrosc.* **1978,** *70,* 432.
13. Jones, R. O. *J. Chem. Phys.* **1985,** *82,* 325.
14. Fairchild, C. E.; Stone, E. J.; Lawrence, G. M. *J. Chem. Phys.* **1978,** *69,* 3632.
15. Okabe, H. *Photochemistry of Small Molecules;* Wiley: New York, 1978.
16. Tanaka, T.; Morino, Y. *J. Mol. Spectrosc.* **1970,** *33,* 538.
17. Herzberg, G. *Spectra of Diatomic Molecules,* 2nd ed.; Van Nostrand Reinhold: New York, 1960.
18. Busch, G. E.; Wilson, K. *J. Chem. Phys.* **1972,** *56,* 3626.
19. Levene, H. B.; Valentini, J. J. *J. Chem. Phys.* **1987,** *87,* 2594.
20. Hughes, R. H. *J. Chem. Phys.* **1956,** *24,* 131.

21. World Meteorological Organization Global Ozone Research and Monitoring Project, Report No. 16, 1985.
22. Jet Propulsion Laboratory, California Institute of Technology, Publication No. 85-37, 1985.
23. Prinn, R. G.; Alyea, F. N.; Cunnold, D. M. *Annu. Rev. Earth Planet Sci.* **1978,** *6,* 43.
24. Schiff, H. I. *Ann. Geophys.* **1972,** *28,* 67.
25. Hay, P. J.; Pack, R. T.; Walker, R. B.; Heller, E. J. *J. Phys. Chem.* **1982,** *86,* 862.
26. Valentini, J. J. *Chem. Phys. Lett.* **1983,** *96,* 395.
27. Valentini, J. J.; Gerrity, D. J.; Phillips, D. L.; Nieh, J. -C.; Tabor, K. D. *J. Chem. Phys.* **1987,** *86,* 6745.
28. Brock, J. C.; Watson, R. T. *Chem. Phys. Lett.* **1980,** *71,* 371.
29. Bigeleisen, J.; Wolfsberg, M. *Adv. Chem. Phys.* **1958,** *1,* 16.
30. Audouze, J. *Comments Astrophys. Space Phys.* **1984,** *10,* 105.
31. Schramm, D. N. In *Nucleosynthesis;* Arnet, W. D.; Truran, J. W., Eds.; University of Chicago: Chicago, 1985.
32. Clayton, R. N.; Grossman, L.; Mayeda, T. K. *Science* **1973,** *182,* 485.
33. Heidenreich III, J. E.; Thiemens, M. H. *J. Chem. Phys.* **1983,** *78,* 892.
34. Heidenreich III, J. E.; Thiemens, M. H. *Geochim. Cosmochim. Acta* **1985,** *49,* 1303.
35. Heidenreich III, J. E.; Thiemens, M. H. *J. Chem. Phys.* **1986,** *84,* 2129.
36. Valentini, J. J. *J. Chem. Phys.* **1987,** *86,* 6757.
37. Aker, P. M.; Germann, G. J.; Valentini, J. J. *J. Chem. Phys.* **1989,** *90,* 4795.
38. Aker, P. M.; Germann, G. J.; Tabor, K. D.; Valentini, J. J. *J. Chem. Phys.* **1989,** *90,* 4809.
39. Aker, P. M.; Valentini, J. J. *Israel J. Chem.* **1990,** *30,* 157.
40. Levine, R. D.; Bernstein, R. B. *Molecular Reaction Dynamics and Chemical Reactivity;* Oxford: London, 1987.
41. Last, I.; Baer, M. In *Potential Energy Surfaces and Dynamics Calculations;* Truhlar, D. G., Ed.; Plenum: New York, 1981.

THE MEASUREMENT OF THERMAL BIMOLECULAR RATE CONSTANTS BY THE FLASH PHOTOLYSIS-SHOCK TUBE (FP-ST) TECHNIQUE:
COMPARISON OF EXPERIMENT TO THEORY

J. V. Michael

Advances in Chemical Kinetics and Dynamics, Volume 1, pages 47–112

ISBN: 1-55938-161-2

I. INTRODUCTION

The measurement of thermal gas phase reactions has been an active subject of research in physical chemistry for over seventy years. The motivation for doing such work in the 1920s and 1930s was almost entirely theoretically inspired. There had to be fundamental physical reasons for the reactivity between two species in a bimolecular interaction. As the richness of the chemistry was realized in naturally occurring environments such as combustion or atmospheric chemistry, chemical kinetics research was increasingly motivated by practical considerations. Today most research work in thermal gas phase chemical kinetics is carried out in support of these practical systems. At the same time, the interest in chemical interactions has become increasingly detailed. The newer experimental studies have concentrated on state-to-state techniques. In fact, the newer results in some instances are so detailed that the gap between theoretical interpretation and experiment has substantially widened even though the techniques used in theoretical investigations are well advanced over those used just a few years ago. With these marvelous developments in both experiment and theory, research workers simply have been caught up in ever increasing specializations in their particular subdisciplines in the general field of chemical dynamics. The end result is that the field has lacked a general focus. Due to the research support structure, it is possible that experimental dynamicists and/or kineticists may not be clear as to whether any progress has been made in the theoretical understanding of bimolecular interactions, and, as a result, experimentalists might totally neglect theory. Conversely, theoreticians might also see little value in modeling studies of complex chemical kinetics systems and likewise neglect this typc of study as being of trivial importance.

Central to all subdisciplines in chemical dynamics is the thermal rate constant investigation. It is the opinion of the author that this type of investigation should continue to hold this central role in the general field. The importance of this work should not be deemphasized since the results have implications for the more specialized studies. Most state-to-state dynamicists consult the thermal rate constant literature before attempting studies on a given bimolecular reaction. The same is true for theoreticians because sophisticated theoretical descriptions require a great deal of effort that could be wasted without some prior information about a given reaction. And chemical modelers of practical systems simply must include well-determined thermal rate constants in chemical mechanisms if they hope to simulate real systems.

The theme of the present review will be to reiterate the importance of the thermal rate constant study from both the practical and theoretical points of view. The reactions chosen for review have been studied by the Flash Photolysis-Shock Tube (FP-ST) technique, and this technique will be discussed in detail. Then a discussion of each of the chemical reactions that have been studied by the technique will follow. With the exception of two reactions, all of the processes that have been studied are H- or D-atom abstractions. Processes of this type are dominated by quantum mechanical tunneling at lower temperatures. The FP-ST technique can give absolute values for rate constants from $\sim$700–2500 K, and in this region of temperature, quantum mechanical tunneling becomes increasingly less important. Hence, the results that are obtained constitute a sensitive and direct probe of the relative energetics on the potential surface for the bimolecular interaction. Since the reactivity is entirely determined by the potential energy, a discussion will be given in each case of the accuracy of the ab initio electronic calculations that have been made. Then theoretical calculations that are mostly based on transition state theory will be presented and compared to the known data. Two addition-elimination cases will be considered first. These will be followed by six H-atom abstraction reactions, three of which are initiated by reaction with O-atoms and three by reaction with H (or D)-atoms.

II. EXPERIMENTAL FP-ST TECHNIQUE

A. Historical Development

The Flash Photolysis-Shock Tube technique was first used by Burns and Hornig in 1960 in their investigation of the recombination of Br atoms.[1] In this application, Br_2 was flash photolyzed prior to shock heating, and then the temporal reformation of Br_2 molecules was followed spectroscopically in the incident shock wave regime. The idea was not exploited again until the late 1970s when Zellner and coworkers[2] designed an apparatus that utilized flash photolysis in the hot stagnant gas region following reflected shock wave heating. The method of observation was OH resonance absorption spectroscopy, and therefore, some reactions of OH-radicals with molecules were successfully studied. This latter application is simply an extension of an earlier experimental technique called the Flash Photolysis-Resonance Absorption (FP-RA) technique. In the FP-RA technique, the gaseous reaction mixture is generally introduced into a static vessel at a known density and temperature, and then a suitable source molecule of known mole fraction is flash photolyzed thereby producing a species of interest. This species is then followed spectroscopically as it reacts with some added reactant molecule. In the application of Zellner and coworkers, the shock tube in the reflected wave mode serves simply as a fast furnace for heating a gaseous mixture of known initial composition and density to some calculable final thermodynamic state. The only additional information necessary is a measurement of the shock wave velocity.

The extension of the method to atomic absorption occurred relatively recently.[3] To be sure, the technique of atomic absorption and its complement, atomic fluorescence, had been extensively used in chemical kinetic investigations in a large number of laboratories. The earliest uses of the atomic resonance absorption spectroscopic (ARAS) method for monitoring atomic concentrations in a chemical kinetic experiment were by Meyerson et al.[4] and Michael and Weston[5] with the applications being carried out in shock tube and flow tube experiments, respectively. There have been notable further advances in the use of the ARAS technique in shock tube experiments principally by Just and coworkers,[6] Skinner and coworkers,[7] Roth and coworkers,[8] and Yumura and Asaba.[9] However, in all of these shock tube studies, the initiation of chemical reactions has been by thermal decomposition.

B. Shock Tube Methods

A schematic diagram of the FP-ST apparatus is shown in Figure 1. The method utilizes a standard shock tube design[10] in which a high pressure driver section is separated from a low pressure driven section by a diaphragm. This diaphragm can be reproducibly ruptured in order

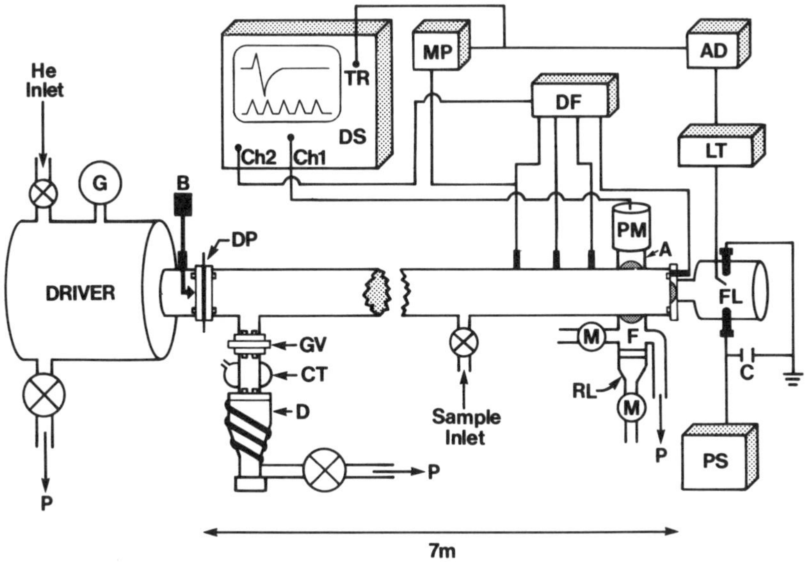

Figure 1. Schematic diagram of the apparatus. P = rotary pump; D = oil diffusion pump; CT = liquid nitrogen baffle; GV = gate valve; G = bourdon gauge; B = breaker; DP = diaphragm; T = pressure transducers; FL = flash lamp; C = capacitor; PS = power supply; M = microwave power supply; F = atomic filter; RL = resonance lamp; A = gas and crystal window filter; PM = photomultiplier; DS = digital oscilloscope; MP = master pulse generator; TR = trigger pulse; DF = differentiator; AD = delayed pulse generator; LT = lamp trigger transformer.

to produce consistent shock strength conditions. In the recent FP-ST applications[3,11] the equipment has been designed with the mechanical diaphragm option (rather than using a pressure break method with a scored diaphragm) because low shock strengths, giving low temperatures, were anticipated. Following the work of Zellner et al,[2] the reflected shock regime is the preferred mode of operation. Therefore, it is important that the diaphragm material does not shatter thereby leaving debris in the shock tube after an experiment. The material chosen for use, a 0.004 in. tempered aluminum sheet (1100-H18), has proven to be ideal in this regard.

The shock tubes in the recent FP-ST applications[3,11] have been fabricated from 304 stainless steel cylindrical tubing. The driver to driven volume ratios are high so that uniform shock waves can be generated over the long length, ~6–7 m, of the driven section. In the newest apparatus[11] the driven section is pumped with a "Santovac 5" charged oil diffusion pump, and the inside surface has been polished to a 0.2 μm finish. In addition, all major flanges are constructed from Varian-type "Conflat" vacuum flanges. These features combine to allow for low ultimate pressures of $<10^{-8}$ torr, and leak plus outgassing rates of $\sim 5 \times 10^{-6}$ torr/min. The turn around time between experiments is ~20 minutes.

Shock velocities are measured with a train of fast response pressure transducers that are spaced at equal intervals along the shock tube. Therefore, acceleration or attenuation in the wave speed can be assessed for a given experiment. In general, if the shock tube is long enough and the driver volume is large enough, the wave velocity is constant well before it reaches the endplate. The values of the velocity from the various intervals are analyzed by least square analysis, and the resultant mean value and standard deviation are used to calculate the thermodynamic state of the hot gas in the stagnant region of the reflected shock wave regime. The typical accuracy of a given shock velocity measurement is ~0.5–1.0% giving a 1–2% uncertainty in the derived temperature.

Since the use of low shock strengths was desired, it has been necessary to characterize the shock wave behavior. Boundary layer effects are a well-recognized complication in shock wave experiments,[12] and correction procedures have been suggested for both incident[13–15] and reflected[16–18] modes of operation. Before the ARAS application of the FP-ST method was used, diagnostic pressure jump experiments were performed.[11,18] These have been compared to a theoretical model[18] that assesses the perturbation from ideal shock wave behavior that is caused by a forming boundary layer in the incident regime as it interacts with the reflected shock wave. The effect, from an hydrodynamic point of view, is negligible; however, failure to include the effect in chemical kinetics experiments can cause errors in activation energies of 5–10%.[19] In all of the cases presented here, the corrections for non-ideal behavior due to boundary layer formation have been applied.

C. Flash Photolysis

The FP-ST technique requires that a suitable source molecule be present in the test gas in order to create a transient species for subsequent observation. As seen in Figure 1, the flash photolysis lamp is located on the endplate of the shock tube, and the photolyzing light is axially collimated along the tube with a MgF_2 lens. The light can additionally be filtered by introducing various windows in the path in order to narrow the spectral distribution of the photolyzing wavelengths. The flash lamp contains N_2 gas at various pressures depending on the flash energy needed for a particular experiment. The original design and characterization of this flash lamp configuration is due to McNesby et al. and Ridley et al.[20] Only the central hot gas following reflected shock wave heating is photolyzed ~200–400 μsec after reflection.

For the H-atom studies to be presented, the selected source molecule must yield a significant H-atom concentration on photolysis. Also, the subsequent hydrogen abstraction process by H-atoms should ideally be negligible. Since the abstraction processes with ammonia and/or water have relatively high activation energies, these two molecules were chosen as the source molecules. On photodecomposition, these species give easily measurable concentrations of atoms. In the D-atom studies, deuterium oxide has been used as the D-atom source. In the O-atoms studies, the source molecule is generally NO; however, in one study[21] SO_2 was used as the source molecule for some experiments. If NO is photolyzed, the products of photodecomposition are $N(^4S)$ or $N(^2D)$ atoms and $O(^3P)$ atoms; for example,

$$NO + h\nu \rightarrow N + O. \qquad (1)$$

Both N-atom states react rapidly with excess NO,

$$N + NO \rightarrow N_2 + O, \qquad (2)$$

producing a second O-atom on a time scale that can be made to be small in comparison to the time for subsequent removal by chemical reaction. In H-, D-, or O-atom studies, the flash duration is ~10μs whereas the typical time for temporal observation is 0.1 to 1.5 ms. Hence, the production rates can always be assumed to be instantaneous in comparison to the time for depletion from bimolecular reaction.

D. Detection by Atomic Absorption

The technique used for the detection of the transient atomic species is atomic resonance absorption spectroscopy (ARAS).[22] As seen in Figure 1, the photometer system is located radially across the shock tube a few centimeters from the endplate. For H-atoms, the photometer configuration has been discussed in detail.[3,23] Unreversed H-atom Lyman-α (hereafter denoted as $Ly_{\alpha H}$) is produced by a 2450 MHz microwave discharge in a 1.7 torr flow of prepurified He. There is sufficient hydrogenous impurity to give measurable radiation. This transition at 121.6 nm is further spectrally isolated by placing a gas filter section[24] that contains 3 cm of dry air at atmospheric pressure in front of the photomultiplier. The absorption coefficient of O_2 is low at 121.6 nm, so that the radiation at this wavelength is only slightly reduced. However, the radiation is strongly attenuated at all other wavelengths within the bandpass of the solar blind photomultiplier. There is always some radiation that is not $Ly_{\alpha H}$, and it is necessary to determine the fraction of this spectral impurity. In order to accomplish this, an H-atom filter section is placed in front of the resonance lamp.[25] This filter section is simply a fast discharge-flow system operating with ~0.2 torr of H_2. With the microwave discharge operating, sufficient H-atom concentration is produced in the optical path to nearly remove all of the $Ly_{\alpha H}$ radiation. A measurement of the fraction of light that is resonance radiation is routinely made before each chemical kinetics experiment.

The configuration used for D-atom studies is exactly the same as that for the H-atom except that a flowing premixture of 5–10 ppm of D_2 in He must be used as the source of $Ly_{\alpha D}$. The light coming from the lamp contains not only the desired $Ly_{\alpha D}$ but also $Ly_{\alpha H}$. Hence, it is necessary to remove $Ly_{\alpha H}$ by operating the H-atom filter section during the kinetics experiments. This configuration has been tested, and it has been shown that D-atoms can be observed in the presence of H-atoms and vice versa.

For O-atom experiments, the source of resonance radiation is also a gaseous microwave discharge. The resonance lamp contains 0.1% O_2 in prepurified He and operates at 1.7 torr. The microwave power is 40 watts. This configuration is the same as that described and characterized by Pamidimukkala et al. (lamp b).[26] For partial spectral isolation, a CaF_2 window is placed between the exit window from the shock tube and the photomultiplier, and the filtering gas is dry N_2. Hence, within the bandpass of the solar blind photomultiplier, there is still the possibility that light other than O-atom resonance light might be detected. It is therefore necessary to determine the fraction that is resonant radiation. In this case, the atomic filter section contains 0.2 torr of O_2, and thus, sufficient O-atoms are produced in the microwave discharge flow system to remove all resonant radiation.[25] As in the H-atom case, the measurement of the fraction of resonant light is made before a kinetics run. The atomic filter is not operating during the experiment. If the transmittance levels are greater than ~0.8, then Beer's law holds even though there is considerable reversal in the lamp.[26] Therefore, in all kinetics experiments, the transmittance levels over which the O-atoms are observed range between 0.85 and 0.99.

E. Analysis of Data

Beer's law holds for H- (D-) or O-atoms provided that the transmittance levels are greater than 0.5[23] or 0.8,[26] respectively. Therefore, atom concentration is equal to $(ABS)_t/\sigma l$, where $(ABS)_t$ is defined as $-\ln(I_t/I_o)$ (I_t is the temporally measured transmitted intensity, and I_o is the incident resonant light intensity), σ is the effective absorption cross section for the atom of interest, and l is the absorption path length. In all cases, the reactant molecule concentration is in large excess over the transient atomic concentration. Therefore, little reactant is consumed, and the rate of depletion of atoms is equal to the product of the thermal rate constant, the concentration of reactant, and the concentration of the atom. After integration and with the assumption of constant reactant concentration, the time dependence of the atomic concentration follows pseudo-first-order kinetics. Since atomic concentration is proportional to $(ABS)_t$, a plot of $\ln(ABS)_t$ against time will be linear with the first-order decay constant being given by the product of the thermal rate constant times the reactant concentration. The top panel of Figure 2 shows a typical experimental record of the transmittance change after flash photolysis as a function of time, and the bottom panel shows the first-order decay plot derived from the experimental record. The bimolecular thermal rate constant can be calculated by dividing the first-order decay constant by the density times the mole fraction of reactant (i. e., concentration of reactant) where the density and temperature in the reflected regime are determined from the measured incident shock wave velocity.

III. RESULTS

A. O + C_2H_2 and O + C_2D_2 → Products

The reaction of O(^{3}P)-atoms with acetylene has long been considered to be important in the oxidation reactions of hydrocarbons. Almost all hydrocarbons are degraded by thermal decomposition produc-

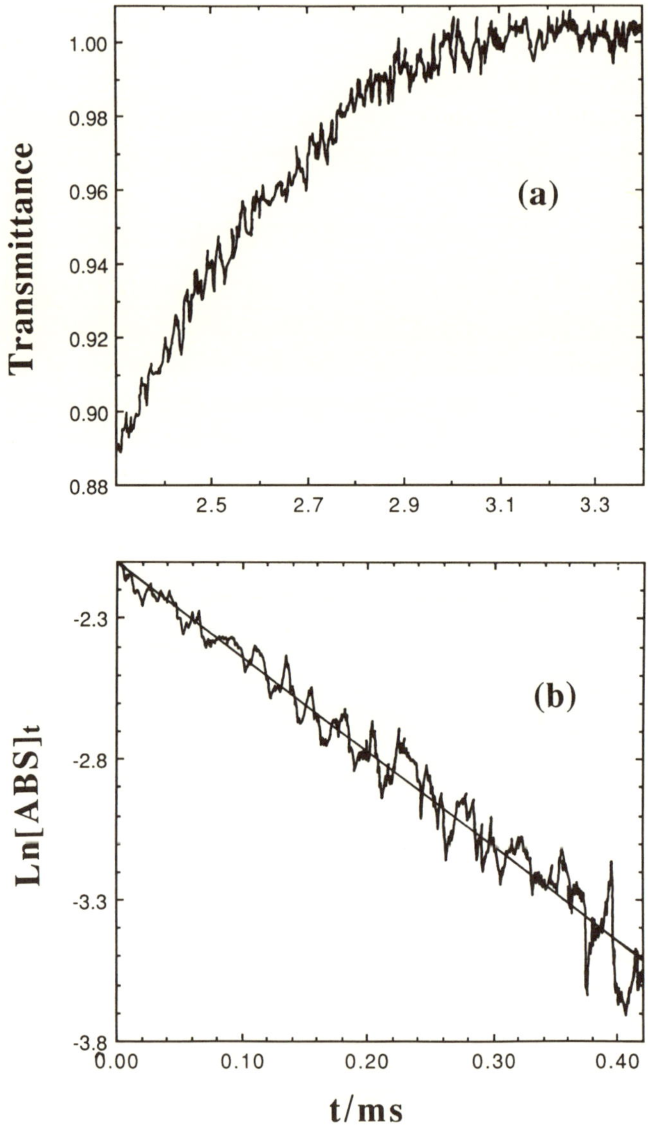

Figure 2. (a) Transmittance from O-atom attenuation after flash photolysis in the reflected shock wave region. (b) First-order plot of $\ln[\mathrm{ABS}]_t$ against time that is obtained from the record in panel (a); $k_{1st} = 3306\ s^{-1}$ in an O + D_2 experiment where $[D_2] = 2.00 \times 10^{15}\ cm^{-3}$ at 1379 K giving k_{O+D_2} (1379 K) = $1.65 \times 10^{-12}\ cm^3\ molecule^{-1}\ s^{-1}$. (Figure 3 from J. V. Michael, J. Chem. Phys., **1989,** 90, 189. Reproduced with the permission of the American Institute of Physics.)

ing species that can give acetylene either through bimolecular reactions or by radical pyrolysis. Therefore, acetylene is an intermediate in these important oxidation reactions.[27]

The original suggestion[28] was that acetylene would be further degraded by reactions with OH-radicals and O-atoms. The O-atom reaction was presumed to be,

$$O + C_2H_2 \rightarrow OH + C_2H, \tag{3}$$

the abstraction process. However, this was soon questioned by Fenimore and Jones,[29] who suggested that the O-atom reaction proceeded through the addition-elimination pathways,

$$O + C_2H_2 \rightarrow H + HCCO, \tag{4}$$

and,

$$O + C_2H_2 \rightarrow CH_2 + CO, \tag{5}$$

This early interest has prompted a considerable number of experimental rate constant studies using a variety of experimental techniques.[30–43] There have been additional studies that have concentrated on the branching ratios for the three pathways, reactions (3), (4), and (5).[33,34,38,40,41,44–51] The reactive system has also been the subject of a theoretical rationalization[52,53] using ab initio electronic structure calculations for both the $^3A'$ and $^3A''$ potential energy surfaces.

Rate constants for both the protonated and deuterated reactions have been measured from ~850–1950 K with the FP-ST technique.[21] The source molecule was NO for the reaction rate experiments; however, SO_2 was used as the source molecule for branching ratio studies. The rate constant results are shown in Figure 3 along with linear least squares lines that are derived from Arrhenius analysis. The rate behavior can be well represented by the Arrhenius equations:

$$k_{O+C2H2} = (1.78 \pm 0.18) \times 10^{-10} \exp(-2714 \pm 118\text{K/T}) \text{ cm}^3 \text{ molecule}^{-1} \text{ s}^{-1}, \tag{6}$$

$$k_{O+C2D2} = (1.73 \pm 0.24) \times 10^{-10} \exp(-2740 \pm 167 \text{ K/T}) \text{ cm}^3 \text{ molecule}^{-1} \text{ s}^{-1}, \tag{7}$$

over the temperature range. From a comparison of equations (6) and (7), there is no observable isotope effect. This fact and the high CH bond strength in acetylene,[54] has shown that direct abstraction, reaction (3), is unimportant over the experimental range. The branching ratio to form H-atoms, reaction (4), was estimated[21] by comparing overall

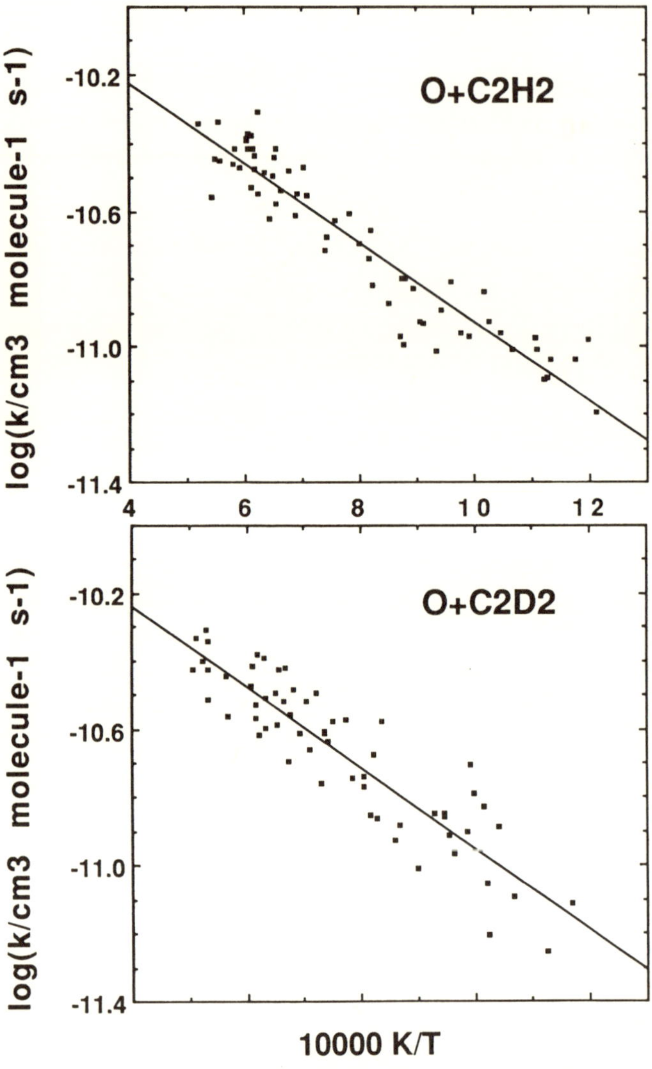

Figure 3. Arrhenius plots of the FP-ST results for O + C_2H_2 and C_2D_2 from ref. 21.

yields of atoms to those formed under identical conditions in the O + $H_2 \rightarrow$ OH + H reaction. The result for the temperature range, 900 to 1200 K, was that 80 ± 15% of the interactions give H-atoms in reaction (4), implying that ~20% of the reaction occurs to produce methylene and CO, reaction (5).

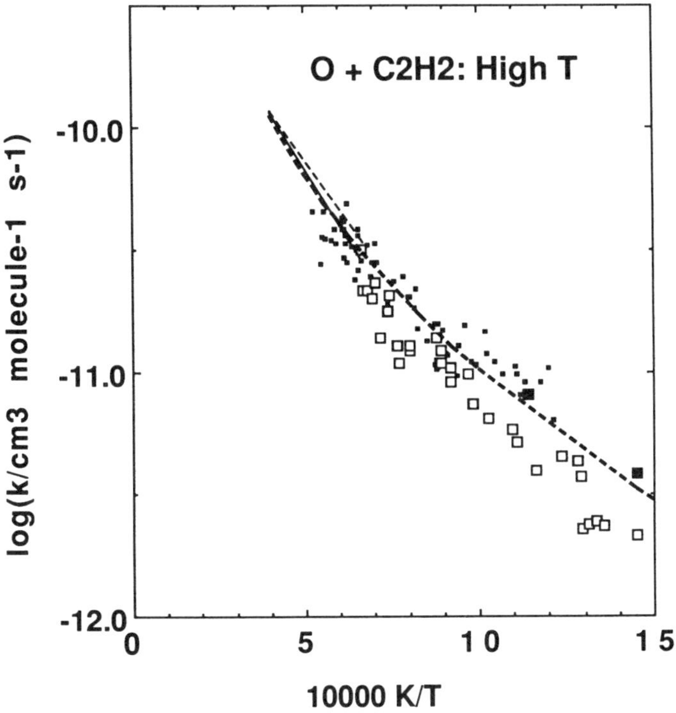

Figure 4. Comparison of the FP-ST results for O + C_2H_2 from ref. 21, ▪, to higher temperature determinations. Shock tube studies; ----- ref. 40 and —— ref. 41. □ - ref. 42. ■ - ref. 43. The bold faced dashed line is calculated from the evaluation, equation (8), as described in the text.

The results from the FP-ST study can be compared to earlier work in which overall rate constants were measured. Figure 4 shows an Arrhenius plot of the FP-ST results for the protonated reaction in comparison to those from earlier studies.[40-43] In general, there is excellent agreement between the FP-ST results and the earlier shock tube results of both Lohr and Roth[40] and Frank et al.[41] Figure 5 shows all lower temperature results.[35-39,42,43] Using all existing data, at least squares expression from the form, $\ln k = \ln A + n \ln T - B/T$, has been derived for the temperature range, 195–2500 K:

$$k_{O+C2H2} = 1.2 \times 10^{-17}\ T^{2.09} \exp(-786\ K/T)\ cm^3\ molecule^{-1}\ s^{-1}. \quad (8)$$

The line calculated from this expression is shown as the long dashed line in both Figures 4 and 5.

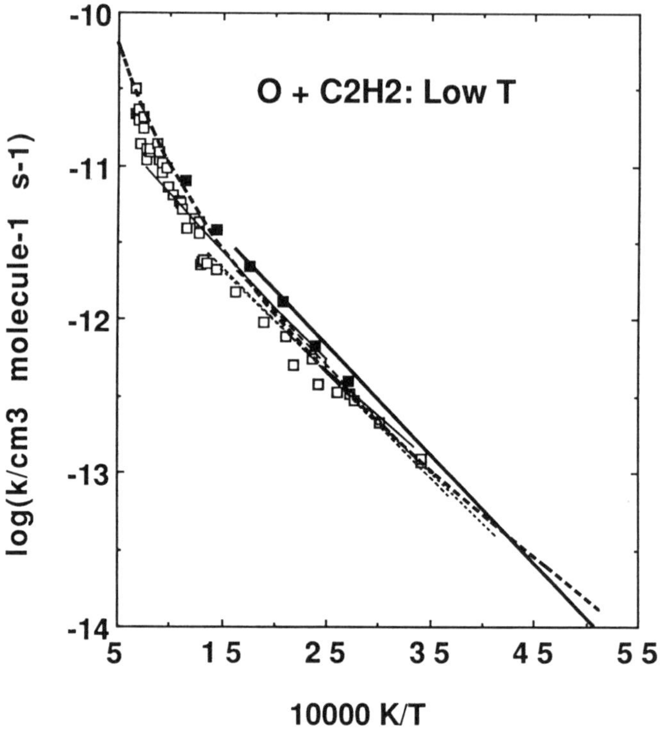

Figure 5. Comparison of the lower temperature results for O + C_2H_2 to one another and to the evaluation, equation (8), shown in the figure as the bold face dashed line. ——— ref. 35; ------ ref. 36; ---- ref. 37; ---- ref. 38; ——— ref. 39; □ - ref. 42; and ■ - ref. 43.

The branching ratio result between 900 and 1200 K of 80% through reaction (4) and 20% through reaction (5), compares favorably to recent results that have been obtained in both shock tube experiments and flow tube experiments. Frank et al.[41] reports 65% through reaction (4) between 1500 and 2500 K, and Peeters et al.[50b] report 59 ± 20% (287 K) and 64 ± 19% (535 K). In a more recent flow tube study, Peeters and coworkers[50c] have revised their 287 K value upward to 81 ± 3%. Hence, a composite temperature independent value of 0.7 ± 0.1 for reaction (4) would be consistent with all determinations over the temperature range, 287 to 2500 K. A recent molecular beam measurement[51] at an average collision energy of 6 kcal/mole gives a somewhat lower value of 42 ± 10% for reaction (4).

The experimental results have been compared to theoretical calculations that are based on an energy scaled ab initio electronic structure calculation by Harding and Wagner.[52,53] These calculations characterize both the $^3A'$ and $^3A''$ potential energy surfaces and then utilize RRKM theory to determine the dynamics of the reaction pathways. In addition, an abstraction transition state model was also considered. The calculations were successful in predicting that abstraction, reaction (3), is totally unimportant below ~2500 K. Also, the calculations explained the branching ratio results showing that the temperature independent value, 0.7 ± 0.1, for reaction (4) is in agreement with theory. The absolute magnitude of the overall rate constant was less successfully predicted. Even with (1) rather large changes in the vibration frequencies of the transition states, (2) an assumption that there is facile crossing between the two surfaces, and (3) additional changes in the scaling parameters for the electronic energy, the low and high temperature rate data cannot be perfectly rationalized.[21] In general, the low temperature results can be explained; however, the high temperature theoretical results lie lower than the error limits of the experimental data. This relationship is shown in Figure 6 for the "best" calculation in comparison to the evaluated experimental result, equation (8).

B. $H + O_2 \rightarrow OH + O$ and $H + O_2 + M \rightarrow HO_2 + M$

The reactions between H-atoms and O_2 are important reactions in gas phase combustion.[55] Not only are they important in the oxidation of H_2, but they also occur in the oxidation mechanisms of all hydrocarbons.[56] Both reactions proceed through the vibrationally hot HO_2* intermediate, and, therefore, the overall process to form O-atoms and OH-radicals is considered as an addition-elimination type reaction. It is well recognized that this overall process is the most important branching reaction in combustion, and it is the competition between branching and recombination (i. e., pressure quenching of HO_2* to form HO_2-radicals) that explains the explosion limit behavior in the H_2/O_2 system.[55]

The reaction to form OH and O has already been extensively studied.[57-62] For several years the rate constant determinations by Schott[60] were considered to be the most reliable. However, Frank and Just[62] have questioned these results. Their experiments indicate that the Schott values are low by about a factor of two in the high temperature region from ~2000–2500 K. Because of its acknowledged impor-

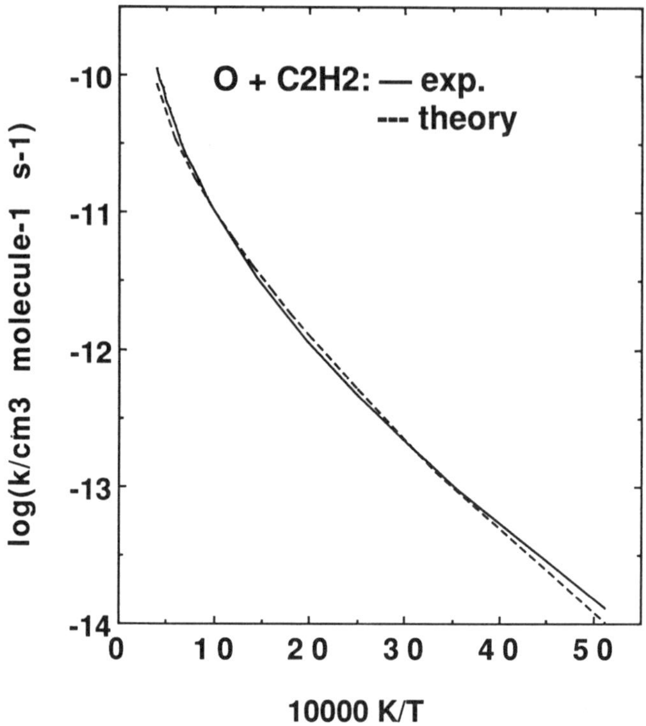

Figure 6. Comparison of the evaluated result, equation (8), to the "best" theoretical prediction for O + C_2H_2 (theory from ref. 21 is based on earlier theoretical calculations from refs. 52 and 53).

tance, the reaction has again been studied an additional five times by four groups in the past three years.[63–66] One of these studies is the FP-ST work that is reviewed here.[66]

There has also been extensive theoretical work on this reaction. For the O with OH interaction there is no electronic energy barrier as shown by the electronic structure calculations of Melius and Blint.[67] With the derived potential energy surface,[67] Miller[68] has performed trajectory calculations and has predicted the T-dependent rate constants for O + OH. Estimates for the H + O_2 reaction have then been made through the equilibrium constant. This calculation has suggested that there is substantial recrossing, and the recrossing becomes more pronounced at higher temperatures. Therefore, the thermal rate behavior shows an increasing negative temperature deviation with increasing temperature that parallels the experimental results of Schott.[60] With the same potential energy surface[67] as a basis, Rai and Truhlar[69] have performed canonical variational transition state theory (CVT) calculations. Since both studies have used the Melius–Blint potential energy surface for O + OH, and this surface does not asymptotically approach a realistic quadrupole-dipole dependence (i. e., r^{-4}) at large separations, it is difficult to assess the significance of the results. On the other hand, Troe[70] and, subsequently, Graff and Wagner,[71] have taken care to empirically describe the long range part of the potential energy surface in a more realistic way, and theoretical rate constants have been calculated using the adiabatic channel[70] and with the ACIOSA (adiabatic capture infinite order sudden approximation)[71] methods. These results either assume no recrossing[71] or treat the phenomenon as a purely statistical effect.[70] They therefore give higher values than the trajectory results of Miller[68] and subsequently agree more closely with the experimental results of Frank and Just.[62] Newer trajectory results with a more realistic potential energy surface have recently been presented by Varandas and coworkers and also by Nyman and Davidsson.[72] The extensive recrossing indicated by Miller has been corroborated in the Varandas et al. calculations;[72a] however, the calculations by Nyman and Davidsson[72b] have questioned these conclusions because the recrossing trajectories apparently violate zero point energy constraints. Hence, there is no clear theoretical agreement as to the significance of the recrossing effect. Unfortunately, the competing experimental data, Schott[60] and Frank and Just,[62] which disagree by the same amount as suggested for the recrossing effect, cannot give any advice as to whether recrossing is a real effect.

Baulch et al. have reviewed the earliest work[57a] on the recombination reaction, H + O_2 + M. There are recent studies of note[66,73,74] one of which is the FP-ST study[66] to be reviewed here. The most notable theoretical prediction is that of Cobos et al.[75]

The FP-ST results have been obtained using both H_2O and NH_3 as source molecules for H-atoms.[66] In order to separate the effects of the addition-elimination path from that due to recombination, experiments have been performed with varying total pressures at relatively low temperatures. The third-order rate constant with Ar as the heat bath gas has been determined for the temperature range, 746–987 K, to be constant within experimental error[66] at $(7.1 \pm 1.9) \times 10^{-33}$ cm^6 $molecule^{-2}$ s^{-1}. The Baulch et al.[57a] recommendation is,

$$k_{H+O_2+Ar} = 4.1 \times 10^{-33} \exp(+500\ K/T)\ cm^6\ molecule^{-2}\ s^{-1}, \quad (9)$$

and the recent flow tube study between 298 and 639 K with He as the heat bath gas by Hsu et al.[74] gives,

$$k_{H+O_2+He} = (4.0 \pm 1.2) \times 10^{-33} \exp(+560 \text{ K/T}) \text{ cm}^6 \text{ molecule}^{-2} \text{ s}^{-1}. \quad (10)$$

It is generally agreed that He and Ar have the same effective collision efficiency, and the FP-ST result can be directly compared to both equations (9) and (10). At the mean temperature of the experiments, 850 K, equations (9) and (10) give respective values of 7.4×10^{-33} and 7.7×10^{-33} to be compared to the FP-ST result of 7.1×10^{-33} cm^6 molecule^{-2} s^{-1}. These results combine to give excellent agreement for the recombination reaction over the temperature range, 300 to 1000 K. The Troe model theoretical calculation of Cobos et al.[75] can be rationalized with the experimental data.

The FP-ST results for the addition-elimination reaction to give O + OH are shown in Figure 7 for the temperature range, 962 to 1705 K.[66] The line shown in the figure is determined from linear least squares analysis,

$$k_{H+O_2} = (2.79 \pm 0.32) \times 10^{-10} \exp(-8119 \pm 139 \text{ K/T}) \text{ cm}^3 \text{ molecule}^{-1} \text{ s}^{-1}, \quad (11)$$

and is accurate to within ± 16%. At the time of the publication of Ref. 66, there was only one relatively direct higher temperature measurement by Frank and Just.[62] The FP-ST results were combined with these values to obtain a composite Arrhenius expression for the temperature range, 962–2577 K. The combined results was quite similar to equation (11), and it agreed better with the theoretical analysis of Troe[70] than with that of Millcr[68] or Varandas and coworkers.[72a] In the past two years, four new studies have been completed[63a,b,64,65] that agree with the FP-ST results up to 1700 K, but they do question the accuracy of the higher temperature study by Frank and Just.[62] These results, expressed in Arrhenius form are:

$$k_{H+O_2} = 9.96 \times 10^{-10} \exp(-11426 \text{ K/T}) \text{ cm}^3 \text{ molecule}^{-1} \text{ s}^{-1} \text{ (1900–2600 K)},^{63a} \quad (12)$$

$$k_{H+O_2} = 6.14 \times 10^{-10} \exp(-10344 \text{ K/T}) \text{ cm}^3 \text{ molecule}^{-1} \text{ s}^{-1} \text{ (1700–2700 K)},^{63b} \quad (13)$$

$$k_{H+O_2} = 1.04 \times 10^{-10} \exp(-6950 \text{ K/T}) \text{ cm}^3 \text{ molecule}^{-1} \text{ s}^{-1} \text{ (1050–2700 K)},^{64} \quad (14)$$

and,

$$k_{H+O_2} = (1.55 \pm 0.07) \times 10^{-10} \exp(-7448 \pm 86 \text{ K/T}) \text{ cm}^3 \text{ molecule}^{-1} \text{ s}^{-1} \text{(1450–3370)}.^{65} \quad (15)$$

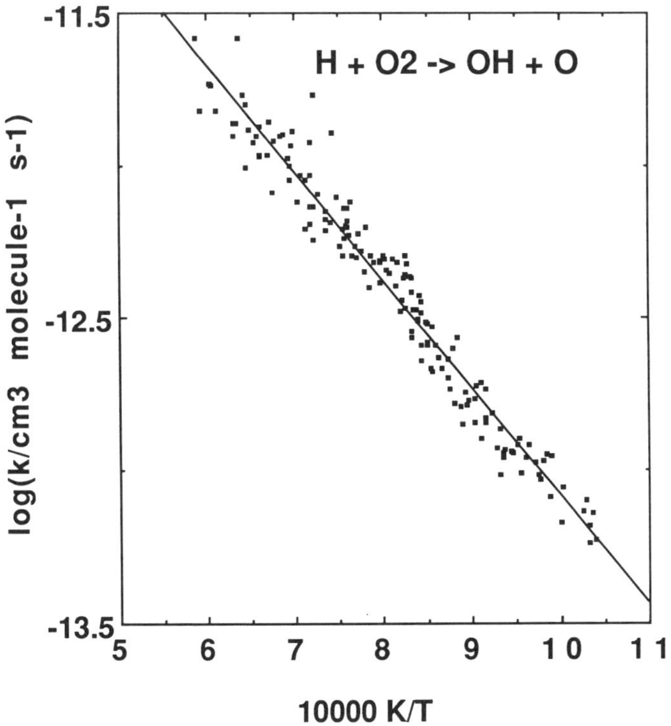

Figure 7. Arrhenius plot of the FP-ST results for H + O_2 from ref. 66. The line is calculated from equation (11).

The conclusion to be reached from the comparisons of the four higher temperature studies[63–65] is that the experimental data agree with one another and with the FP-ST results[66] below about ~1700–1800 K; however, above 2000 K the spread in values increases to about a factor of two as temperature increases to ~2500 K. One of the new studies[64] agrees more closely with the work of Schott,[60] whereas another[63a] is closer to Frank and Just.[62] The two remaining studies lie somewhere between the two extremes. The magnitude of the recrossing effect predicted by Miller[68] and Varandas and coworkers[72a] at ~2500 K is of about the same as these discrepancies. If recrossing is not important, then the three theoretical studies, Troe,[70] Graff and Wagner,[71] and Nyman and Davidsson[72b] agree better with the results of Frank and Just.[62] At the present time, there is still confusion in the high temperature rate behavior, and therefore, there is no way for the experimental data to be used as a confirmation or denial for the phenomenon of recrossing in the $H + O_2 \rightarrow O + OH$ reaction.

C. $O + CH_4 \rightarrow OH + CH_3$

The reaction between $O(^3P)$-atoms and CH_4 is important in the oxidation of methane.[76] It has therefore been studied a number of times with a variety of techniques,[77–83] and one of these is the FP-ST study[83] that will be reviewed here. This case is representative of a simple H-atom abstraction process that can be considered from a theoretical point of view. The ab initio potential energy surface at the POL-CI level of calculation has been reported by Walch and Dunning.[84] Subsequently, Wagner, Dunning, Walch, and Schatz[85] have predicted the rate behavior with conventional transition state theory. These theoretical calculations supersede earlier estimates that are based on LEPS[86] or BEBO[87] potential energy surfaces. There are two additional theoretical studies by Golden[88] and by Cohen[89] that assess the rate behavior from the thermochemical kinetics TST point of view.

The FP-ST rate constant results[83] have been obtained with NO as the source of O-atoms. These results are shown as an Arrhenuis plot in Figure 8 for the temperature range, 763 to 1755 K. A definite curvature is indicated in the figure, and a linear least squares analysis has been carried out on an equation of the form, $\ln k = \ln A + n \ln T - B/T$. The expression obtained is,

$$k_{O+CH_4} = 9.20 \times 10^{-19}\ T^{2.4} \exp(-2838\ K/T)\ cm^3\ molecule^{-1}\ s^{-1}. \tag{16}$$

The FP-ST results have been combined with other studies[77, 79–82] and an evaluation of the rate behavior has been made for the temperature range, 400 to 2250 K. The resulting three-parameter expression,

$$k_{O+CH_4} = 1.15 \times 10^{-15}\ T^{1.56} \exp(-4270\ K/T)\ cm^3\ molecule^{-1}\ s^{-1}, \tag{17}$$

is generally within $\sim \pm 50\%$ of all the studies used in the evaluation. Even though equations (16) and (17) look to be quite different, the actual differences are small, being $< 27\%$ over the temperature range of the FP-ST study. The line calculated from equation (17) is shown along with the FP-ST data in Figure 8.

The theoretical calculation of rate constants by Wagner et al.[85] has used the ab initio potential energy surface of Walch and Dunning.[84] The

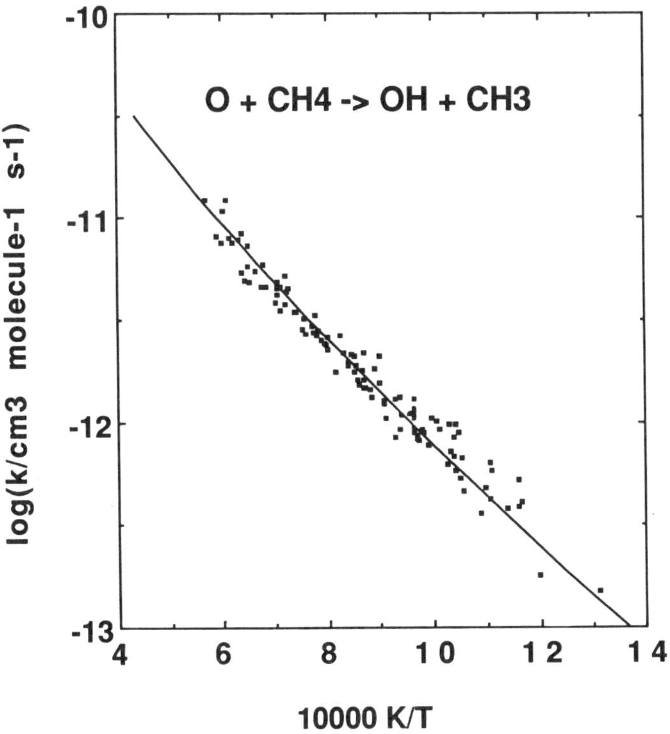

Figure 8. Arrhenius plot of the FP-ST results for O + CH_4 from ref. 83. The line is calculated from equation (17).

computed barrier height at the saddle point was 14.4 kcal/mole; however, it was scaled down by 2.4 kcal/mole to 12.0 kcal/mole. The justification for this scaling was that a similar correction was needed for the O + H_2 reaction when a similarly sized basis set was used in the ab initio calculation. When the basis set size was increased, the calculated energy at the saddle point was 2.4 kcal/mole lower than in the simpler calculation. With this 2.4 kcal/mole decrease as the only adjustment, O + CH_4 rate constant values were then calculated with conventional transition state theory for both $^3A'$ and $^3A''$ surfaces, and they were then combined to give the overall rate constant prediction. Their results have been expressed[83] by the three parameter equation,

$$k^{th}_{O+CH_4} = 3.44 \times 10^{-18}\ T^{2.32} \exp(-3713\ K/T)\ cm^3\ molecule^{-1}\ s^{-1}, \quad (18)$$

over the temperature range, 400 to 2000 K. This "best" theoretical estimate is within a few percent of both the original theoretical data from which it was determined and the experimental results as expressed by equation (17). This can be seen in Figure 9 where equations (17) and (18) are explicitly compared to one another over the respective ranges of temperature. The earlier theoretical calculations[87–89] are also high quality estimates, being well within $\sim\pm 30\%$ over the temperature range. Hence, there is a good theoretical basis for understanding the rate behavior for the O + CH_4 reaction.

D. O + $H_2O \rightarrow$ OH + OH and the Reverse Reaction

The reaction between $O(^3P)$ atoms and water is an important reaction in the oxidation of H_2 and small hydrocarbons particularly in the later stages of reaction when both OH-radical and H_2O concentrations are high.[56c] In fact, this reaction and its reverse are important in the

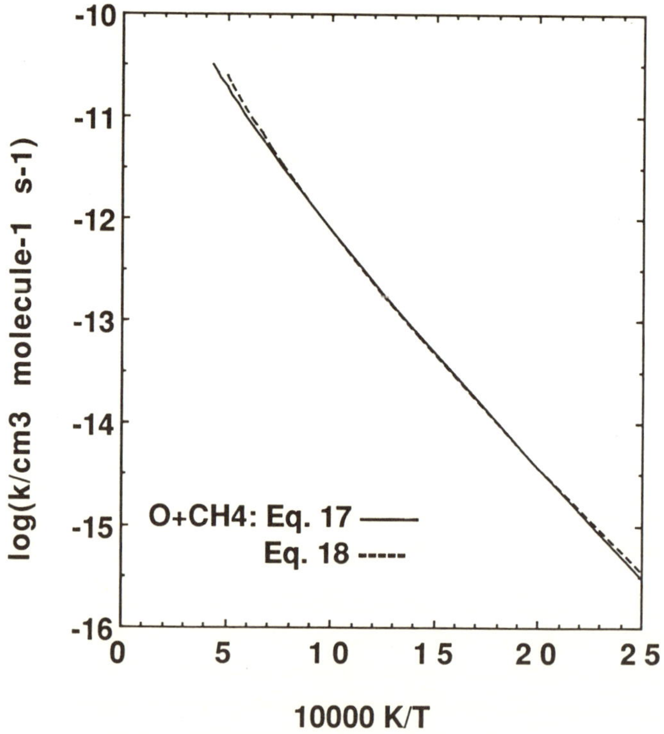

Figure 9. Comparison of the evaluated result from ref. 83 [equation (17)] for O + CH_4 to the theoretical calculation from ref. 85 [equation (18)].

establishment of the well-known pseudo-equilibrium in post-flame gases.[55c]

There have been only two direct determinations of the thermal rate constants for the O + H_2O reaction,[90,91] and one of these is the FP-ST study that will be reviewed here. However, there have been several direct determinations[92–103] of the reverse action, OH + OH. Two of these studies were performed with the shock tube technique[92,93] and the remaining studies were carried out by the discharge-flow method.[94–103] In this latter group, all studies except one[103] were performed at room temperature. Rate constants have been measured by Wagner and Zellner[103] over the temperature range, 250–580K. All data have been reviewed in detail previously,[57] the exception being the new FP-ST study.[91]

This reaction system has also been the subject of theoretical analysis. Harding and Wagner[104] have predicted the thermal rate behavior with the POLYRATE[105] version of variational transition state theory. The basis of the calculation is an ab initio potential energy surface for both $^3A''$ and $^3A'$ states of the interaction. At the saddle points, the $^3A''$ state is lower lying by 2.4 kcal/mole than the $^3A'$ state, and therefore, the bulk of the reactivity occurs through the $^3A''$ state. Rate constants have been individually calculated for both states, and the final predicted values have been taken as the sum.

Thermal rate constants have been measured by the FP-ST technique over the temperature range, 1500–2400 K.[91] The results are shown in Figure 10 along with the least squares line calculated from the Arrhenius expression,

$$k_{O+H_2O} = (1.12 \pm 0.20) \times 10^{-10} \exp(-9115 \pm 304 \text{ K/T}) \text{ cm}^3 \text{ molecule}^{-1} \text{ s}^{-1}. \tag{19}$$

These results are in excellent agreement with the discharge-flow results of Albers et al.,[90] and an Arrhenius expression can be derived for the combined temperature range, 753–2400 K,

$$k_{O+H_2O} = (1.20 \pm 0.04) \times 10^{-10} \exp(-9282 \pm 37 \text{ K/T}) \text{ cm}^3 \text{ molecule}^{-1} \text{ s}^{-1} \tag{20}$$

If the lower temperature results for the back reaction, OH + OH, are transformed to k_{O+H_2O} through the equilibrium constants as given in the JANAF Tables,[106] then a three parameter expression can be evaluated from data for both forward and reverse processes. This evaluation gives,

$$k_{O+H_2O} = 8.44 \times 10^{-14} \text{ T}^{0.946} \exp(-8571 \text{ K/T}) \text{ cm}^3 \text{ molecule}^{-1} \text{ s}^{-1}. \tag{21}$$

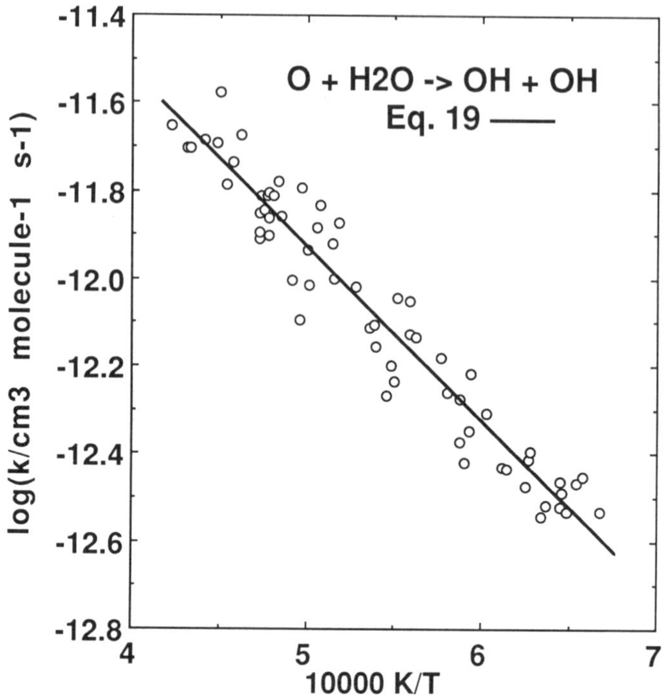

Figure 10. Arrhenius plot of the FP-ST results for O + H_2O from ref. 91. The line is calculated from equation (19).

Equation (21) is plotted along with the direct and transformed data in Figure 11, and it gives values that are within ±50% of all the studies[90-103] over the entire temperature range, 253–2400 K. The theoretical prediction of Harding and Wagner[104] is also shown in Figure 11 as the dashed line.

Rate constants for the back reaction, OH + OH, can be obtained from the results of Albers et al.[90] and Lifshitz and Michael[91] by transformation through the JANAF equilibrium constants,[106] and these can be directly compared to the direct discharge-flow and shock tube experiments.[93-103] This comparison is shown in Figure 12 along with the transformed expression obtained from equation (21) and the equilibrium constants. The theoretical prediction of Harding and Wagner[104] is shown as the dashed line. Also shown is a conventional transition state theoretical calculation[107] that is based on the same potential surfaces as given by Harding and Wagner. In this latter calculation the barrier heights for both states have been decreased by 0.5 kcal/mole, relative

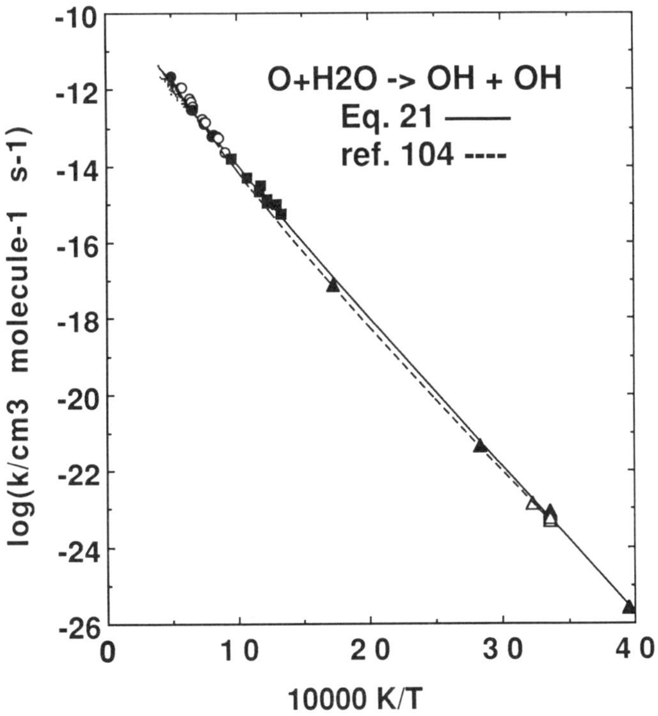

Figure 11. O + H_2O Arrhenius plot comparison of the direct and transformed data to the data evaluation given in ref. 91 as equation (21). The dashed line is a variational transition state theoretical calcuation from ref. 104. · - FP-ST results, ref. 91; ■ - ref. 90; Transformed data from: ● - ref. 92; ○ - ref. 93; △ - ref. 94–96, 98–102; and ▲ - ref. 103.

to both products and reactants, from that used by Harding and Wagner. However, the constant difference of 2.4 kcal/mole between the states has not been changed. Hence, the barrier heights at the saddle points of the forward processes are 21.20 and 23.60 kcal/mole for the $^3A''$ and $^3A'$ surfaces, respectively. The respective values for the back processes are 2.50 and 4.90 kcal/mole. These small adjustments are well within the accuracy of the ab initio calculations.[104] Considering the scatter of the data, it is impossible to choose between the theoretical methods, but such a choice might be possible if the data were more accurate, particularly in the intermediate temperature regime, 500–1200 K. It is clear that further experimental studies in this regime on both the

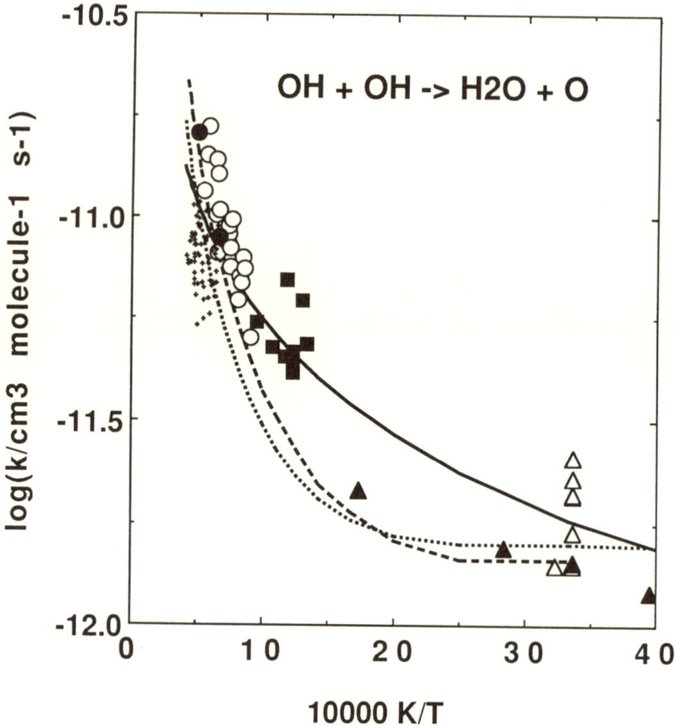

Figure 12. OH + OH Arrhenius plot comparison of the direct and transformed data to theoretical calculations. The symbols are the same as given in Figure 11 except that the refs. 90 and 91 are transformed and the other data are direct. —— - calculated from equation (21) divided by equilibrium constants from ref. 106; ---- - variational transition state theoretical prediction from ref. 104; ------ conventional transition state theoretical prediction from ref. 107.

forward and reverse processes would be helpful in order to confirm the shapes of the non-Arrhenius behavior for both reactions.

E. O + H_2 and O + D_2

The reaction between $O(^3P)$ atoms and hydrogen is an important propagation reaction in the combustion of H_2. Because of this prior interest, the rate constant data for this process are numerous as can be seen in the reviews of Baulch, et al.; Warnatz; Cohen and Westberg,[57] and Tsang and Hampson.[108] The individual studies are additionally reviewed in detail in the recent FP-ST study that will be considered here.[109] Even though data for the related O + D_2 reaction are less extensive, there are earlier studies,[11,110–115] one of which is the FP-ST study[11] that will be reviewed here.

Theoretical predictions of thermal rate constants for both the protonated and deuterated reactions are also extensive. This particular system has served as a test case for theoretical development, and this progression has been thoroughly reviewed.[115,116] The latest theoretical work has utilized a modified RMOS-POLCI (rotated Morse oscillator spline fit to the ab initio POLCI calculations) potential energy surface by Lee et al.[117] The original ab initio calculations were by Walch et al.[118] Some energy adjustments were necessary in order to reproduce the correct endoergicity. With these adjustments, two theoretical predictions have been made with different dynamical methods. Garrett and Truhlar[119] have used variational transition state theory whereas Bowman et al.[120] have used the CEQB method. In a subsequent paper,

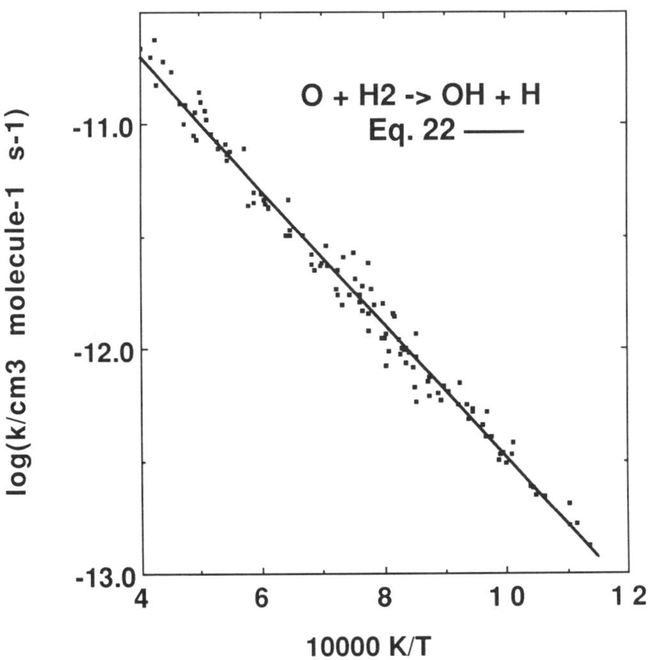

Figure 13. Arrhenius plot of the FP-ST results for O + H_2 from ref. 109. The line is calculated from equation (22).

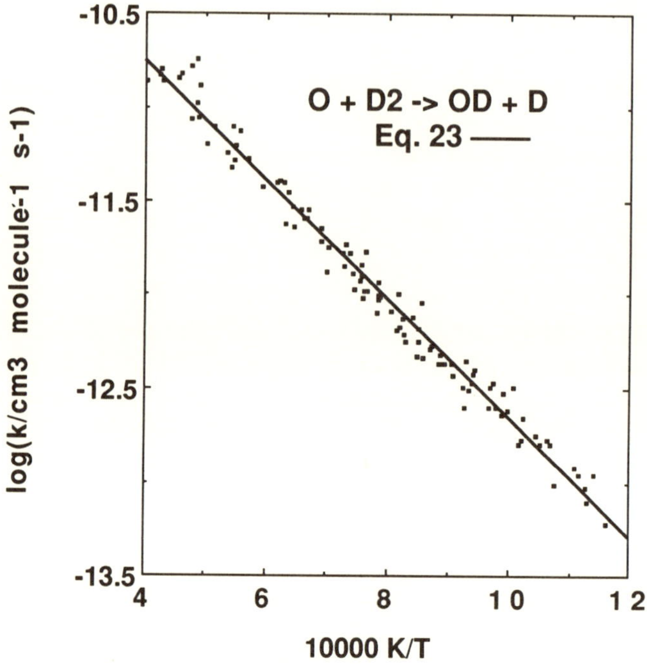

Figure 14. Arrhenius plot of the FP-ST results for O + D_2 from ref. 11. The line is calculated from equation (23).

Joseph et al.[121] have modified the surface slightly and have applied their latest version of variational transition state theory, ICVT/LAG, to both the protonated and deuterated cases. Hence, the best estimates to date are the CEQB (collinear exact quantum with an adiabatic treatment of the bending vibration) method[120] (based on a more general treatment of the reaction dynamics that is labeled the reduced dimensionality theory of reaction rates[116,122]) and the ICVT/LAG method.[121] It should be noted in this latter calculation[121] that the barrier height has been increased by 0.45 kcal/mole over the ab initio value,[118] and some discontinuities have been removed in the RMOS fit. In view of the accuracy of the original ab initio calculation,[118] such minor adjustments are allowed.

The data obtained for the O + H_2 and O + D_2 reactions by the FP-ST technique are shown in Figures 13[109] and 14,[11] respectively. The Arrhenius equation lines shown in these figures are linear least squares representations and are given explicitly by,

$$k_{O+H_2} = (3.10 \pm 0.20) \times 10^{-10} \exp(-6854 \pm 84 \text{ K/T}) \text{ cm}^3 \text{ molecule}^{-1} \text{ s}^{-1}, \tag{22}$$

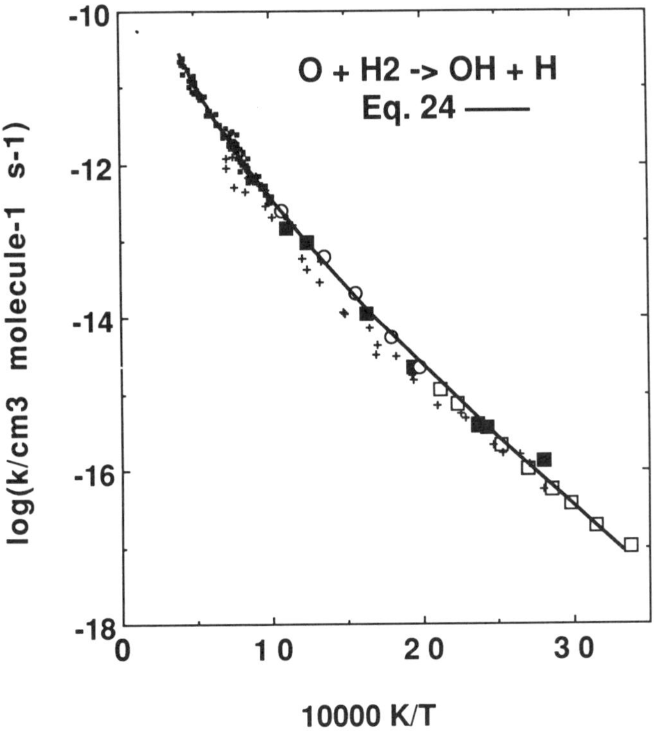

Figure 15. Comparison of the evaluated rate constant behavior for O + H_2, equation (24), to the experimental data. ▪ - FP-ST results, ref. 109; ○ - ref. 109 (FP-RF results); ■ - ref. 77; □ - ref. 113; + - ref. 114.

and,

$$k_{O+D_2} = (3.22 \pm 0.25) \times 10^{-10} \exp(-7293 \pm 98 \text{ K/T}) \text{ cm}^3 \text{ molecule}^{-1} \text{ s}^{-1}. \quad (23)$$

The temperature range of applicability for both equations is 800–2500 K.

The FP-ST data for O + H_2[109] have been combined with flash photolysis-resonance fluorescence (FP-RF) data from two studies,[109,113] and a three parameter expression has been derived for the entire temperature range of all experiments, 300–2500 K. The resulting expression is,

$$k_{O+H_2} = 8.44 \times 10^{-20} \text{ T}^{2.67} \exp(-3167 \text{ K/T}) \text{ cm}^3 \text{ molecule}^{-1} \text{ s}^{-1}. \quad (24)$$

Figure 15 shows the line calculated from equation (24) in relation to

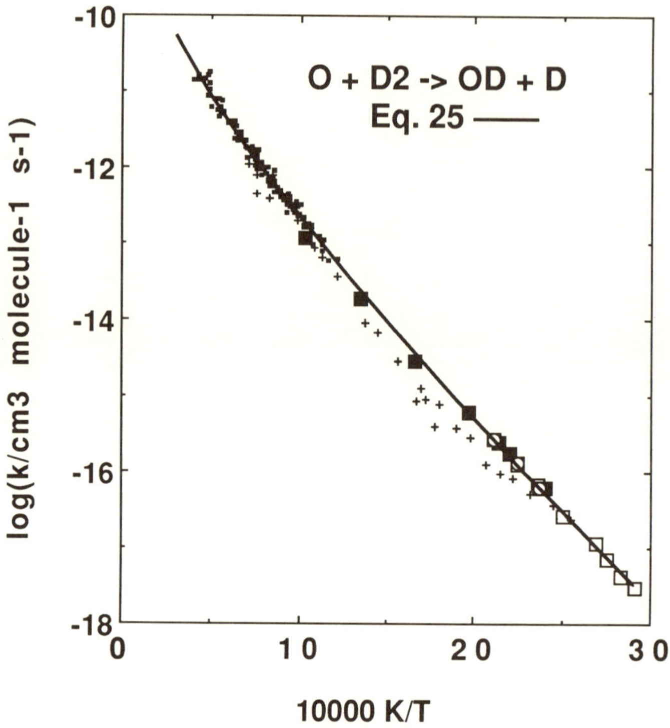

Figure 16. Comparison of the evaluated rate constant behavior for O + D_2, equation (25), to the experimental data. ▪ - FP-ST results, ref. 11; ■ - ref. 111; ○ - ref. 113; + - ref. 114; □ - ref. 115.

the FP-ST work,[109] the FP-RF work,[109,113,114] and the discharge-flow work of Westenberg and deHaas.[77] As can be seen, the equation gives values that are within ±20% of the measurements with the exception of the recent work of Marshall and Fontijn[114] where the values are 30–40% lower, particularly in the intermediate temperature range, 500–1000 K. Similarly, the FP-ST data for O + D_2[11] have been combined with the FP-RF data of Presser and Gordon,[113] and Zhu et al.,[115] and the resulting three parameter representation of these data is,

$$k_{O+D_2} = 2.43 \times 10^{-16} T^{1.70} \exp(-4911 \text{ K/T}) \text{ cm}^3 \text{ molecule}^{-1} \text{ s}^{-1}, \quad (25)$$

for the temperature range, 343–2487 K. Figure 16 shows the line calculated from equation (25) in comparison to studies by Westenberg and deHaas,[111] Marshall and Fontijn,[114] and the abovementioned work by Gordon and coworkers.[113,115] The equation agrees to within ~±30% with the values obtained by Westenberg and de Haas; however, the

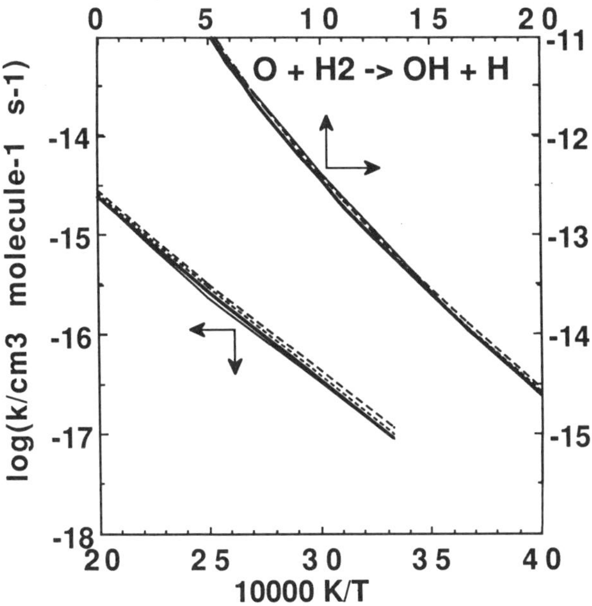

Figure 17. Theoretical to experimental comparison of rate constants for O + H_2. —— - experimental evaluation, equation (24); -- - ref. 119; ---- ref. 120; —— - ref. 121.

discrepancy between this combined result and the results of Marshall and Fontijn is about the same as in the O + H_2 reaction comparison; that is, ~35–45% lower than equation (25). The results of Pamidimukkala and Skinner[112] are not shown in the figure, but their inferred expression from chemical modeling calculations is only ~20% lower than the combined result.

The most modern theoretical estimates of the rate constants for k_{O+H_2} and k_{O+D_2} are shown in Figures 17 and 18 in comparison to the experimental evaluations, equations (24) and (25), respectively. The calculations shown in both figures have been carried out with the CEQB method[120] and also with variational transition state theory.[119,121] The latter variational TST calculation required energy scaling of the ab initio PES by 0.45 kcal/mole, a value that is well within the uncertainty of the relative energetics in the original calculation.[118] Therefore, this slight adjustments is not in contradiction to the ab initio PES. Inspection of Figure 17 shows that all three calculations give excellent predictions of the rate behavior, being higher by <18% than experiment. The calculation by Joseph et al.[121] is in best agreement with equation (24). The agreement for the O + D_2 case is similarly excel-

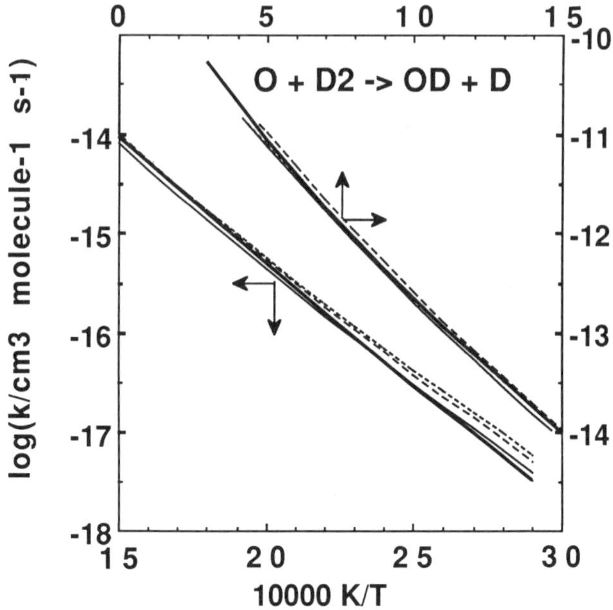

Figure 18. Theoretical to experimental comparison of rate constants for O + D_2. ——- experimental evaluation, equation (25); ——- ref. 119; ----- ref. 120; —— - ref. 121.

lent as shown in Figure 18. The best representation is again the scaled variational TST calculation of Joseph et al.;[121] however, the earlier VTST and CEQB calculations are equally good over most of the temperature range, being higher than the evaluation, equation (25), at lower temperatures due to an overestimation of quantum mechanical tunneling. This conclusion is further illustrated in Figure 19 where the experimental (equation (24) divided by (25)) and theoretical kinetic isotope effects are compared. A similar overestimation can be seen in the theory to experimental comparison of the branching ratio for O + HD → OD + H or OH + D.[124, 125]

Lastly, it is useful to compare these successful theoretical calculations to similar conventional transition state theory calculations that are based on the same ab initio PES.[107] The conventional TST method (not shown) gives poor predictions for both reactions if quantum mechanical tunneling is not included. Use of the Wigner formula results in predictions that are in good agreement with experiment over part of the temperature range; however, this simplified method fails to predict the extent of upward curvature in the isotope effect, Figure 19, because

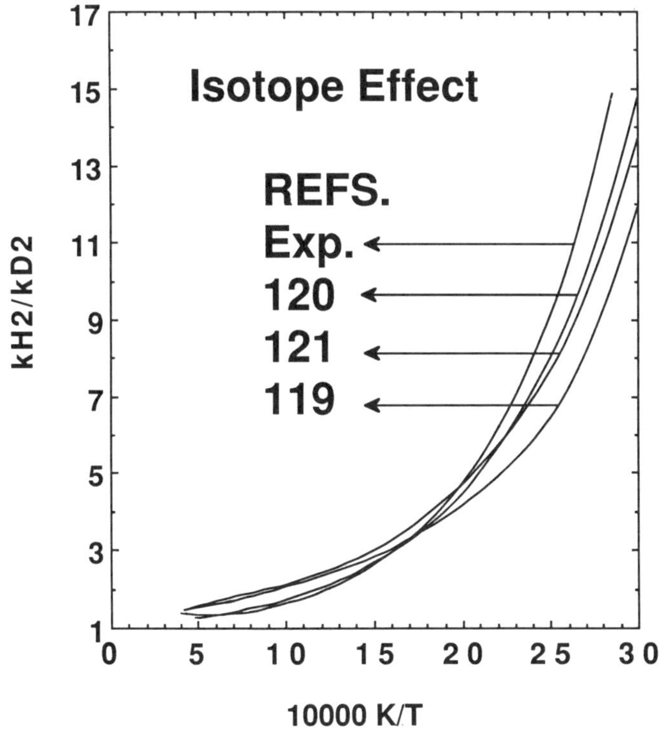

Figure 19. Theoretical to experimental comparison of the kinetic isotope effect calculated from equation (24) divided by equation (25) (labeled Exp. in the figure). The theoretical predictions are from refs. 119–121 as labeled.

the ultra-simple Wigner formula becomes increasingly inadequate at lower temperatures. Two conclusions arise from this comparison: (1) Quantum mechanical tunneling in gas phase H-atom abstraction reactions is a real phenomenon, and (2) the methods for describing it in the more modern theories must be used in order to account for the magnitude observed in experiments. These two conclusions and the accurate predictability of these modern calculations demonstrate the significant advances that have been made in theoretical chemical kinetics.

F. $H + NH_3 \rightarrow NH_2 + H_2$ and the Reverse Reaction

The reaction between H-atoms and ammonia is a propagation process in reactive systems of practical importance. It has been included in the kinetics (a) of pyrolysis[9,126], (b) of NH_3 oxidation,[127] and (c) of the "thermal de-NO_x" process for removing nitrogen oxides from smoke stack gases.[128] The process has therefore been studied previously by a number of techniques[3,9,129–133] including the presently reviewed FP-ST method.[130]

The interaction has been studied theoretically by Melius[134] and has been significantly extended by Garrett et al.[135] The saddle point properties have been determined with ab initio methods at several levels of approximation.[135] At all levels, the saddle point structure is the boat form. In general, the calculated energy differences are higher than allowed from thermodynamic considerations. However, these have been corrected by the BAC (bond additivity correction) method. It is the BAC corrected potential energy surfaces that are subsequently used in thermal rate constant calculations.

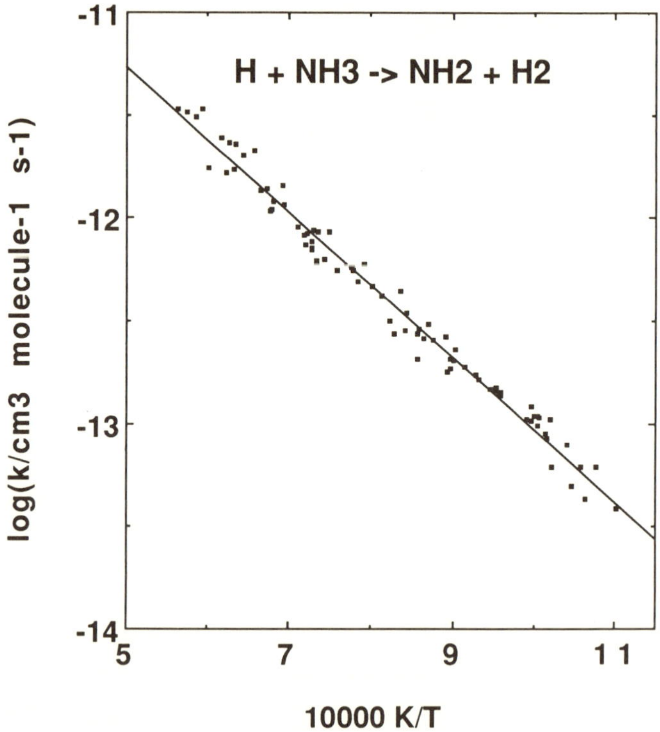

Figure 20. Arrhenius plot of the FP-ST results for H + NH_3 from ref. 130. The line is calculated from equation (26).

The FP-ST results between 908 and 1777 K[130] are shown in Figure 20 along with the line calculated from the linear least squares Arrhenius expression,

$$k_{H+NH_3} = (3.02 \pm 0.30) \times 10^{-10} \exp(-8067 \pm 117 \text{ K/T}) \text{ cm}^3 \text{ molecule}^{-1} \text{ s}^{-1}. \quad (26)$$

Subsequent to this study, three additional determinations of the rate constant have appeared,[131-133] and these have extended the temperature range down to 490 K. One of these[132] preempts the results of an earlier study.[131] When all results are included,[130,132,133] a three parameter expression can be derived that accurately expresses the temperature dependence over the range, 490–1780 K. The least squares result,[132]

$$k_{H+NH_3} = 9.0 \times 10^{-19} \text{ T}^{2.40} \exp(-4990 \text{ K/T}) \text{ cm}^3 \text{ molecule}^{-1} \text{ s}^{-1}, \quad (27)$$

is plotted in Figure 21 along with the data from the FP-ST and lower temperature studies.

A variation of the FP-ST method has been used to determine the equilibrium constants between the title reaction and its reverse over the temperature range, 900–1620 K.[136] These data have been used to estimate the heat of formation of NH_2-radicals. The value obtained, ΔH_{f0}^0 = (46.0 ± 0.3) kcal/mole, is in good agreement with the photoionization mass spectrometric determination (45.8 ± 0.3) of Gibson et al.[137] and also with the value (46.2 ± 1.5) recommended in the JANAF tables.[106] The experimentally determined equilibrium constants coupled to the rate constants for the forward reaction, equation (26), have then been used to transform the equilibrium data into values for $k_{NH_2+H_2}$, and they have been expressed by the Arrhenius equation,

$$k_{NH_2+H_2} = 5.38 \times 10^{-11} \exp(-6492 \text{ K/T}) \text{ cm}^3 \text{ molecule}^{-1} \text{ s}^{-1}, \quad (28)$$

over the temperature range, 900–1620 K.[136] In addition, there are two direct studies of the back reaction, the first being a discharge-flow study[133] and the second being a flash-photolysis study.[138] In both studies, NH_2-radicals were monitored by spectroscopic techniques. The results of these studies and the FP-ST study are plotted in Figure 22. The agreement between the FP-ST results and those of Hack et al. is excellent over the region of temperature overlap. The two sets of data can then be analyzed together giving the Arrhenius expression,

$$k_{NH_2+H_2} = (3.0 \pm 0.6) \times 10^{-11} \exp(-5813 \pm 180 \text{ K/T}) \text{ cm}^3 \text{ molecule}^{-1} \text{ s}^{-1}, \quad (29)$$

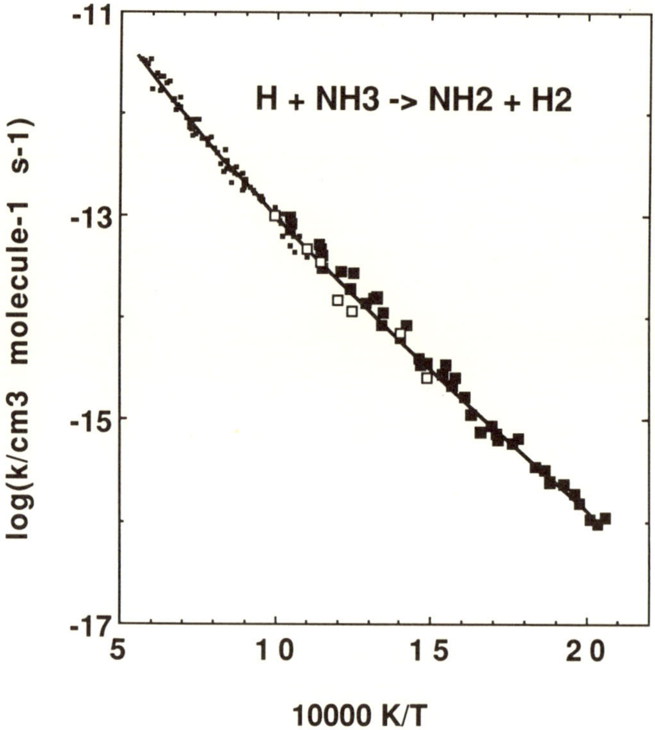

Figure 21. Comparison of the evaluated rate constant behavior for H + NH_3 to the experimental data. ▪ - FP-ST results, ref. 130; ■ - ref. 132; □ - ref. 133. The line is calculated from equation (27).

for 670≤T≤1620 K. The data of Demissy and Lesclaux[138] are summarized by the Arrhenius expression,

$$k_{NH_2+H_2} = 2.16 \times 10^{-12} \exp(-4330 \text{ K/T}) \text{ cm}^3 \text{ molecule}^{-1} \text{ s}^{-1}, \quad (30)$$

over the temperature range, 370–520 K. Inspection of Figure 22 and the equations shows that there is no obvious contradiction between the extrapolated values from equation (29) and those from equation (30). However, the values from equation (30) are not compatible with the evaluation for the forward process, equation (27). If an experimental equilibrium constant at 500 K (a common temperature of applicability) is calculated from equation (27) divided by (30), a value of 0.33 is obtained. The JANAF calculated value is 0.113.[106] Since the JANAF derived values appear to be accurate over the high[136] and intermediate[133] temperature ranges, acceptance of the high value at low temperature

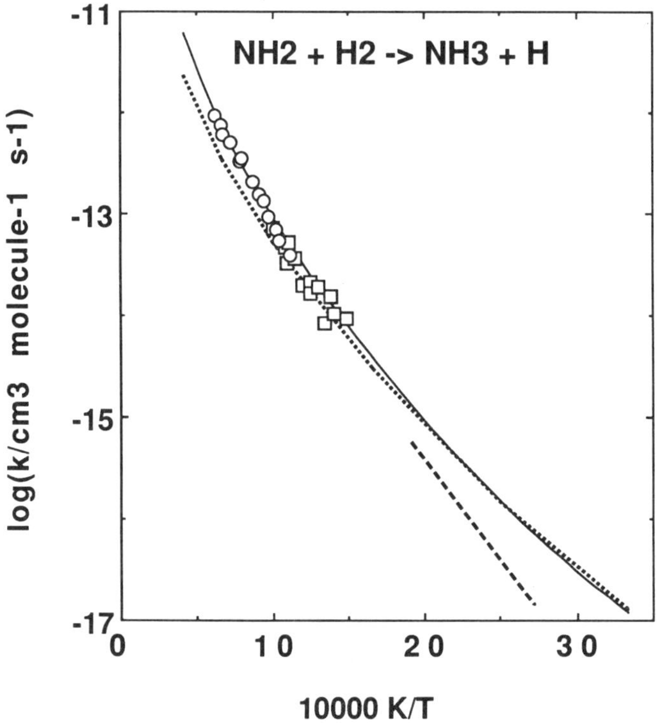

Figure 22. Theoretical to experimental comparison of rate constants for $NH_2 + H_2$. ○ - FP-ST results, ref. 136; □ - ref. 133; ——— - ref. 138; Theory: ---- - ref. 135 (variational TST scaled down by 1 kcal/mole); —— - ref. 135 (variational TST from energy scaled (1 kcal/mole) and frequency scaled calculation for the forward rate, combined with equilibrium constants (see text)).

would force a reconsideration of the pre-exponential factor in the statistical mechanically calculated JANAF inferences. This in turn would suggest the improbable situation that errors exist in the spectroscopic values used in the calculation. Hence, it has to be concluded that either the data for the forward reaction from Ko et al.[132] are high, or the data for the back reaction from Demissy and Lesclaux are low, either one by a factor of two to three. Since the former experiment involves fewer assumptions in the analysis of the data, it is probable that the latter data are systematically low. This would be consistent with the theoretical analysis that follows because increasing the Ko et al. data[132] by a factor of three would then make them difficult to reconcile with any theoretical method.

Melius has calculated the ab initio potential energy surface at the HF/BAC level of approximation,[134] and the saddle point properties have been used by Michael, et al.[130] to estimate with conventional transition state theory, the rate behavior for H + NH_3 over the temperature range, 600–1800 K. The results of this calculation are shown in Figure 23 along with the line calculated from the evaluation, equation (27). Marshall and Fontijn[131] have used the same model with Eckart tunneling. However, in the work of Ko et al.,[132] the value calculated at 950 K with equation (27) has been used to fix the conventional transition state T-independent multiplier. Then, the barrier height has been varied with (a) no tunneling, (b) Wigner tunneling, and (c) Eckart tunneling models, in order to fit the experimental data. The Wigner tunneling model requires a barrier height of 15.18 kcal/mole, whereas the Eckart tunneling model inference is $V^\ddagger = 15.54$ kcal/mole. Both calculations are shown in Figure 23 where it is seen that the Eckart model almost perfectly reproduces the evaluation. Even though the agreement is good, the method allows for two free parameters. The most sophisticated estimate comes from the work of Garrett et al.[135] who have expanded the basis set in the ab initio calculations and have then applied the BAC method. Their best calculation, HF/BAC-MP4, has subsequently been used in a variational transition state theory calculation (POLYRATE code[105]). It should be noted that these workers did not scale the calculated vibration frequencies in their preferred estimate. The agreement with experiment is excellent if the BAC potential energy surface is additionally decreased, relative to H + NH_3, by 1 kcal/mole at the maximum of the adiabatic ground state barrier. These results are shown in Figure 23. In view of the accuracy of the method, this small adjustment is not serious.

Garrett et al.[135] have also estimated rate constants for the back reaction, NH_2 + H_2 from the same scaled surface (i. e., 1 kcal/mole). These calculations are shown in Figure 22 along with the three sets of data cited previously.[133,136,138] The variational transition state calculation predicts values that are ~40% lower than the higher temperature data. This disagreement can be traced to the inability of the model to reproduce the correct temperature dependence of the equilibrium constants. Garrett et al. have scaled only the electronic energy barrier. However, in one set of calculations, they have examined the effect of vibrational frequency scaling on the forward rate constant, k_{H+NH_3}. Without electronic energy scaling, this calculation (not shown) is only slightly higher than the conventional TST calculation[130] that is shown in

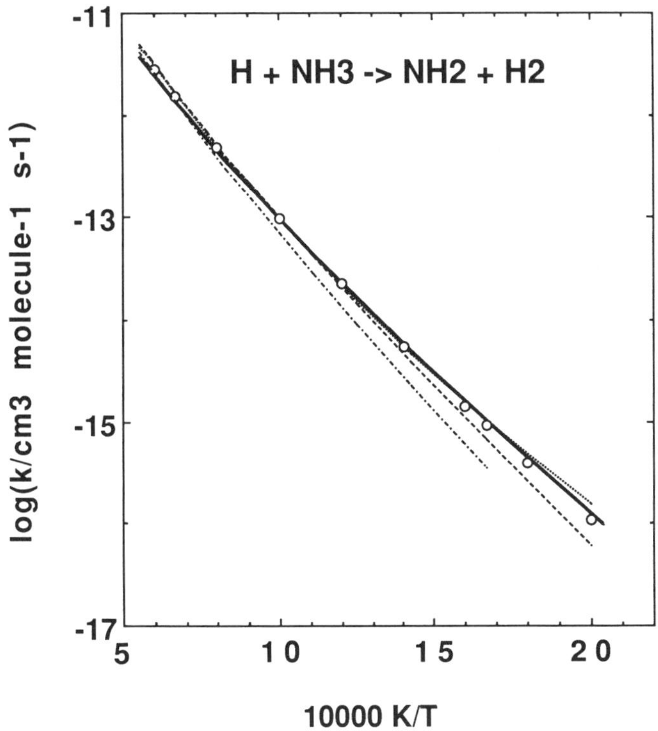

Figure 23. Theoretical to experimental comparison of rate constants for H + NH_3. The evaluation, equation (27) is the bold line. - · - · - ref. 130, conventional TST/W; ---- ref. 131, two parameter TST/W fit; ..... ref. 131, two parameter conventional TST with Eckart tunneling (TST/E); ○ - ref. 135, variational TST calculation energy scaled down by 1 kcal/mole.

Figure 23. If the Garrett et al. frequency scaled calculation is additionally energy scaled so that the barrier is reduced by 0.7 kcal/mole (in contrast to the earlier adjustment of 1 kcal/mole), then values are obtained that deviate less than 15% from the preferred Garrett et al. calculation shown in Figure 23. Within the accuracy of the ab initio potential energy surface, both adjustments are allowed.

The scaling of both vibration frequencies and electronic energy then supplies results that are compatible with the known equilibrium constants and the rates of forward and reverse reactions. Since the vibrational scaling factor is always derived by comparing ab initio results to the experimental values for the stable products and/or reactants, the scaled values will always reproduce the correct equilibrium constants within a few percent if the ergicity for the overall reaction is also correct. Hence, the scaled results (both vibrational and electronic) for k_{H+NH_3} by Garrett et al.[135] can be transformed to $k_{NH_2+H_2}$ with the correct[136] JANAF values.[106] The results of this transformation are shown in Figure 22 as the thin solid line.

The comparison of theory to experiment that is illustrated in Figures 22 and 23 shows that this system is microscopically reversible. All data for both forward and reverse reactions are in excellent agreement with the theoretical calculations, as described, with the exception of those of Demissy and Lesclaux.[138] As mentioned previously, these data seem to be too low by about a factor of two or three.

G. H + H_2O, D + D_2O, and the Reverse Reactions

The reaction between H-atoms and water is important in the combustion mechanism for the oxidation of hydrogen.[55] Since the H_2/O_2 mechanism is included as a subset in the oxidation mechanisms of hydrocarbons,[56e] the reaction will occur in these oxidations also. In general, the reaction acts as an inhibition reaction in the later stages of combustion, contributing to the well known pseudo-equilibrium in the post-combustion regime. The reverse process, OH + H_2, is a crucial chain propagation step in hydrogen combustion, and as such, it has been evaluated several times as evidenced by the large number of reviews on the reaction.[56b,57,139–143]

Since the forward and reverse reactions are so important in combustion, there has been a concurrent interest in the theoretical description of the system. An ab initio potential energy surface calculation has been presented by Walch and Dunning.[144] This surface was used by Schatz and Walch to calculate thermal rate constants for the reverse protonated reaction, OH + H_2.[145] Subsequently, Schatz and Elgersma[146] constructed a parameterized fit to the ab initio potential energy surface and performed a trajectory calculation study. This same parameterized surface was then used by Isaacson and Truhlar[147] in variational transition state theoretical estimates (CVT/SCSAG) of the thermal rate behavior. Even though the agreement with experiment was moderately successful, the success may be fortuitous since the parameterized fit[146] is known to be inaccurate in certain regions of potential energy space. This has prompted a new more accurate ab initio calculation by Kraka and Dunning.[148]

Even though there are indirect determinations of rate constants for the forward reaction,[149–151] there have only been two direct studies,[152,153] and both of these have been carried out with similar shock tube techniques. One of these is the FP-ST study that will be reviewed here,[152] and the other is closely related. Davidson et al.[153] have utilized excimer laser photolysis and laser absorption techniques for observing the OH-radical production rate in the presence of reactant in order to measure rate constants for the reaction. These developments are indicative of the increasing flexibility of the transient photolysis-absorption spectrophotometric method in shock tube experiments that will be continually evolving in the next decade. With this flexibility, it will soon be routinely possible to measure thermal rate constants for radical-molecule and radical-radical reactions.

Figure 24 shows the results from both direct studies[152,153] of the forward reaction. The results of Michael and Sutherland[152] have yielded the linear least squares Arrhenius expression,

$$k_{H+H_2O} = (4.58 \pm 0.61) \times 10^{-10} \exp(-11558 \pm 243 \text{ K/T}) \text{ cm}^3 \text{ molecule}^{-1} \text{ s}^{-1}, \quad (31)$$

which is applicable over the temperature range, 1246–2297 K. Equation (31) is plotted in Figure 24. The results of Davidson et al.[153] give about the same A-factor, but the experimental activation energy is about 7.5% lower than that in equation (31). Therefore the two studies agree quite well with one another being within 20–40% over the common temperature range of applicability.

Figure 25 shows the results for the deuterated forward reaction, D + D_2O, that have been obtained by the FP-ST technique.[154] There are no earlier direct studies of this reaction. These data, between 1285 and 2261 K, have been analyzed by linear least squares methods giving the Arrhenius equation,

$$k_{D+D_2O} = (2.90 \pm 0.73) \times 10^{-10} \exp(-10815 \pm 356 \text{ K/T}) \text{ cm}^3 \text{ molecule}^{-1} \text{ s}^{-1}, \quad (32)$$

Equation (32) is plotted in Figure 25. Even though equations (31) and (32) appear to differ, the results from the two studies are closely similar, giving a kinetic isotope effect of unity. Since the data bases for both protonated and deuterated cases have precisions of about $\pm 25\%$, the calculated isotope effect for the forward process is accurate to $\sim \pm 35\%$.

In order to assess whether an isotope effect of unity is reasonable,

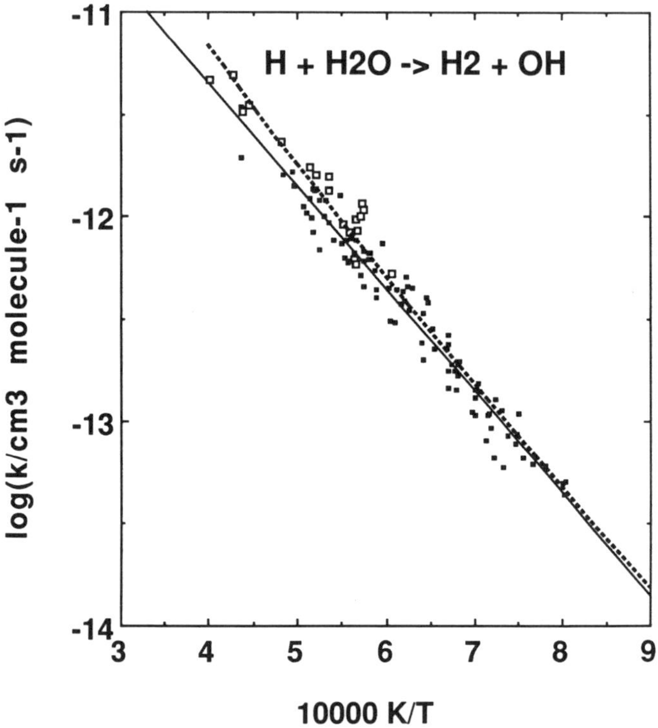

Figure 24. Arrhenius plot of the results for H + H_2O from ref. 152 (FP-ST), ▪, and ref. 153 (LP-ST), □. The solid line is calculated from equation (31), and the dashed line is the CTST/W calculation, described in the text and based on ref. 148.

conventional transition state theory predictions with Wigner tunneling (CTST/W) have been carried out[154] with the new ab initio potential energy surface calculation by Kraka and Dunning[148] as the basis. This calculation is sufficiently accurate that scaling of the vibration frequencies in reactants, products, and transition states has not been required. However, it has been necessary to increase the overall endoergicity by 0.95 kcal/mole (to ΔH^0_{0K} = 16.406 kcal/mole) in order to reproduce the equilibrium constants for both protonated and deuterated cases as reported in the JANAF tables.[106] This energy deficit of 0.95 kcal/mole has been additionally partitioned so that the forward barrier is increased by 0.30 kcal/mole while the reverse barrier has been decreased by 0.65 kcal/mole; that is, $V\ddagger_f$ = 21.766 kcal/mole and $V\ddagger_r$ = 5.360 kcal/mole. This partitioning scheme has been chosen so as to give a

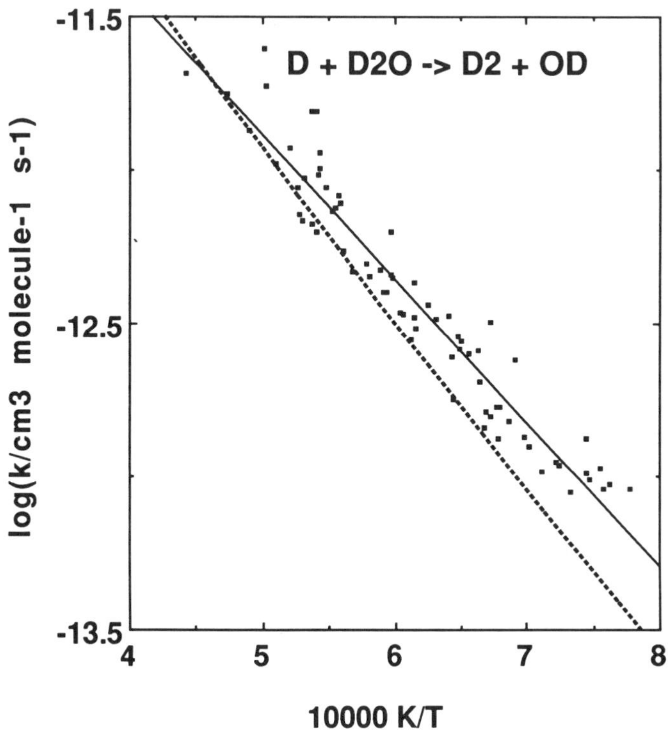

Figure 25. Arrhenius plot of the results for D + D_2O from ref. 153 (FP-ST), ▪. The solid line is calculated from equation (32), and the dashed line is the CTST/W calculation, described in the text and based on ref. 148.

good representation for the protonated reverse reaction where the data are the most extensive. It should be emphasized that these adjustments in electronic energy are well within the estimated absolute error of ±2 kcal/mole in the ab initio calculation[148] and are therefore allowed. The results of the CTST/W calculations are shown in Figures 24 and 25 as the dashed lines. In both cases, the calculation overlaps the scatter of the data; however, it is obvious that the calculation is higher for H + H_2O, and lower for D + D_2O, than the linear least square lines from equations (31) and (32), respectively. The theoretically predicted isotope effect (k_H/k_D) ranges from 1.8 to 1.5 whereas the data give 1.0 ± 0.5. It should be noted that including the data of Davidson et al.[153] in an evaluation for the forward process does not appreciably improve this comparison. Considering the accuracy of the ab initio calculations and, particularly, the simple transition state theory, the agreement between theory and experiment is quite good for both forward processes. The experimental values are lower than theory for the protonated case by ≤30%; however, the same comparison for the deuterated case gives experimental values higher by up to 45% than theory in the lower temperature range of the data.

The theoretical predictions of the equilibrium constants for both protonated and deuterated reactions[154] agree within ~1% with the experimental values as given in the JANAF tables.[106] Therefore, a purely theoretical estimate of the rate behavior for the reverse reactions can be derived from the CTST/W calculations for the forward reac-

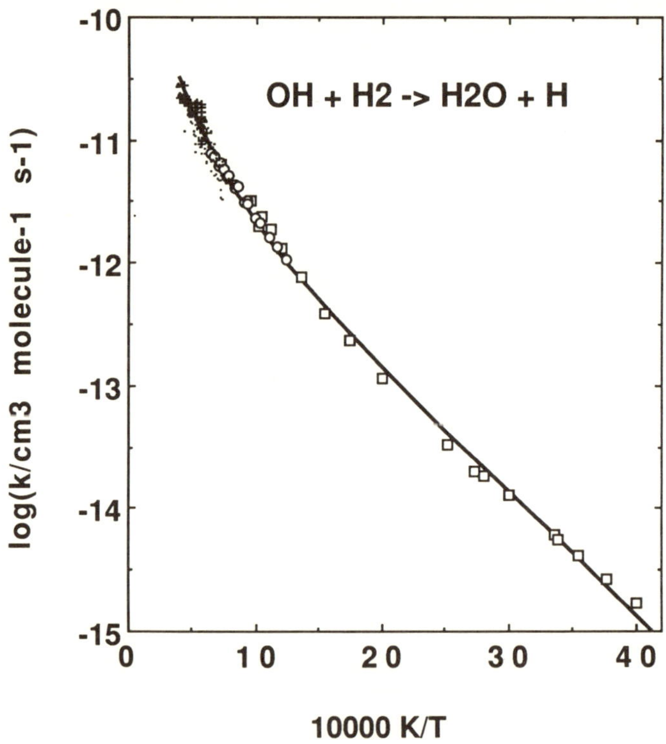

Figure 26. OH + H_2 Arrhenius plot comparison of the direct and transformed data to the CTST/W theoretical calculation as described in the text. Direct results from: □ - ref. 156, 157; ▲ - ref. 158; and ○ - ref. 159. Transformed data from: · - ref. 152 (FP-ST results); and + - ref. 153 (LP-ST results).

tions. Likewise, experimentally based estimates for the reverse reactions can be derived from the values for the forward reaction[152–154] by dividing them by the experimental equilibrium constants.[106] The results of these operations are shown in Figures 26 and 27 for the protonated and deuterated reverse reactions, respectively. In these figures the small dots are the transformed FP-ST points and the plus symbols are the transformed data of Davidson et al. The lines are from the CTST/W theoretical calculations.

The additional data shown in Figure 26 come from direct studies of the reverse reaction. Even though there are several earlier studies,[155] the data shown in the figure come from Ravishankara et al.,[156] Tully and Ravishankara,[157] Frank and Just,[158] and Oldenborg et al.[159] Including all

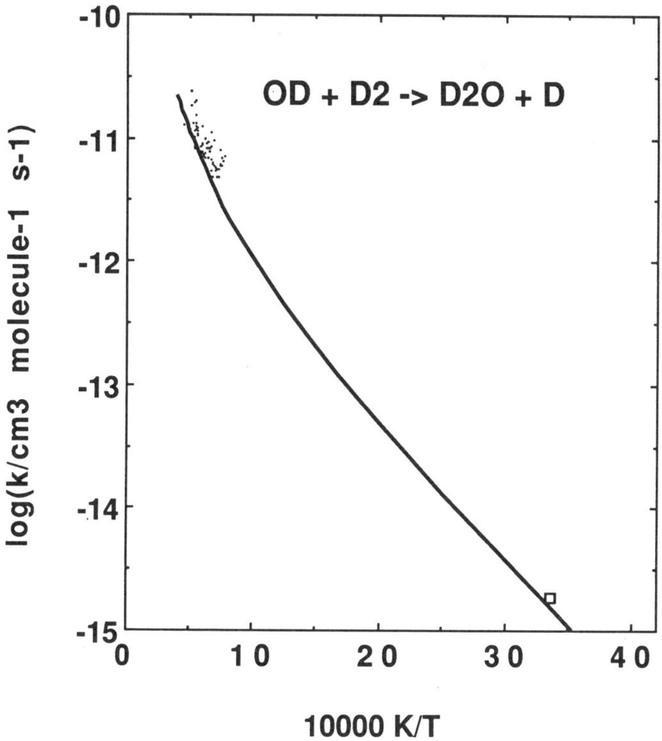

Figure 27. OD + D_2 Arrhenius plot comparison of the direct and transformed data to the CTST/W theoretical calculation as described in the text. Direct results from: □ - ref. 160. Transformed data from: · - ref. 154 (FP-ST results).

the data, these latter workers have derived a three parameter expression that is applicable over the temperature range, 250–2581 K.

$$k_{OH+H_2} = 3.56 \times 10^{-16}\ T^{1.52} \exp(-1736\ K/T)\ cm^3\ molecule^{-1}\ s^{-1} \quad (33)$$

The earlier results[155] agree well with equation (33). The CTST/W theoretical result also shown in Figure 26 cannot be expressed by one simple three parameter expression. Rather, it can be expressed to within ±10% by two and three parameter expressions,

$$k_{OH+H_2}^{theory} = 1.57 \times 10^{-11} \exp(-2346\ K/T)\ cm^2\ molecule^{-1}\ s^{-1},\ (250\text{–}600K) \quad (34a)$$

$$k_{OH+H_2}^{theory} = 1.75 \times 10^{-16}\ T^{1.63} \exp(-1753\ K/T)\ cm^3\ molecule^{-1}\ s^{-1},\ (600\text{–}2500K) \quad (34b)$$

over the indicated temperature ranges. Equations (34a,b) are within ≤±15% of equation (33) over the entire temperature range, 250–2600 K, showing that the CTST/W theoretical description of this reaction is excellent. The variational transition state theoretical calculations by Isaacson and Truhlar[147] are not nearly as good a representation of the rate behavior, possibly because they used the less accurate potential energy surface[144] and fit.[146]

Figure 27 shows the transformed data from the FP-ST study[154] and one point from a recent room temperature study by Vaghjiani et al.[160] There is obviously not enough data for an evaluation; however the FP-ST data taken alone can be expressed between 1285–2261K by the Arrhenius equation,

$$k_{OD+D_2} = (6.6 \pm 1.7) \times 10^{-11} \exp(-3320 \pm 356\ K/T)\ cm^3\ molecule^{-1}\ s^{-1}. \quad (35)$$

The theoretical CTST/W calculation can be represented to within ±10% by the three parameter expression,[154]

$$k_{OD+D_2}^{theory} = 7.94 \times 10^{-18}\ T^{1.98} \exp(-1820\ K/T)\ cm^3\ molecule^{-1}\ s^{-1},\ (200\text{–}2500K). \quad (36)$$

Because the equilibrium constants are theoretically well understood, the relationship between the data and theory is the same for both forward and reverse reactions. Therefore, the experimental values, as expressed by equation (35), are 45% higher than theory, equation (36), at 1300 K. However at 2300 K, theory and experiment agree well with one another. The theoretical result, equation (36), predicts a value at 298 K of 1.40×10^{-15}, to be compared to the experimental value[160] of

$(1.88 \pm 0.30) \times 10^{-15}$, both in units of cm^3 $molecule^{-1}$ s^{-1}. It is probable that equation (36) will turn out to be accurate to within a factor of two over the temperature range, 300–2200 K.

In conclusion, the simple CTST/W calculation,[154] that is based on the slightly adjusted ab initio potential energy surface,[148] predicts the rate behavior quite well. However, it fails to predict the experimental isotope effect of unity. One possibility is that a further refinement of theory may be necessary. This may be a case where reactive bottlenecks cause the observed effect, and, therefore, the phenomenon might turn out to be a dynamical effect rather than one that is based solely on the potential energy surface. This possibility can be explored within the framework of variational transition state theory.[105,147] However another possibility may be that the experimental data are not quite correct. It should be pointed out that the FP-ST data sets for $H + H_2O$ and $D + D_2O$ were obtained with two different apparatuses. Even though care has been taken to eliminate systematic error, such errors may be present in the data. To corroborate this possibility, it should be noted that Ravishankara et al.[156] have reported values for k_{OH+H_2}/k_{OH+D_2}. Presuming that the H(D)-atom in the OH(D)-radical acts as a spectator, the reported isotope effect should be closely similar to that reported here. Their kinetic isotope effect value at 1000 K is 1.6 ± 0.5 and can be compared to the FP-ST derived value of 1.0 ± 0.5. To be sure, these values are not contradictory within combined errors; however, the rather large errors do suggest the necessity for further experimental work with a more accurate technique. Lastly, it is interesting to note that the simple CTST/W calculations (equation (34) divided by (36)) give predictions that within experimental error are nearly equal to those reported in the mixed isotope case of Ravishankara et al.[156] Hence, the improved ab initio potential energy surface at Kraka and Dunning[148] with the indicated adjustments[154] should certainly be used as the basis for further theoretical predictions on the microscopically reversible $H + H_2O$ (and isotopically substituted) system.

H. $D + H_2$ and $H + D_2$

Theoretical understanding is the only motivation for rate constant studies of these two isotopic modifications of the simplest chemical reaction. Experimental research on these reactions has been ongoing for over sixty years,[161,162] culminating in the classic work during the 1960s and 1970s of Le Roy and coworkers[163] and Westenberg and de

Haas.[164] There is a new low temperature study on the H + D_2 reaction by Jayaweera and Pacey.[165] All earlier determinations of thermal rate constants have been carried out with the discharge flow tube technique where the highest attainable temperature has been 745 K. Recently, the FP-ST technique has been used to determine rate constants for both reactions up to ~2000 K,[166,167] and these higher temperature data will be reviewed here. For D + H_2 and H + D_2, data now exist over the temperature ranges, 167–1979 K and 256–2160 K, respectively.

The possibility of theoretically predicting the rate behavior of gas phase reactions actually preceded experimentation and undoubtedly supplied the original motivation for such work. It was supposed that if the potential energy of interaction could be known with high accuracy, then a statistical theory could be devised to explain the thermal reactivity for such reactions.[168] The early theoretical work and its relation to the experiments has been thoroughly reviewed[162] a number of times.

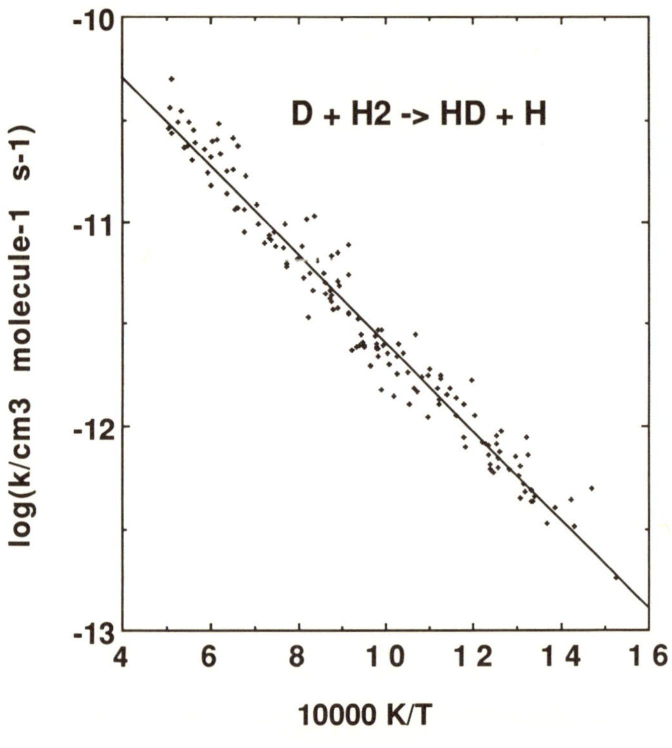

Figure 28. Arrhenius plot of the FP-ST results for D + H_2 from ref. 166. The line is calculated from equation (37).

The main impediment to an exact understanding of gas phase reactions has been the accuracies of potential energy surfaces.[169] Hence, with increasingly complex electronic structure calculational methods, there has been a continuing effort to describe the potential energy surface for the H + H_2 interaction. From the work of Liu and Siegbahn,[170] it can now be claimed that the potential energy surface for this simplest three electron case is known with higher accuracy than in any other reactive system. An analytic fit to these ab initio results has subsequently been given by Truhlar and Horowitz,[171] and this potential energy surface is commonly referred to as the LSTH surface. With newer ab initio calculations and better fitting techniques, a new and presumably more accurate potential energy surface by Varandas et al. has appeared.[172] These workers have used a double many body expansion technique, and therefore, this new potential surface is called the DMBE surface.

Figures 28 and 29 show Arrhenius plots of the FP-ST results for D

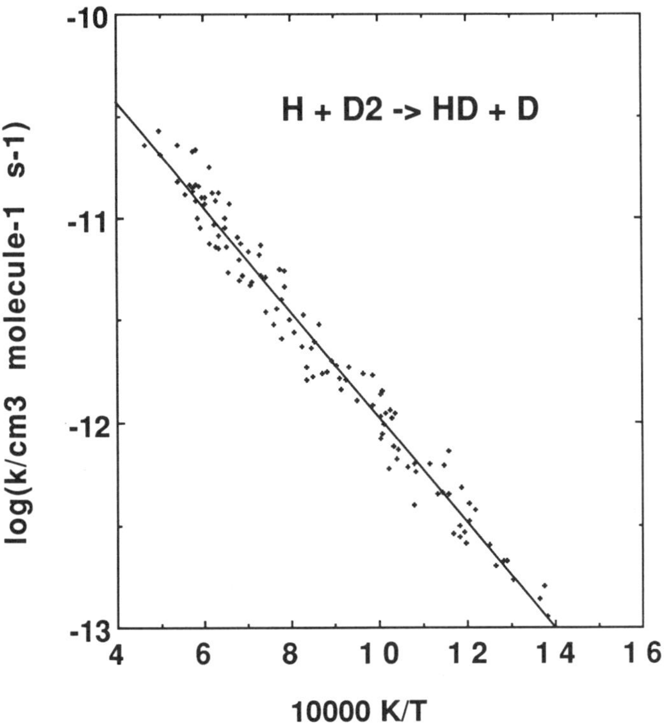

Figure 29. Arrhenius plot of the FP-ST results for H + D_2 from ref. 167. The line is calculated from equation (38).

+ H_2,[166] and H + D_2,[167] respectively. Though there is curvature in the plots, two parameter Arrhenius equations can be derived that describe the data over their respective temperature ranges, 655–1979 K and 724–2061 K,

$$k_{D+H_2} = (3.76 \pm 0.31) \times 10^{-10} \exp(-4985 \pm 80 \text{ K/T}) \text{ cm}^3 \text{ molecule}^{-1} \text{ s}^{-1}, \quad (37)$$

and

$$k_{H+D_2} = (3.95 \pm 0.35) \times 10^{-10} \exp(-5919 \pm 95 \text{ K/T}) \text{ cm}^3 \text{ molecule}^{-1} \text{ s}^{-1}. \quad (38)$$

Both sets can be combined with the lower temperature results of Le Roy and coworkers,[163] Westenberg and de Haas,[164] and Jayaweera and

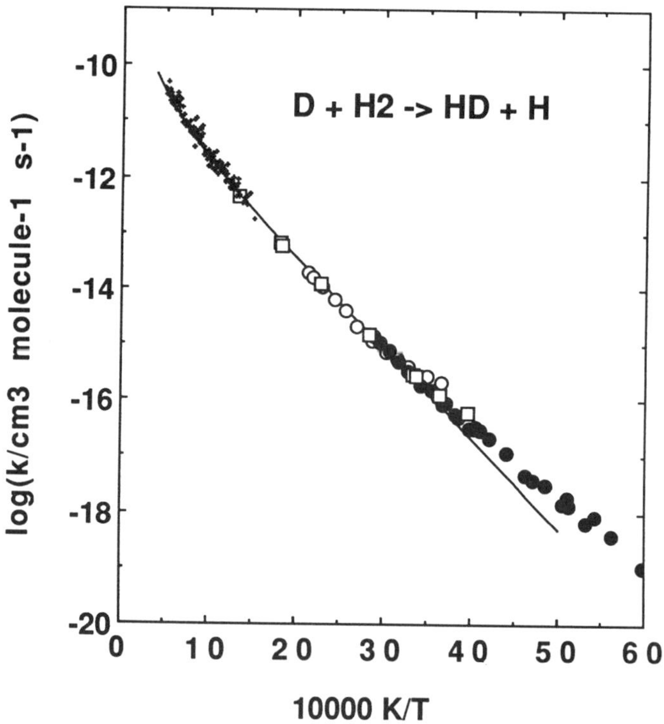

Figure 30. D + H_2 Arrhenius plot comparison of the direct data to the DMBE potential energy surface based CTST/W theoretical calculation as described in the text. Direct results from: ○ - ref. 163b; ● - ref. 163e; □ - ref. 164; and ▪ - ref. 166 (FP-ST).

Pacey,[165] and the resulting Arrhenius plots are shown in Figures 30 and 31. The highest temperature attained in these earlier studies is 745K,[164] and equations (37) and (38) give values at that are within 8% of the earlier measurements[164] showing the excellent continuity of the data sets. Using all existing data, least squares expressions of the form, $\ln k = \ln A + n \ln T - B/T$, have been derived from the combined data sets.[166,167]

$$k_{D+H_2} = 4.0 \times 10^{-18}\ T^{2.29} \exp(-2627\ K/T)\ cm^3\ molecule^{-1}\ s^{-1} \quad (167\text{–}1979K), \qquad (39)$$

and,

$$k_{H+D_2} = 1.69 \times 10^{-17}\ T^{2.10} \exp(-3527\ K/T)\ cm^3\ molecule^{-1}\ s^{-1} \quad (256\text{–}2160K). \qquad (40)$$

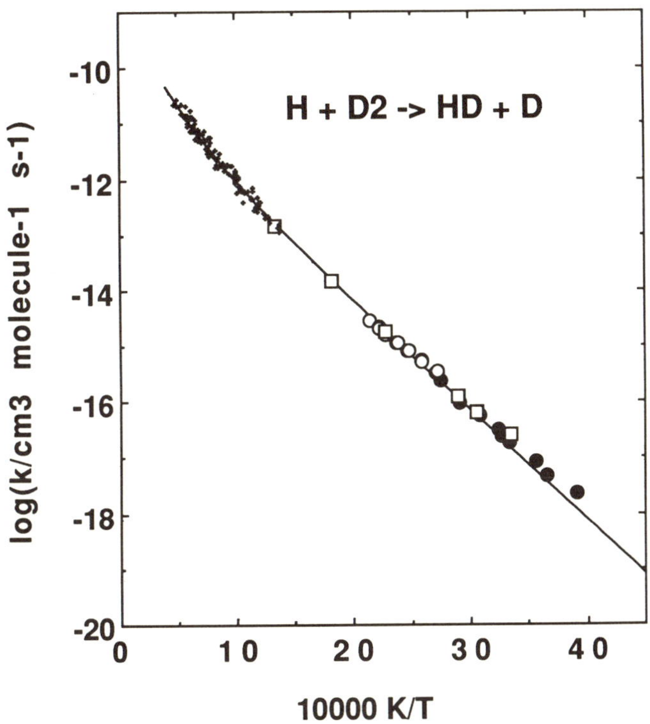

Figure 31. H + D_2 Arrhenius plot comparison of the direct data to the DMBE potential energy surface based CTST/W theoretical calculation as described in the text. Direct results from: ○ - ref. 163a; □ - ref. 164; ● - ref. 165; and ▪ - ref. 167 (FP-ST).

Equation (39) gives values above 250 K that are within a few percent of all the data; however, between 167–250 K the equation is ~20–40% lower than the data of Mitchell and Le Roy.[163c] Equation (40) agrees with all of the data to within ±23% over the entire temperature range.

Since the potential energy surface for this case is known with high accuracy,[170-172] the prediction of the thermal rate behavior for these two cases then constitutes a stringent test of the statistical theories used in chemical kinetics. The simplest prediction can be made with conventional transition state theory with Wigner tunneling corrections (CTST/W) where all DMBE PES frequencies are taken to be harmonic.[166,167] This type of calculation is shown for both reactions as solid lines in Figures 30 and 31. In both cases, the agreement with the data is excellent above ~300 K, deviating by at most, 25%, from the data as expressed by equations (39) and (40). At lower temperatures the simple Wigner formula is not sufficiently accurate to explain the extent of tunneling. The point should be made for these cases that formulations of the rate behavior must include tunneling corrections. If this phenomenon is not included, the theoretical prediction will fail badly. This strongly suggests that tunneling corrections should routinely be applied in all H-atom abstraction reactions involving first and second row non-metallic elements.

Variational transition state theoretical calculations by Truhlar and coworkers have also been carried out for both of the present reactions.[173,174] The results of these calculations are summarized in Figures 32 and 33 where comparisons to the data representations, equations (39) and (40), are shown. The earlier calculations[173] are generally lower than experiment by 20–40% over most of the temperature range. These calculations were based on the original LSTH potential energy surface.[170,171] For the D + H_2 case, a more recent variational TST calculation,[174] that is based on the more accurate DMBE potential energy surface,[172] has been presented. These results are shown in Figure 32 as the dash-dotted line. This new calculation substantially improves the comparison to the lower temperature data, being well within the combined error of the experiments.[163c] Even though the figure would indicate that there is a low temperature divergence from equation (39), this equation, as noted above, gives values that are ~20–40% lower than the data. Since the reactivity in this lower temperature region is almost entirely due to quantum mechanical tunneling, one can conclude that the tunneling model used by Truhlar and coworkers is accurate for these cases. Above ~300 K the two variational TST

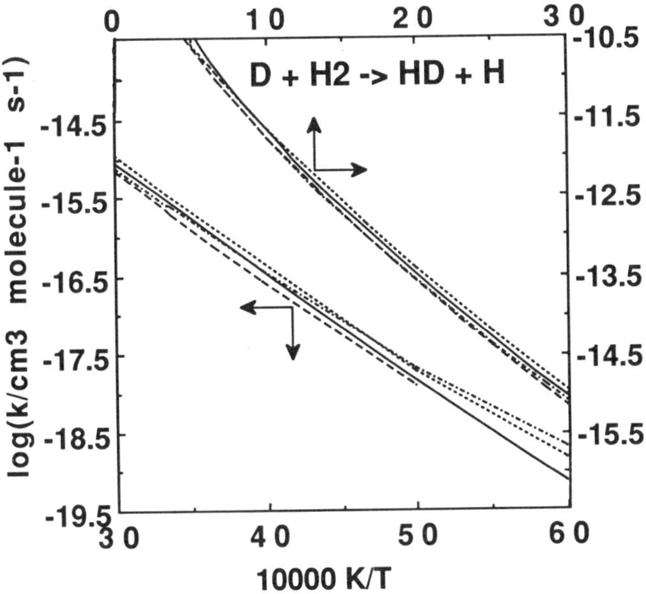

Figure 32. Theoretical to experimental comparison of rate constants for D + H_2. —— - experimental evaluation, equation (39); ----- - ref. 173 (anharmonic VTST based on LSTH potential energy surface); - · - · - · ref. 174 (anharmonic VTST based on DMBE potential energy surface); --- ref. 175 (harmonic CEQB based on DMBE potential energy surface).

calculations for the D + H_2 case (i. e., the ones based on the LSTH and DMBE potential energy surfaces) are virtually identical. In the temperature region above ~1000 K, the results for both cases lie uniformly lower than the data by 20–40%. The reason for this discrepancy probably lies in the use of anharmonic terms for both stretching and bending vibrations in the transition states. Truhlar and coworkers have included only positive quartic terms in these vibrations, and it may be necessary to determine (from the potential energy surface) higher order terms in the description of these modes. Even without this addition, it should be emphasized that the anharmonic variational TST calculations are giving high quality predictions over the entire temperature range being at most 40% lower than the experiments.

Figures 32 and 33 show the results of recent reduced dimensionality[116,122] theoretical estimates for the two reactions. The CEQB version

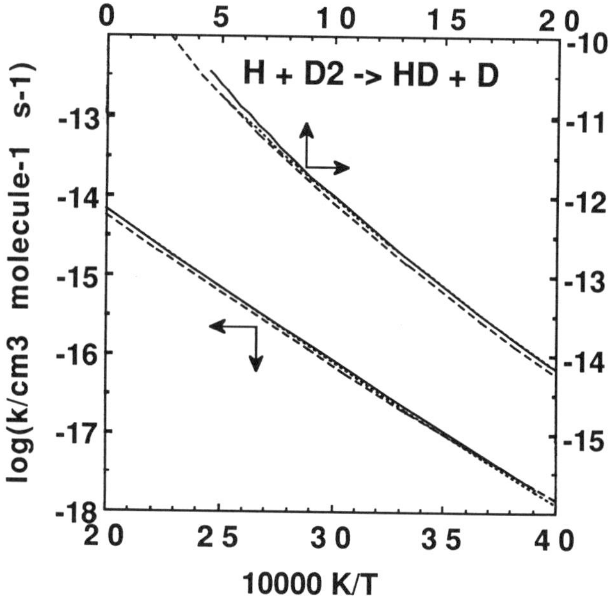

Figure 33. Theoretical to experimental comparison of rate constants for H + D_2. —— - experimental evaluation, equation (40); --- ref. 173 (anharmonic VTST based on LSTH potential energy surface); ---- ref. 175 (harmonic CEQB based on DMBE potential energy surface).

of the theory has been used with both the LSTH[175] and DMBE[176] potential energy surfaces. The calculations, that are shown in the figures as short dashed lines, are based on the harmonic DMBE potential energy surface.[176] They are in excellent agreement with the experiments when the attendant error of the experiments is considered. There is a slight divergence between the experiments and the calculations near the top of the temperature range; however, this discrepancy is not serious. In addition, the agreement for both reactions with the abovementioned variational TST calculations in the lower temperature region is excellent showing that the CEQB method for describing tunneling is also accurate.

A question arises as to why a discrepancy of 20–40% is considered to be a disagreement in these cases, whereas such a discrepancy is not considered to be serious in all of the other cases considered in this review. The answer lies in the accuracy of the potential energy surface. In this H + H_2 cases, no options are available for adjusting the parameters derived from the ab initio calculations. In all of the other cases (with the possible exception of O + H_2), the absolute accuracies of the energetics, including the barrier heights at the saddle points, are too high by 2–4 kcal/mole. The numerical methods associated with these calculations can contribute an additional random error of ±1 kcal/mole. Also, the values of the vibration frequencies that are derived from the potential energy surfaces can be high by as much as 10–20%. Hence, some adjustments are required in order to make the models agree with experimentally derived quantities (i. e., spectroscopically observed frequencies and/or measured equilibrium constants and/or measured activation energies). However, it should be emphasized that these adjustments are almost always well within the accuracies of the potential energy surface calculations and are therefore totally allowed because they are not in contradiction to the ab initio calculations. Since the potential energy is accurately determined for these two H + H_2 cases, no such adjustments are allowed. Hence, any discrepancies with the data are undoubtedly due to the approximations in the dynamical theories for calculating thermal rate constants.

Lastly, the kinetic isotope effect, KIE, for ~250–2200 K, can be determined from equation (39) divided by (40),

$$\text{KIE} = 0.237\ T^{0.19} \exp(900\ \text{K}/T). \tag{41}$$

Figure 34 shows a plot of equation (41) along with predictions from the various theoretical estimations; that is, harmonic conventional TST,[166,167] variational TST,[171,172] and CEQB[176] methods. All three methods agree quite well with equation (41) within experimental error. It is interesting to note that the conventional TST and CEQB methods are almost superimposable. This is not surprising because the individual rate constant estimates from the two methods (cf. Figures 30 and 31 to 32 and 33) are nearly identical over the 300–2000 K temperature range. Reiterating again the importance of quantum mechanical tunneling, if the phenomenon is not included in these descriptions then the prediction of the kinetic isotope effect shown in Figure 34 would not agree at all with experiment. Similar calculations for the exchange rate, $(k_{D+H_2}k_{H+D_2})/(k_{D+H_2} + k_{H+D_2})$, have been given,[167] and the predictions are compared to older experimental values.[161c,d,e, 162c] The conclusion from this comparison is that the older data are not accurate enough to help with the choice of theoretical method because all predictions for the exchange rate lie quite close to one another and also lie between the earlier experimental data.

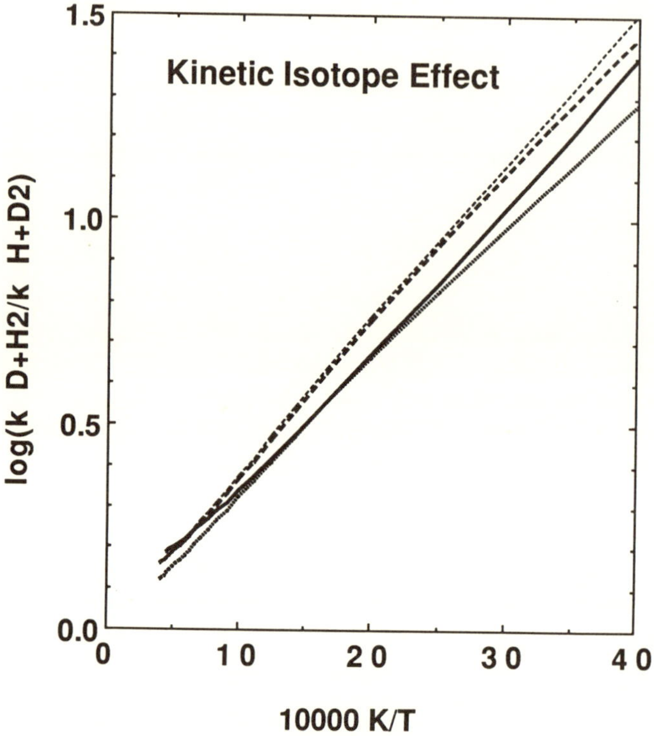

Figure 34. Theoretical to experimental comparison of the kinetic isotope effect, k_{D+H_2}/k_{H+D_2}. Bold line is the experimental result, equation (39) divided by (40); bold dotted line is from ref. 173 (anharmonic VTST based on LSTH potential energy surface); thin dashed line is from ref. 175 (harmonic CEQB based on DMBE potential energy surface); and bold dashed line is from refs. 166 and 167 (CTST/W based on DMBE potential energy surface).

IV. CONCLUSIONS

The Flash Photolysis-Shock Tube (FP-ST) method with atomic resonance absorption spectroscopic detection is a special case of the more general transient photolysis-shock tube method with spectroscopic detection of transient species. Various new modifications (e.g., laser photolysis with laser absorption and/or fluorescence) of this general method will undoubtedly continue to be developed in the next few years. As has already been emphasized,[177] the reason for these developments will primarily be to support combustion modeling. It is clear that the present FP-ST method has already contributed significantly to the understanding of the simplest combustion mechanism; that is, that for hydrogen oxidation. The important steps in this mechanism including all reverse processes are,

$$H + O_2 \rightleftharpoons OH + O$$
$$O + H_2 \rightleftharpoons OH + H$$
$$OH + H_2 \rightleftharpoons H_2O + H$$
$$O + H_2O \rightleftharpoons OH + OH$$
$$H + O_2 + M \rightleftharpoons HO_2 + M.$$

As described in the Results section, the absolute rate constants in this oxidation have now been measured, and therefore, the rate behavior for the thermalized combustion that exists in dilute shock tube experiments[58–62] can be completely described. This claim must be qualified due to the continuing controversy surrounding the rate behavior of the $H + O_2$ reaction above 1700 K. However, this case illustrates the role that the technique, and the expected derivative techniques, will be having in the general field of fuel combustion modeling.

The FP-ST technique also has a role in the basic theoretical understanding of bimolecular reactions. As mentioned in the Introduction and illustrated with several cases in the Results section, the technique can be used in the temperature region above ~1000 K where tunneling in H-atom abstraction becomes relatively unimportant. This means that the temperature dependence of directly determined rate constants gives a direct experimental probe of the relative energetics of the potential energy of interaction. Also, the magnitude of preexponential factors has implications regarding intermediate configurations in an interaction. The claim can therefore be made that a well-determined analytical representation of an absolute rate constant, without any other information, is in itself a very significant theoretical statement. Though this claim is not original, it is forgotten from time to time and should periodically be restated.

The theoretical calculations that have been presented in the Results section show that theoretical understanding is really quite well advanced. In fact, the most significant advance in the entire field of chemical kinetics and dynamics in the past decade has been in theory. Even though there continue to be difficulties in the addition-elimination reactions (probably due to both the ab initio potential energy surface calculations and the dynamical methods), the calculations presented on the H-atom abstraction reactions have shown that transition state theoretical methods are describing reality quite well. To be sure, there are competing dynamical methods; however, these methods, regardless of their sophistication, give quite similar results. This is an important observation. The most sophisticated method to date[178] has recently been used to describe dynamical results[179] on the H + D_2 reaction. This method, quantum mechanical scattering on the LSTH (or DMBE) potential energy surface, is being extended to the thermalized system. Such calculations will certainly agree with TST and/or CEQB methods since both of these methods agree with the data, and this will be an important confirmation of the approximate TST with the tunneling correction method.

The abovementioned confirmation of the TST approximation is important because the uncertainties involved in theoretical analysis continue to be in the accuracies of the relative energetics in ab initio electronic structure calculations of potential energy surfaces. At the present time, these methods are most accurate in calculations involving first and second row non-metallic elements, and, in the best calculations, accuracies of ± 2 kcal/mole have been claimed. Within a few years, the widespread availability of computational resources will become common, and more complete calculations with larger and larger basis sets will routinely be performed on systems with three to five atoms. These should rival the accuracy of the DMBE potential energy surface for H + H_2; however, even with this expected higher accuracy, a complete description of a potential energy surface will not be a trivial exercise.[180] Therefore, less rigorous electronic structure potential energy calculations will continue to be presented, particularly for larger molecules. As in the past, the thermal rate data will dictate energy scaling factors, and the experimental values of vibration frequencies will continue to be used to determine vibrational scaling factors. With scaling, or even with less rigorous methods of potential energy determination (guesses are even better than nothing), it will be particularly important to continue to apply TST theoretical methods to thermal bimolecular interactions, especially in cases that have either been only partially studied or have not been studied at all (e. g., reactions of species with molecules in an homologous series). This less rigorous approach serves modelers who are trying to understand the complex chemistry that exists in natural environments. Such calculations also continue to be important to theoreticians because they, as a group, are advised on what to expect when further computational advances are made. There is no doubt but that the theoretical advances in both potential energy determination and dynamics within the past decade have given an increasingly clearer understanding of bimolecular interactions, and this fundamental understanding is an important goal of Science.

ACKNOWLEDGMENTS

The author wishes to thank Drs. A. F. Wagner and J. P. Hessler for a thorough reading of the manuscript and comments. In addition, the author thanks Dr. L. B. Harding, Dr. R. L. Shepard, and Professor G. C. Schatz for helpful suggestions. This work was supported by the U. S. Department of Energy, Office of Basic Energy Sciences, Division of Chemical Sciences, under Contract No. W-31-109-Eng-38.

REFERENCES

1. Burns, G.; Hornig, D. F. *Can. J. Chem.* **1960,** *38,* 1702.
2. (a) Ernst, J.; Wagner, H. Gg.; Zellner, R. *Ber. Bunsen-Ges. Phys. Chem.* **1978,** *82,* 409. (b) Niemitz, K. J.; Wagner, H. Gg.; Zellner, R. *Z. Phys. Chem.* (Frankfurt am Main) **1981,** *124,* 155.
3. Michael, J. V.; Sutherland, J. W.; Klemm, R. B. *Int. J. Chem. Kin.* **1985,** *17,* 315.
4. (a) Meyerson, A. L.; Thompson, H. M.; Watt, W. S. *J. Chem. Phys.* **1965,** *42,* 3331. (b) Meyerson, A. L.; Watt, W. S. *J. Chem. Phys.* **1968,** *49,* 425.
5. Michael, J. V.; Weston, Jr., R. E. *J. Chem. Phys.* **1966,** *45,* 3632.
6. Roth, P.; Just, Th. *Ber. Bunsen-Ges. Phys. Chem.* **1975,** *79,* 682; Braun-Unkhoff, M.; Frank, P.; Just, Th. *Symp. (Int.) Combust., [Proc.]* **1988,** *22,* 1053; and earlier references cited therein.
7. Bernfeld, D.; Skinner, G. B. *J. Phys. Chem.* **1983,** *87,* 3732; and earlier references cited therein.
8. Roth, P.; Barner, U.; Lohr, R. *Ber. Bunsen-Ges. Phys. Chem.* **1979,** *83,* 929; Thielen, K.; Roth, P. *AIAA Journal* **1986,** *24,* 1102; Burmeister, M.; Gulati, S. K.; Natarajan, K.; Thielen, K.; and Roth, P. *Symp. (Int.) Combust., [Proc.]* **1988,** *22,* 1083; and earlier references cited therein.
9. Yumura, M.; Asaba, T. *Symp. (Int.) Combust., [Proc.]* **1980,** *18,* 863.
10. Bradley, J. N. *Shock Waves in Chemistry and Physics;* Wiley: New York, 1962.
11. Michael, J. V. *J. Chem. Phys.* **1989,** *90,* 189.
12. (a) Mirels, H. *Eighth International Symposium on Shock Waves and Shock Tubes;* 1971; paper 6; (b) *Phys. Fluids* **1966,** *9,* 1907; (c) *Am. Inst. Aeronaut. Astronaut. J.* **1964,** *2,* 84; (d) *Phys. Fluids* **1963,** *6,* 1201; and references and reports (particularly those from NACA) cited therein.

13. Strehlow, R. A.; Belford, R. L. *AAE Tech. Report No. 69-1;* University of Illinois, Urbana, 1969; (b) Belford, R. L.; Strehlow, R. A.; *Annu. Rev. Phys. Chem.* **1969,** *20,* 247.
14. Belles, F. E.; Brabbs, T. A. *Symp. (Int.) Combust., [Proc.]* **1970,** *13,* 165.
15. Fujii, N.; Koshi, M.; Ando, H.; Asaba, T. *Int. J. Chem. Kin.* **1979,** *11,* 285.
16. (a) Skinner, G. B. *J. Chem. Phys.* **1959,** *31,* 268; (b) Skinner, G. B.; Sweet, R. C.; Davis, S. K. *J. Phys. Chem.* **1971,** *75,* 1; (c) Skinner, G. B.; Rogers, D.; Patel, K. B. *Int. J. Chem. Kin.* **1981,** *13,* 481.
17. Brabbs, T. A.; Zlatarich, S. A.; Belles, F. A. *J. Chem. Phys.* **1960,** *33,* 307.
18. Michael, J. V.; Sutherland, J. W. *Int. J. Chem. Kin.* **1986,** *18,* 409.
19. Michael, J. V.; Fisher, J. R. *AIP Conference Proceedings 208, Current Topics in Shock Waves, Seventeenth International Symposium on Shock Waves and Shock Tubes* **1989,** *17,* 210.
20. (a) McNesby, J. R.; Braun, W.; Ball, J. in *Creation and Detection of the Excited State;* Lamola, Ed.; Dekker: New York, 1971, Vol. 1, part B, p. 503; (b) Ridley, B. A.; Davenport, J. A.; Stief, L. J.; Welge, K. H. *J. Chem. Phys.* **1972,** *57,* 520.
21. Michael, J. V.; Wagner, A. F. *J. Phys. Chem.* **1990,** *94,* 2453.
22. Mitchell, A. C. G.; Zemansky, M. W. *Resonance Radiation and Excited Atoms;* Cambridge, London, 1934.
23. Maki, R. G.; Michael, J. V.; Sutherland, J. W. *J. Phys. Chem.* **1985,** *89,* 4815; and references cited therein.
24. Barker, J. R.; Michael, J. V. *J. Opt. Soc. Am.* **1968,** *58,* 1615.
25. The method for O- and N-atoms is described in Lee, J. H.; Michael J. V.; Payne, W. A.; Stief, L. J. *J. Chem. Phys.* **1978,** *69,* 3069; and is used for H-atoms by Miller, J. C.; Gordon, R. J. *J. Chem. Phys.* **1983,** *78,* 3713.
26. Pamidimukkala, K. M.; Lifshitz, A.; Skinner, G. B.; Wood, D. R. *J. Chem. Phys.* **1981,** *75,* 1116.
27. (a) Bittner, J. D.; Howard, J. B. *Prog. Astronaut Aeronaut.* **1978,** *62,* 335; (b) Warnatz, J.; Bockhorn, H.; Moser, A.; Wenz, H. W. *Symp. (Int.) Combust., [Proc.]* **1982,** *19,* 197.
28. (a) Bradley, J. N.; Kistiakowsky, G. B. *J. Chem. Phys.* **1961,** *35,* 264; (b) Kistiakowsky, G. B.; Richards, L. W., *J. Chem. Phys.* **1962,** *36,* 1707; (c) Hand, C. W.; Kistiakowsky, G. B. *J. Chem. Phys.* **1962,** *37,* 1239; (d) Glass, G. P.; Kistiakowsky, G. B.; Michael, J. V.; Niki, H. *J. Chem. Phys.* **1965,** *42,* 608; (e) Glass, G. P.; Kistiakowsky, G. B.; Michael J. V.; Niki, H. *Symp. (Int.) Combust., [Proc.]* **1964,** *10,* 513.
29. Fenimore, C. P.; Jones, G. W. *J. Chem. Phys.* **1963,** *39,* 1514.
30. Jonathan, N.; Marmo, F. F.; Padur, J. *J. Chem. Phys.* **1965,** *42,* 1463.
31. Arrington, C. A.; Brennen, W.; Glass, G. P.; Michael, J. V.; Niki, H., *J. Chem. Phys.* **1967,** *43,* 525.
32. Brown, J. M.; Thrush, B. A. *Trans. Faraday Soc.* **1967,** *63,* 630.
33. Williamson, D. G.; Bayes, K. D. *J. Phys. Chem.* **1969,** *73,* 1232.
34. (a) Jones, I. T. N.; Bayes, K. D. *Symp. (Int.) Combust., [Proc.]* **1972,** *14,* 277; (b) *Proc. Roy. Soc. London, A* **1973,** *335,* 547.
35. Westenberg, A. A.; deHaas, N. *J. Chem. Phys.* **1969,** *73,* 1181.
36. Hoyermann, K.; Wagner, H. Gg.; Wolfrum, J. *Z. Phys. Chem. (Frankfurt am Main)* **1969,** *63,* 193.

37. James, G. S.; Glass, G. P. *J. Chem. Phys.* **1969,** *50,* 2268.
38. Aleksandrov, E. N.; Arutyunov, V. S.; Koslov, S. N. *Kinet. Catal.* **1981,** *22* 391.
39. Homann, K. H.; Wellmann, Ch. *Ber. Bunsen-Ges. Phys. Chem.* **1983,** *87,* 527.
40. Lohr, R.; Roth, P. *Ber. Bunsen-Ges. Phys. Chem.* **1981,** *85,* 153.
41. Frank, P.; Bhaskaran, K. A.; Just, Th. *Symp. (Int.) Combust., [Proc.]* **1986,** *21,* 885.
42. Mahmud, K.; Fontijn, A. *J. Phys. Chem.* **1987,** *91,* 1918.
43. Russell, J. J.; Senkan, S. M.; Seetula, J. A.; Gutman, D. *Symp. (Int.) Combust., [Proc.]* **1988,** *22* 1007.
44. Williamson, D. G. *J. Phys. Chem.* **1971,** *75,* 4053.
45. Kanofsky, J. R.; Lucas, D.; Pruss, F.; Gutman, D. *J. Phys. Chem.* **1974,** *78,* 311.
46. Blumenberg, B.; Hoyermann, K.; Sievert, R. *Symp. (Int.) Combust., [Proc.]* **1976,** *16,* 841.
47. Homann, K. H.; Schweinfurth, H. *Ber. Bunsen-Ges. Phys. Chem.* **1981,** *85,* 569.
48. Clemo, A. R.; Duncan, G. L.; Grice, R. *J. Chem. Soc., Faraday Trans. 2* **1982,** *78,* 1231; **1983,** *79,* 637.
49. Homann, K. H.; Wellmann, Ch. *Ber. Bunsen-Ges. Phys. Chem.* **1983,** *87,* 609.
50. (a) Vinckier, C.; Schaekers, M.; Peeters, J. *J. Phys. Chem.* **1985,** *89,* 508; (b) Peeters, J.; Schaekers, M.; Vinckier, C. *J. Phys. Chem.,* **1986,** *90,* 6552; (c) Peeters, J.; Boullart, W.; Van de Ven, P.; Vinckier, C. Joint Meeting of the French and Italian Sections of the Combustion Institute, Amalfi, 1987, paper 2.12; *XIIIth International Conference on Photochemistry,* Budapest, 1987, p. 69.
51. Schmoltner, A. M.; Chu, P. M.; Lee, Y. T. *J. Chem. Phys.* **1989,** *91,* 5365.
52. Harding, L. B. *J. Phys. Chem.* **1981,** *85,* 10.
53. Harding, L. B.; Wagner, A. F. *J. Phys. Chem., 90,* 2974.
54. Green, P. G.; Kinsey, J. L.; Field, R. W. *J. Chem. Phys.* **1989,** *91,* 5160; and references cited therein.
55. (a) Dainton, F. S. *Chain Reactions: An Introduction;* Methuen: London, 1956; p. 122; (b) Bradley, J. N. *Flame and Combustion Phenomena;* Methuen: London, 1969; (c) Glassman, I. *Combustion;* Academic Press: New York, 1987; p. 56.
56. For earlier reviews see: (a) Wagner, H. Gg. *Symp. (Int.) Combust., [Proc.]* **1972,** *14,* 27; (b) Khandewal, S. C.; Skinner, G. B. In Lifshitz, Ed. *Shock Waves in Chemistry;* Dekker: New York, 1981; (c) Westbrook, C. K.; Dryer, F. L. *Symp. (Int.) Combust., [Proc.]* **1980,** *18,* 749; (d) Bittner, J. D.; Howard, J. B. *Symp. (Int.) Combust., [Proc.]* **1982,** *19,* 211; (e) Westbrook, C. K.; Dryer, F. L. *Prog. Energy Combust. Sci.* **1984**, *10,* 1.
57. For reviews of earlier work see: (a) Baulch, D. L.; Drysdale, D. D.; Horne, D. G.; Lloyd, A. C. *Evaluated Data for High Temperature Reactions,* Butterworths, London, 1972; Vol. 1, (b) Warnatz, J. In *Combustion Chemistry;* Gardiner, Ed.; Springer-Verlag: New York, 1984; (c) Cohen, N.; Westberg, K. R. *J. Phys. Chem. Ref. Data* **1983,** *12,* 531.

58. Gutman, D.; Schott, G. L. *J. Chem. Phys.* **1967,** *46,* 4576.
59. Gutman, D.; Hardwidge, E. A.; Dougherty, F. A.; and Lutz, R. W. *J. Chem. Phys.* **1967,** *47,* 4400.
60. Schott, G. L. *Symp. (Int.) Combust., [Proc.]* **1968,** *12,* 569; Schott, G. L. *Combust. Flame* **1973,** *13,* 357.
61. Pamidimukkala, K. M.; Skinner, G. B. *Thirteenth Int. Symp. on Shock Waves and Shock Tubes;* SUNY Press: Albany, NY, 1981; p. 585.
62. Frank, P.; Just, Th. *Ber. Bunsen-Ges. Phys. Chem.* **1985,** *89,* 181.
63. (a) Fujii, N.; Shin, K. *Chem. Phys. Lett.* **1988,** *151,* 461; (b) Fujii, N.; Sato, T.; Miyama, H.; Shin, K. S.; Gardiner, W. C., Jr., *AIP Conference Proceedings 208, Current Topics in Shock Waves, Seventeenth International Symposium on Shock Waves and Shock Tubes,* **1989,** *17,* 456.
64. Yuan, T.; Wang, C.; Yu, C.-L.; Frenklach, M.; Rabinowitz, M. J. *J. Phys. Chem.,* **1991,** *95,* 1258. The three parameter expression given for k_{H+O_2} has been transformed into the approximate Arrhenius expression given as equation (14).
65. Masten, D. A.; Hanson, R. K.; Bowman, C. T. *J. Phys. Chem.* **1990,** *94,* 7119.
66. Pirraglia, A. N.; Michael, J. V.; Sutherland, J. W.; Klemm, R. B. *J. Phys. Chem.* **1989,** *93,* 282.
67. Melius, C. F.; Blint, R. J. *Chem. Phys. Lett.* **1979,** *64,* 183.
68. Miller, J. A. *J. Chem. Phys.* **1981,** *74,* 5120; *J. Chem. Phys.* **1986,** *84,* 6170.
69. Rai, S. N.; Truhlar, D. G. *J. Chem. Phys.* **1983,** *79,* 6046.
70. Troe, J. *J. Phys. Chem.* **1986,** *90,* 3485; *Symp. (Int.) Combust., [Proc.]* **1988,** *22,* 843, and comments on 862.
71. Graff, M. M.; Wagner, A. F. *J. Chem. Phys.* **1990,** *92,* 2423.
72. (a) Varandas, A. J. C.; Brandao, J.; Quintales, L. A. M. *J. Phys. Chem.* **1988,** *92,* 3732; Quintales, L. A. M.; Varandas, A. J. C.; Alvarino, J. M., *J. Phys. Chem.,* **1988,** *92,* 4552. (b) Davidsson, J.; Nyman, G., *J. Chem. Phys.* **1990,** *92,* 2407; Nyman, G.; Davidsson, J. *J. Chem. Phys.* **1990,** *92,* 2415.
73. Slack, M. W. *Combust. Flame* **1978,** *28,* 241.
74. (a) Hsu, K. -J.; Durant, J. L.; Kaufman, F. *J. Phys. Chem.* **1987,** *91,* 1985; (b) Hsu, K. -J.; Anderson, S. M.; Durant, J. L.; Kaufman, F. *J. Phys. Chem.* **1989,** *93,* 1018.
75. Cobos, C. J.; Hippler, H.; and Troe, J. *J. Phys. Chem.* **1985,** *89,* 342.
76. (a) Herron, J. T.; Huie, R. E. *J. Phys. Chem. Ref. Data* **1973,** *2,* 467; (b) Tsang, W. *J. Phys. Chem. Ref. Data* **1986,** *15,* 1087.
77. Westenberg, A. A.; deHaas, N. *J. Chem. Phys.* **1967,** *46,* 490; *J. Chem. Phys.* **1969,** *50,* 2512.
78. Barassin, J.; Combourieu, J. *Bull. Soc. Chim. France,* **1974,** p. 1.
79. Brabbs, T. A.; Brokaw, R. S. *Symp. (Int.) Combust., [Proc.]* **1974,** *15,* 893.
80. Roth, P.; Just, Th. *Ber. Bunsen-Ges. Phys. Chem.* **1977,** *81,* 572.
81. Klemm, R. B.; Tanzawa, T.; Skolnik, E. G.; Michael, J. V. *Symp. (Int.) Combust., [Proc.]* **1980,** *18,* 785.
82. (a) Felder, W.; Fontijn, A. *Chem. Phys. Lett.* **1979,** *67,* 53; (b) Fontijn, A. *Symp. (Int.) Combust., [Proc.]* **1980,** *18,* 797.
83. Sutherland, J. W.; Michael, J. V.; Klemm, R. B. *J. Phys. Chem.* **1986,** *90,* 5941.

84. Walch, S. P.; Dunning, T. H. Jr. *J. Chem. Phys.*, **1980,** *72,* 3221.
85. Wagner, A. F.; Dunning, T. H., Jr.; Walch, S. P.; Schatz, G. C. Private communication.
86. Luntz, A. C.; Andresen, P. *J. Chem. Phys.* **1980,** *72,* 5851.
87. Michael, J. V.; Keil, D. G.; Klemm, R. B. *Int. J. Chem. Kin.* **1983,** *15,* 705.
88. Golden, D. M. *J. Phys. Chem.* **1979,** *83,* 108; and see also Figure 2, Barker, J. R. *Symp. (Int.) Combust., [Proc.]* **1980,** *18,* 798.
89. Cohen, N. *Int. J. Chem. Kin.* **1986,** *18,* 59.
90. Albers, E. A.; Hoyermann, K.; Wagner, H. Gg.; Wolfrum, J. *Symp. (Int.) Combust., [Proc.]* **1970,** *13,* 81.
91. Lifshitz, A.; Michael, J. V. *Symp. (Int.) Combust. [Proc.]* **1990,** *23,* 59.
92. Rawlins, W. T.; Gardiner, W. C., Jr. *J. Chem. Phys.* **1974,** *60,* 4676.
93. Ernst, J.; Wagner, H. Gg.; Zellner, R. *Ber. Bunsenges. Phys. Chem.* **1977,** *81,* 1270.
94. del Greco, F. P.; Kaufman, F. *Disc. Faraday Soc.* **1962,** *33,* 128; Kaufman, F. *Ann. Geophys.* **1964,** *20,* 106.
95. Dixon-Lewis, G.; Wilson, W. E.; Westenberg, A. A. *J. Chem. Phys.* **1966,** *44,* 2877.
96. Wilson, W. E.; O'Donovan, J. T. *J. Chem. Phys.* **1967,** *47,* 5455.
97. Breen, J. E.; Glass, G. P. *J. Chem. Phys.* **1970,** *52,* 1082.
98. Westenberg, A. A.; deHaas, N. *J. Chem. Phys.* **1973,** *58,* 4066.
99. McKenzie, A.; Mulcahy, M. F. R.; Steven, J. P. *J. Chem. Phys.* **1973,** *59,* 3244.
100. Clyne, M. A. A.; Down, S. *J. Chem. Soc. Faraday Trans. 2,* **1974,** *70,* 253.
101. Trainor, D. W.; von Rosenberg, C. W., Jr. *J. Phys. Chem.* **1974,** *61,* 1010.
102. Farquharson, G. K.; Smith, R. H. *Aust. J. Chem.* **1980,** *33,* 1425.
103. Wagner, G.; Zellner, R. *Ber. Bunsenges. Phys. Chem.* **1981,** *85,* 1122.
104. Harding, L. B.; Wagner, A. F. *Symp. (Int.) Combust., [Proc.]* **1988,** *22,* 983.
105. Issacson, A. D.; Truhlar, D. G.; Rai, S. N.; Steckler, R.; Hancock, G. C.; Garrett, B. C.; Redmon, M. J. *Computer Physics Communications* **1987,** *47,* 91.
106. Chase, M. W., Jr.; Davies, C. A.; Downey, J. R., Jr.; Frurip, D. J.; McDonald, R. A.; Syverud, A. N. *J. Chem. Ref. Data* **1985,** *14,* Supplement No. 1.
107. Michael, J. V. Unpublished results.
108. Tsang, W.; Hampson, R. F. *J. Phys. Chem. Ref. Data* **1986,** *15,* 1078.
109. Sutherland, J. W.; Michael, J. V.; Pirraglia, A. N.; Nesbitt, F. L.; Klemm, R. B. *Symp. (Int.) Combust., [Proc.]* **1986,** *21,* 929.
110. Clyne, M. A. A.; Thrush, B. A. *Proc. Roy. Soc. London A.* **1963,** *275,* 544.
111. Westenberg, A. A.; deHaas, N. *J. Chem. Phys.* **1967,** *47,* 4241.
112. Pamidimukkala, K. M.; Skinner, G. B. *J. Chem. Phys.* **1982,** *76,* 311.
113. Presser, N.; Gordon, R. J. *J. Chem. Phys.* **1985,** *82,* 1291.
114. Marshall, P., Fontijn, A. *J. Chem. Phys.* **1987,** *87,* 6988.
115. Zhu, Y-F.; Arepalli, S.; Gordon, R. J. *J. Chem. Phys.* **1989,** *90,* 183.
116. Bowman, J. M.; Wagner, A. F. In *The Theory of Chemical Reaction Dynamics;* Clary, Ed.; Reidel: Boston, 1986, p. 47.
117. Lee, K. T.; Bowman, J. M.; Wagner, A. F.; Schatz, G. C. *J. Chem. Phys.* **1982,** *76,* 3563; *J. Chem. Phys.* **1982,** *76,* 3583.

118. Walch, S. P.; Dunning, T. H., Jr.; Raffenetti, R. C.; Bobrowicz, F. W. *J. Chem. Phys.* **1980,** *72,* 406; Walch, S. P.; Wagner, A. F.; Dunning, T. H., Jr.; Schatz, G. C. *J. Chem. Phys.* **1980,** *72,* 2894.
119. Garrett, B. C.; Truhlar, D. G. *J. Chem. Phys.* **1984,** *81,* 309.
120. Bowman, J. M.; Wagner, A. F.; Walch, S. P.; Dunning, T. H., Jr.; *J. Chem. Phys.* **1984,** *81,* 1739.
121. Joseph T.; Truhlar, D. G.; Garrett, B. C. *J. Chem. Phys.* **1988,** *88,* 6982.
122. Bowman, J. M. *Adv. Chem. Phys.* **1985,** *61,* 115.
123. Robie, D. C.; Arepalli, S.; Presser, N.; Kitsopoulos, T.; Gordon, R. J. *Chem. Phys. Lett.* **1987,** *134,* 579.
124. Garrett, B. C.; Truhlar, D. G.; Bowman, J. M.; Wagner, A. F.; Robie. D.; Arepalli, S.; Presser, N.; Gordon, R. J. *J. Am. Chem. Soc.* **1986,** *108,* 3515.
125. Bowman, J. M.; Wagner, A. F. *J. Chem. Phys.* **1987,** *86,* 1967; Wagner, A. F.; Bowman, J. M. *J. Chem. Phys.* **1987,** *86,* 1976.
126. Holzrichter, K.; Wagner, H. Gg. *Symp. (Int.) Combust., [Proc.]* **1980,** *18,* 769.
127. (a) Dean, A. M.; Hardy, J. E.; Lyon, R. K. *Symp. (Int.) Combust., [Proc.]* **1982,** *19,* 97; (b) Miller, J. A.; Smooke, M. D.; Green, R. M.; Kee, R. J. *Comb. Sci. Technol.* **1983,** *34,* 149; (c) Miller, J. A.; Bowman, C. T. *Prog. Energy Combust. Sci.* **1989,** *15,* 287, and reference therein.
128. Lyon, R. K. U. S. Patent 3,900,554, 1975; *Int. J. Chem. Kin.* **1976,** *8,* 315; *Environ. Sci. Technol.* **1987,** *21,* 231.
129. Dove, J. E.; Nip, W. S. *Can. J. Chem.* **1976,** *52,* 1171.
130. Michael, J. V.; Sutherland, J. W.; Klemm, R. B. *J. Phys. Chem.* **1986,** *90,* 497.
131. Marshall, P.; Fontijn, A. *J. Chem. Phys.* **1986,** *85,* 2637.
132. Ko, T.; Marshall, P., Fontijn, A. *J. Phys. Chem.* **1990,** *94,* 1401.
133. Hack, W.; Rouveirolles, P.; Wagner, H. Gg. *J. Phys. Chem.* **1986,** *90,* 2505.
134. Melius, C. F. Private communication, October, 1983.
135. Garrett, B. C., Koszykowski, M. L.; Melius, C. F.; Page, M. *J. Phys. Chem.* **1990,** *94,* 7096.
136. Sutherland, J. W.; Michael, J. V. *J. Chem. Phys.* **1988,** *88,* 830.
137. Gibson, S. T.; Greene, J. P.; Berkowitz, J. *J. Chem. Phys.* **1985,** *83,* 4319.
138. Demissy, M.; Lesclaux, R. *J. Am. Chem. Soc.* **1980,** *102,* 2897.
139. Wilson, W. E., Jr. *J. Phys. Chem. Ref. Data* **1972,** *1,* 535.
140. Dixon-Lewis, G.; Williams, D. J. *Comprehensive Chemical Kinetics;* Elsevier: Amsterdam, 1977; Vol. 17, p. 1.
141. Zellner, R. *J. Phys. Chem.* **1979,** *83,* 18.
142. Kaufman, F. *Annu. Rev. Phys. Chem.* **1979,** *30,* 411.
143. Baulch, D. L.; Cox, R. A.; Hampson, R. F., Jr.; Kerr, J. A.; Troe, J.; Watson, R. T. *J. Phys. Chem. Ref. Data* **1980,** *9,* 295.
144. Walch, S. P.; Dunning, T. H., Jr. *J. Chem. Phys.* **1980,** *72,* 1303.
145. Schatz, G. C.; Walch, S. P. *J. Chem. Phys.* **1980,** *72,* 776.
146. Schatz, G. C.; Elgersma, H. *Chem. Phys. Lett.* **1980,** *73,* 21.
147. Isaacson, A. D.; Truhlar, D. G. *J. Chem. Phys.* **1982,** *76,* 1380.
148. Kraka, E.; Dunning, T. H., Jr. Private communication.
149. Fenimore, C. P.; Jones, G. W. *J. Phys. Chem.* **1958,** *62,* 693.

150. Dixon-Lewis, G.; Sutton, M. M.; Williams, A. *Trans. Faraday Soc.* **1965,** *61,* 255.
151. Madronich, S.; Felder, W. *J. Phys. Chem.* **1984,** *88,* 1857.
152. Michael, J. V.; Sutherland, J. W. *J. Phys. Chem.* **1988,** *92,* 3853.
153. Davidson, D. F.; Chang, A. Y.; Hanson, R. K. *Symp. (Int.) Combust., [Proc.]* **1988,** *22,* 1877.
154. Fisher, J. R.; Michael, J. V. *J. Phys. Chem.* **1990,** *94,* 2465.
155. Kaufman, F.; delGreco, F. P. *Discuss. Faraday Soc.* **1962,** *33,* 128; Dixon-Lewis, G.; Wilson, W. E.; Westenberg, A. A. *J. Chem. Phys.* **1966,** *44,* 2877; Greiner, N. R. *J. Chem. Phys.* **1969,** *51,* 5049; Stuhl, F.; Niki, H. *J. Chem. Phys.* **1972,** *57,* 3671; Westenberg, A. A.; deHaas, N. *J. Chem. Phys.* **1973,** *58,* 4061; Smith, I. W. M.; Zellner, R. J. *J. Chem. Soc. Faraday Trans. 2* **1974,** *70,* 1045; Atkinson, R.; Hansen, D. A.; Pitts, J. N., Jr. *J. Chem. Phys.* **1975,** *63,* 1703; Overend, R. P.; Paraskevopoulos, G.; Cvetanovic, R. J. *Can. J. Chem.* **1975,** *53,* 3374.
156. Ravishankara, A. R.; Nicovich, J. M.; Thompson, R. L.; Tully, F. P. *J. Phys. Chem.* **1981,** *85,* 2498.
157. Tully, F. P., Ravishankara, A. R. *J. Phys. Chem.* **1980,** *84,* 3126.
158. Frank, P.; Just, Th. *Ber. Bunsen-Ges. Phys. Chem.* **1985,** *89,* 181.
159. Oldenborg, R. C.; Harradine, D. M.; Loge, G. W.; Winn, K. R. *J. Phys. Chem.;* forthcoming.
160. Vaghjiani, G. L.; Ravishankara, A. R.; Cohen, N. *J. Phys. Chem.* **1989,** *93,* 7833.
161. (a) Farkas, A. Z. *Elektrochem.* **1930,** *36,* 782; *Z. Physik. Chem.* **1930,** *B10,* 419. (b) Gieb, K. H.; Harteck, P. *Z. Physik. Chem.,* **1931,** *Bodenstein Festband,* 849. (c) Farkas, A.; Farkas, L. *Proc. Roy. Soc. (London),* **1935,** *A152,* 124. (d) van Meersche, M. *Bull. Soc. Chim. Belg.* **1951,** *60,* 99. (e) Boato, G.; Careri, G.; Cimino, A.; Molanari, E.; Volpi, G. G. *J. Chem. Phys.* **1956,** *24,* 783.
162. For reviews of earlier theoretical and experimental work see: (a) Glasstone, S.; Laidler, K. J.; Eyring, H. *The Theory of Rate Processes;* McGraw Hill: New York, 1941. (b) Laidler, K. J. *Chemical Kinetics;* McGraw Hill: New York, 1965. (c) Johnston, H. S. *Gas Phase Reaction Rate Theory;* Ronald: New York, 1966. (d) Kaufman, F. *Annu. Rev. Phys. Chem.* **1969,** *20,* 43. (e) Truhlar, D. G.; Wyatt, R. E. *Annu. Rev. Phys. Chem.* **1979,** *27,* 1.
163. (a) Schulz, W. R.; Le Roy, D. J. *Can. J. Chem.* **1964,** *42,* 2480; *J. Chem. Phys.* **1965,** *42,* 3869. (b) Ridley, B. A.; Schulz, W. R.; Le Roy, D. J. *J. Chem. Phys.* **1964,** *44,* 3344. (c) Le Roy, D. J.; Ridley, B. A.; Quickert, K. A., *Disc. Faraday Soc.* **1967,** *44,* 92. (d) Quickert, K. A.; Le Roy, D. J. *J. Chem. Phys.* **1970,** *53,* 1325; *J. Chem. Phys.* **1971,** *54,* 5444. (e) Mitchell, D. N.; Le Roy, D. J. *J. Chem. Phys.* **1973,** *58,* 3449.
164. Westenberg, A. A.; de Haas, N. *J. Chem. Phys.* **1967,** *47,* 1393.
165. Jayaweera, I. S. Ph. D. Dissertation, Dalhousie University, 1989; Jayaweera, I. S.; Pacey, P. D. *J. Phys. Chem.* **1990,** *94,* 3614.
166. Michael, J. V.; Fisher, J. R. *J. Phys. Chem.* **1990,** *94,* 3318.
167. Michael, J. V. *J. Chem. Phys.* **1990,** *92,* 3394.
168. London, F. *Probleme der Modernen Physik (Sommerfeld Festschrift);* 1928, p.

104; London, F. *Z. Elektrochem.* **1929,** *35,* 552; Pelzer, H.; Wigner, E. *Z. Physik. Chem. B.* **1932,** *58,* 445; Eyring, H. *J. Chem. Phys.* **1935,** 3, 107; Evans, M. G.; Polanyi, M. *Trans. Faraday Soc.,* **1935,** *31,* 875; *Trans. Faraday Soc.* **1937,** *33,* 448; Polanyi, M. *J. Chem. Soc.* **1937,** 627.

169. See references 162d and e for a discussion of earlier semi-empirical and ab initio calculations. See also, Truhlar, D. G.; Wyatt, R. E. *Adv. Chem. Phys.* **1977,** *36,* 141.

170. (a) Liu, B. *J. Chem. Phys.* **1973,** *58,* 1925. (b) Siegbahn, P.; Liu, B. *J. Chem. Phys.* **1978,** *68,* 2457.

171. Truhlar, D. G.; Horowitz, C. J. *J. Chem. Phys.* **1978,** *68,* 2466.

172. Varandas, A. J. C.; Brown, F. B.; Mead, C. A.; Truhlar, D. G.; Blais, N. C. *J. Chem. Phys.* **1987,** *86,* 6258.

173. Garrett, B. C.; Truhlar, D. G. *Proc. Natl. Acad. Sci.* **1979,** *76,* 4755; *J. Chem. Phys.* **1980,** *72,* 3460.

174. Garrett, B. C.; Truhlar, D. G.; Varandas, A. J. C.; Blais, N C. *Int. J. Chem. Kin.* **1986,** *18,* 1065.

175. Sun, Q.; Bowman, J. M. *J. Phys. Chem.* **1990,** *94,* 718.

176. Michael, J. V.; Fisher, J. R.; Bowman, J. M.; Sun, Q. *Science* **1990,** *249,* 269.

177. Miller, J. A.; Fisk, G. A. *Chem. and Eng. News* **1987,** *65*(35), 22.

178. (a) Pack, R. T.; Parker, G. A. *J. Chem. Phys.* **1987,** *87,* 3888; (b) Zhao, M.; Mladenovic, M.; Truhlar, D. G.; Schwenke, D. W.; Sun, Y.; Kouri, D. J.; Blais, N. C. *J. Am. Chem. Soc.* **1989,** *111,* 852; (c) Park, T. J.; Light, J. C. *J. Chem. Phys.* **1989,** *91,* 974; (d) Zhang, J. Z. H.; Miller, W. H. *J. Chem. Phys.* **1989,** *91,* 1528; (e) Manolopoulos, D. E.; Wyatt, R. E. *J. Chem. Phys.* **1990,** *92,* 810; and references cited therein.

179. Kliner, D. A. V.; Rinnen, K.-D.; Zare, R. N. *Chem. Phys. Lett.* **1990,** *166,* 107.

180. See for example, Harding, L. B. *J. Phys. Chem.* **1989,** *93,* 8004.

DIFFUSION CONTROLLED ELEMENTARY REACTIONS IN LOW DIMENSIONS

Raoul Kopelman and Yong-Eun Koo

Advances in Chemical Kinetics and Dynamics, Volume 1, pages 113–138

ISBN: 1-55938-161-2

I. INTRODUCTION

Chemical kinetics in low dimensions usually involves active sites or surfaces which lower significantly the activation energy barriers. At the same time, there is essentially no convective stirring in low dimensions and the diffusion process is less efficient due to the compactness[1] of the Brownian motion. As a result, low-dimensional reactions are very often diffusion limited. Furthermore, the so-called compact Brownian motion leads to anomalous diffusion[2] and to anomalous rate laws.[3] Finally, these anomalous macroscopic rate laws are a manifestation of the unexpected entropic self-ordering of reacting molecules, that is, a dramatic self-organization on a microscopic or mesoscopic (or sometimes macroscopic) length scale.[4]

At the risk of oversimplification, we present here some mental pictures. Random walk or diffusion in three dimensions is largely non-repetitive. A drunk space-pilot is not likely to come back to his Bar-planet using random navigation. On the other hand, a drunk in an alley always comes back to the Bar. In more mathematical terms, a random walker in three dimensions has a finite probability of escaping from his point of origin. However, in one or even two dimensions, the random walker's probability of escaping from the origin is zero (in infinite time!). Thus, the trace of the random walker in the latter case is "compact." It efficiently sweeps out the environment of the origin. Basically it is localized in space. This is no longer true in three (or higher) dimensions, where the trace is non-compact (sometimes called "ramified") and the particle is thus delocalized. This difference is crucial for chemical reaction kinetics. If A and B "annihilate" each other by reacting, one cannot expect a B particle to survive in a region in which one or more A particles are localized. This effective kinetic exclusion principle obviates the possibility of a Poissonian (integrated) particle distribution, in low dimensions. Instead, one finds segregated patches of A and B populations of particles and the reaction proceeds only at the surfaces or interfaces of these patches. Thus the reaction laws differ drastically from the classical laws, which were derived on the assumption of statistically random distributions in space (Maxwell's hypothesis that the most probable distributions are those that maximize the entropy).

Practically every heterogeneous reaction is a low-dimensional reaction, whether it occurs on a surface, inside a pore, at a micelle, or elsewhere. In addition, some homogeneous reactions, such as solid-state reactions, occur preferentially in low-dimensional environments, for example, domain boundaries, defect sites, surface steps, or radiation tracks. Reactions such as polymerization, gelation, and cellular growth also exhibit low-dimensional dynamics. We note that, mechanistically, these low-dimensional reactions may involve molecules, radicals, atoms, electrons, holes, excitons, or solitons.

Experimentally, simple elementary reactions that have been observed to follow strange rate laws range from solution photodimerization inside porous polymeric membranes[5,6] to exciton fusion in molecular alloys.[3] Further evidence for such anomalies comes from excimer kinetics in polymer blends,[7–9] oxygen quenching in porous media,[10] energy transfer in photosynthetic systems,[11] and neuron gating kinetics in sodium channels.[12] There also appear to be many other instances where anomalous rate laws have been fitted (unnecessarily) by complex mechanisms.

There are three major categories of elementary, bimolecular mechanisms:

$$A + A \rightarrow \text{Products} \tag{1}$$

$$A + B \rightarrow \text{Products} \tag{2}$$

$$A + C \rightarrow C + \text{Products} \tag{3}$$

In the above we assume that the products are inert or leave the reaction medium (e.g., desorb). We also assume that in reaction (2) there is no A + A or B + B reaction mechanism. We note that in reaction (3) C is a catalyst or catalytic site. Reactions (1) and (2) are described by textbook bimolecular rate equations and reaction (3) by the pseudo-monomolecular equation. We also note that reaction (1) is "second order" in A concentration while reaction (2) is "second order" overall, that is, first order in A and first order in B. Reaction (3) is actually first order in both A and C (if C is varied from case to case) and thus also "second order" overall, while being pseudo-first order in a given case (where only A varies in time). These three categories (1–3) give rise to extremely different rate laws, all different from the classical law for low dimensions. Within each category there arise various subcategories each with different rate laws which depend upon distinctions such as dimensionality, initial conditions, source (input) condi-

Table 1. Reaction Orders X in One Dimension

Reaction	*Steady State*	$t \to \infty$
Random $A + A \to A_2 \uparrow$	3	3
$\downarrow A_2 \to A + A \to A_2 \uparrow$	2	3
Classical biomolecular	2	2
Random $A + B \to AB \uparrow$	4*	5
$\downarrow AB \to A + B \to AB \uparrow$	2	3
If $A + A \to A_2 \uparrow$ and $B + B \to B_2 \uparrow$	3	3
Examples: hydrogen on Pt Wire		
$\downarrow H_2 \to H + H \to H_2 \uparrow$	2	3
$\downarrow D_2 \to D + D \to D_2 \uparrow$	2	3
$H + D \to HD \uparrow$	3	3

Note: *For hard-sphere particles and vertical reactions; otherwise $X = 2$.

tions, and other subtle constraints which do not affect the textbook rate laws. Even the achievement of a steady-state is no longer guaranteed by a steady input of particles. We find that a catalyst may be poisoned "mathematically," that is, as a result of the mathematical description of fluctuations (note that the lower the dimension, the more serious the fluctuations). Furthermore, oscillatory and even chaotic behaviors become possible for elementary binary reactions. The above is very different from the Maxwell hypothesis underlying the kinetic theory of gases, the hypothesis of randomly even (Poissonian) spatial distribution of all particles. In chemical reactions with no stirring or convection, this hypothesis is no longer of universal validity.

The new "zoo" of rate laws is demonstrated by a few examples of reactions constrained to one dimension (Table 1). While some early theoretical work and some early experiments may have indicated rate law anomalies,[13,14] it was the development of computer simulations that became the major force behind the recent developments in this area.

The importance of short-time fluctuations has already been pointed out by Einstein in his 1905 papers on Brownian motion. However, the possibility of drastic long-time fluctuations was probably first pointed out by the famous Soviet chemist, physicist, and astrophysicist, Zeldovich.[15] Similar long-time fluctuation phenomena have been pointed out in the context of solid-state kinetics.[16] It is the occurrence of such long-time and long-range fluctuations that leads to reactant self-ordering and to the strange new rate laws for elementary reaction mechanisms.

We first review the theory and simulations from the standpoint of an experimentalist, emphasizing the nature and form of the novel laws. We also attempt to give physicochemical rationalizations for and insights into this new paradigm. We then describe some selected experiments in order to demonstrate these laws and the consequences and opportunities which they create.

II. THEORY AND SIMULATIONS

A. Batch Reactions

Batch reactions or "big bang" reactions were the first to be treated with respect to the long-time, long-range "fluctuations" that give rise to the anomalous asymptotic (long-time) behavior. For instance,[17,18] the "trapping reaction" and the "target reaction" are both special cases of reaction (3):

$$A + C \rightarrow C. \tag{3'}$$

In the trapping reaction C is fixed and A moves, while in the target (quenching) reaction A is fixed and C is mobile. Classically, one expects the survival function C_A (A concentration) to decrease exponentially in time:

$$C_A = C_A^\circ \exp(-kt) \tag{4}$$

where C_A° is the "big bang" ($t = 0$) reactant concentration and k the rate constant. In the Smoluchowski treatment of diffusion-limited reactions[17] $k \sim D$ where D is the diffusion constant. However, this is only valid in three (or higher) dimensions. For instance, the Smoluchowski approach in one dimension leads to a different asymptotic solution:

$$C_A \sim \exp(-k' t^{1/2}). \tag{5}$$

Taking into account long-range fluctuations leads to[16]

$$C_A \sim \exp(-k'' t^{1/3}). \tag{6}$$

Many such new relations have been derived[16,18] for low dimensions (including fractal dimensions).

Replacing equations (1–3) by

$$A + i = P \qquad i = A, B, C \tag{7}$$

where P is any resulting products (including A and/or i), one can consider the *differential rate equation* (rather than the integrated forms, involving the "survival function" C_A):

$$-(dC_A/dt) = k(t)C_AC_i. \qquad i = A,B,C \tag{8}$$

The non-classical results can now be written (for long times, $t \rightarrow \infty$)

$$k(t) \sim t^{-h} \qquad h < 1 \tag{9}$$

with the special case $h = 0$ for the classical (textbook) case. In particular, for low dimensions ($d < 2$), one finds[19] for certain A + A reactions:

$$h = 1 - d_s/2, \tag{10}$$

while for the same A + B → 0 reactions,[20]

$$h = 1 - d_s/4 \tag{11}$$

whenever the spectral (fracton)[21] dimension (d_s) is $d_s < 2$.

We note that in a non-Euclidean medium, where different dimensions are invoked, the spectral dimension is the one closely associated with dynamic phenomena, such as vibrations or reaction kinetics. This is discussed in more detail below.

Another simple way of describing many of the results, and particularly those for steady-state reactions, is:

$$-dC_A/dt = kC_A^yC_i^z \tag{12}$$

where k is a time-independent constant and the partial orders y and z may be *non-integers*, and the same is true for the overall order

$$x = y + z. \tag{13}$$

We emphasize that the above is for an elementary bimolecular reaction, where the molecularity of the reaction is described by equation (8). Obviously, by the classical (textbook) reaction laws, x, y, and z are all integers ($y = 1$, $z = 1$, $x = 2$). For instance, for A + A reactions on a critical percolation cluster $x = 5/2$, for a Sierpinski gasket it is $x = 2.47$ and, in general, $x = 1 + 2/d_s$. Similarly, this empirical form [equations (12),(13)] accounts for the strange results listed in Table 1.

In the last few years there has been a bewildering explosion of new results. It would be difficult to discuss all the problems and contradictions of these results here. Instead, we list a few papers that include partial reviews[4,22,23] and give below what we consider to be the main problems of interest.

The beauty of classical kinetics lies in the universality of its formulation. The *functional form* of the rate laws does not depend on the dimension or dimensionality of the medium, initial conditions (for batch reactions), details of source (for steady-state reactions) nature of motion (ballistic, diffusion), relative mobility and reaction of the components, and so on. Such factors only affect the values of the parameters. It is also implied that global and local rate laws are the same and that the particle distribution functions depend only on the concentrations, but not on whether the ensemble is at equilibrium or far from it, or at steady-state or far from it.

When deviations from classical behavior are studied, it is important to distinguish the various factors behind such deviations and the interactions among them. Often it is only with a combination of two such factors that one finds deviations from the classical picture, for example, with a combination of low-dimensionality and a random source term (but not with low-dimensionality and a "geminate" source term).

Non-classical diffusion-controlled reaction kinetics do retain scaling and universality relations.[24–26] However, they differ in detail for different kinds of reactions (A + A vs. A + B), for different modes of reactions (steady-state vs. batch), for different conditions (source term structure, initial distributions), for different conservation laws (exact vs. statistical equality of reactant species), and so on. To get a better insight into the class structure of such scaling laws is one problem of interest to us.

Global rate laws have been of much interest. Indeed, they were the first targets of modern investigations. However, it has become apparent that there are instances where precise global laws do not exist. Can they be substituted by local laws? Can one always define a local law? What is the relation between a local and a global law? This is another problem of interest to us.

Distribution functions have been the linch-pins of statistical mechanics. They have also been recognized as the basic information required for the qualitative and quantitative understanding of reaction kinetics.[4,27] Probably the most stringent test of any kinetic theory is the accuracy of the distribution functions predicted by it. Computer simulations are adept at testing distribution functions with almost as much ease as testing rate laws.[27,28] Examples[29] are given in Figures 1 and 2.

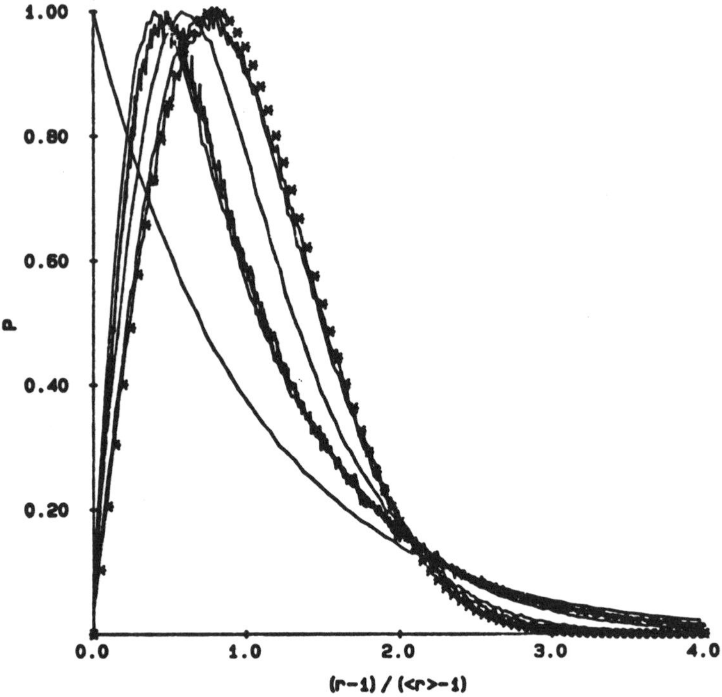

Figure 1. Normalized distribution of gap distances, r, for one-dimensional kinetics.[29] The normalization of r is done by calculating the quantity $(r - 1)/(<r> - 1)$, where $<r>$ is the average gap distance for all particles at any given time. Lattice size is $L = 10000$ sites, initial density is 0.05. The exponential single curve is the initial ($t = 0$) distribution, the two groups of curves are for the $A + A \rightarrow 0$ reaction (left) and $A + A \rightarrow A$ reaction (right); in each group the distribution shown for different reactions times: $t = 100$, 500, and 1000. All curves are averages of 5000 runs. The stars are solution of the analytical equation of Doering and Ben Avraham, for $A + A \rightarrow A$.

The distribution functions can also serve as a bridge between local and global rate laws (to the extent that they exist). Furthermore, on an empirical basis, computer simulations of distribution functions show trends and clues that may be useful for the design of improved theories. Such trends may also be of much aid to experimentalists, providing new qualitative or intuitive model pictures. For instance, the build-up of depletion zones around reactants or catalytic sites depends on

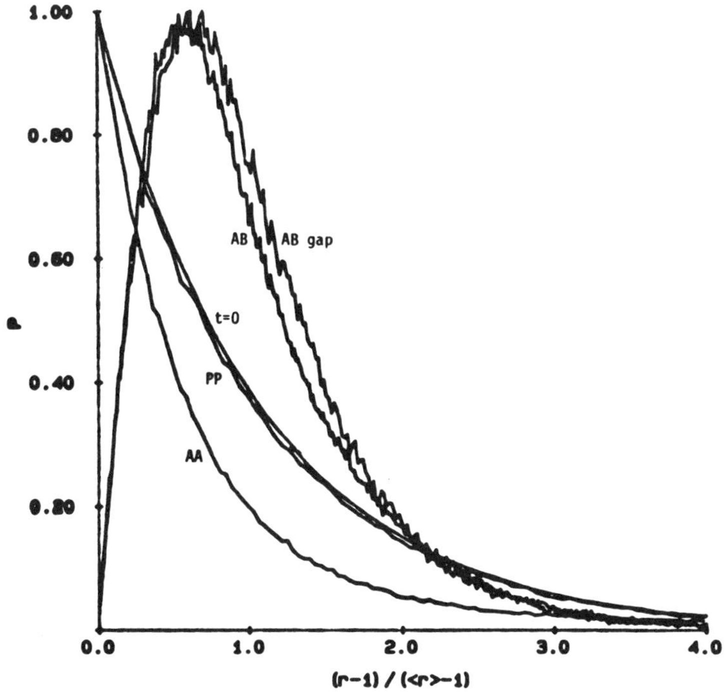

Figure 2. NNDD (Nearest-neighbor distance distributions) for one-dimensional A + B → 0 reactions.[29] From the left, the three exponential-type curves are NNDD for t = 1000 like-like (AA), for t = 1000 particle-particle (PP), and for t = 0 (same for all cases, i.e., AA, BB, and PP), respectively. The two curves with a well-developed maximum are t = 1000 like–unlike (AB) NNDD (left) and like-unlike (AB) gap distribution (right). All NNDD curves are averages of 2000 runs and the gap curve is that of 5000 runs. Total number of lattice sites is 10000 and the initial (t = 0) concentration is 0.05 for A and 0.05 for B (0.1 for P).

specific parameters such as medium dimensionality and reactant mobility.[30,31] Recently there has been much progress in the area of analytical models, for diffusion-controlled reaction kinetics.[32,33]

A + A Reactions

This is apparently the simplest and best understood class of reaction. Somewhat surprisingly, the *coagulation* reaction

$$A + A \rightarrow A \tag{14}$$

is better understood than the *annihilation* reaction,

$$A + A \rightarrow 0. \tag{15}$$

Even more surprising are the drastic differences between these two reactions, in particular the difference in the particle distribution functions. This difference first became apparent from our simulations on interparticle distribution functions.[4,28] It was elegantly proven in the recent analytical work of Doering and Ben Avraham.[32] Specifically, in one dimension, the *interparticle distance distribution probability* $P(r)$ was found to have the form:

$$P(r) = (\pi/2)Cr \exp[-(\pi/2)C^2r^2/2] \tag{16}$$

for long times, where $C = C(t)$ is the instantaneous density and r is the interparticle distance. (This is shown in Figure 1 together with simulation results.) The first important observation is that this distribution differs drastically from the Poisson distribution (in one dimension):

$$P(r) = Cexp(-Cr). \tag{17}$$

The "maximum" of the Poisson distribution is at $r = 0$. The kinetic distribution has a minimum there (with $P(0) = 0$). The maximum of the kinetic distribution is at $r = (2/\pi)^{1/2} C^{-1}$. This maximum *increases* with time, that is, as the density decreases (this is obviously not so for the Poissonian distribution). Furthermore, for large values of r, the kinetic distribution has a Gaussian tail (while the Poissonian has an exponential one). We also note that for small values of r, the kinetic distribution *increases* linearly with r, while the Poissonian one *decreases* exponentially. As the reaction kinetics is mainly determined by the small r range, the resulting rate law is *cubic* in C, while with a Poissonian distribution it is *quadratic* in C (the classical result). This is consistent with the older Torney and McConnell[34] result (with the Anacker et al.[35] transformation).

We now turn to the *annihilation* reaction ($A + A \rightarrow 0$). There is no analytical result, so far, for one dimension. However, there is an *asymptotic analytical* result,[36] for $r \rightarrow \infty$: The tail of the distribution has an *exponential*, rather than Gaussian form. (This is also seen from the simulation results in Figure 1.) This is a major distinction. However, the small r part of the distribution is linear, as in the coagulation ($A + A \rightarrow A$) case, thus giving a similar *cubic* rate law. On the other hand, the central part (intermediate r) already differs from the coagulation case, and the maximum falls at about 0.4 C^{-1} (half the value of the coagulation case). A *rough* approximation[28,29] is

$$P(r) \quad Cr \quad \exp[-\alpha Cr], \tag{18}$$

where α is a coefficient that has not been determined from simulations, due to the poor quality of the functional fit.[29]

A + B → 0 Reactions

We assume that there occur no A + A or B + B reactions (otherwise this case reverts to the A + A → 0 reaction). This is the Zeldovich[15] case, which essentially was the start of modern diffusion-reaction theory. This model assumes a random "big bang" ($t = 0$) distribution, with random fluctuations. Globally $C_A = C_B = C/2$ at $t = 0$ and at every other instant in time. However, locally, $\Delta C = C_A - C_B$ fluctuates around zero at $t = 0$. His formalism shows that $2|\Delta C|/C$ *increases* in time and becomes of the order of *unity* at $t \to \infty$, provided that the dimension of the reaction space is $d < 4$. This means that a segregation of reactants proceeds in time. A very small segregation ($2|\Delta C|/C << 1$) at $t = 0$ turns into a large one (1) at $t \to \infty$. Concomitantly, the classical survival probability

$$C = 2C_A = 2C_B \qquad t^{-1} \qquad t \to \infty \tag{19}$$

is valid only for high dimensions ($d > 4$) but is replaced by

$$C \qquad t^{-d/4} \qquad t \to \infty,\ d < 4 \tag{20}$$

for all "practical" dimensions ($d \leq 3$). This formalism was generalized somewhat (and demonstrated numerically) by Toussaint and Wilczek,[37] and generalized to fractals by Kang and Redner[20] (replacing d by d_s, the "spectral" dimension). The latter generalization is somewhat problematic.

In the above discussions it was implied that A and B have identical mobilities. The other extreme case is where one of them (say B) is immobile ("sitters"). This case, as well as the intermediate cases, has been studied by Zumofen et al.[38,22] with some interesting results. It is intuitively obvious that complete spatial segregation is harder to achieve and will take longer. Quantitatively, it is not yet clear how the particle distributions depend on the relative mobilities of the two species.

Toussaint and Wilczek[37] have also argued that for a correlated $t = 0$

distribution (A and B land with a specified distance between them), the asymptotic rate law is different:

$$C \quad t^{-d/2}, \qquad d \leq 2,\ t \rightarrow \infty \tag{21}$$

that is, the same as for A + A reactions,[3,20] and that there is *no* segregation. It turns out[39] that the situation is more complex and equation (21) is only correct for strictly geminate conditions (where the distance between particles equals the particle size). For correlated but non-geminate initial conditions (where the correlation length is much larger than the particle size):

$$C \quad t^{-(d+2)/4} \qquad d \leq 2,\ t \rightarrow \infty \tag{21a}$$

$$C \quad t^{-1}, \qquad d \geq 2,\ t \rightarrow \infty \tag{21b}$$

Finally we note that there are no published experiments showing self-organization or anomalous kinetics for A + B reactions.

A + C → C Reactions

The early work for this case has centered around "trapping" reactions, that is, where A diffuses and C is fixed in space. This work also originated in the Soviet Union and is summarized, in part, in the monograph of Agranovich and Galanin.[16] The typical rate law in one dimension was given in equation (6) and was rationalized by a long-time distribution of particles *away* from the traps. The first exact (analytical) work on particle distributions appears to be that by Weiss et al.[30] It is for a *single* trap C in three- and one-dimensional continua of particle A (with density C_A). Again the nearest neighbor (C-A) distance distribution is far from random in one dimension (while essentially random in d = 3). Actually, the distribution function is quite similar to that for A + A → A [equation (16)] in one dimension. Similarly, also, the global rate law is, in one dimension,

$$|dC_A/dt| \quad t^{-1/2}C_A \tag{22}$$

while in three dimensions the classical result is recovered:

$$|dC_A/dt| \quad C_A \tag{23}$$

This is again similar to the A + A case. We note that equations (22) and (23) are the first exact results for the original Smoluchowski problem[17] of accretion of molecules by a colloidal particle (or a crystallization seed). Smoluchowski's solution for the three-dimensional case is indeed equivalent to equation (23).

Self-Ordering

In each of the three above cases one finds deviations from uniformly random particle distributions, deviations that increase with time. Often it requires only a few time "steps" (in the simulation or the discrete model) to attain a drastically different (e.g., "Wigner-like"[4,28]) distribution. This spontaneous self-ordering, based totally on diffusion and reaction (but no long-range forces) is of major interest. There are similarities[29] in the "patterns" of different reaction classes (A + A vs. A + B) and dissimilarities[4,29] in those of very similar reactions (A + A → 0 vs. A + A → A). Some of the patterns take much longer to form than others. Dimensionality and initial conditions are also of prime importance. The investigation of such patterns requires some new tools (e.g., order parameters). For instance, we have defined[27,40] about a dozen different "segregation parameters." The relative usefulness of such order parameters in different situations is also of some importance. Particle distributions or partial particle distributions, as well as quantities derived from them, can also be used as such tools.

A + B → C Where A and B are Segregated throughout the Reaction

Recently, analytical work on an A + B → C type reaction-diffusion process in an effective one-dimensional system has been done by Galfi and Racz.[41] The reactant A of constant density a_0 and B of constant density b_0 are initially separated. They meet at time 0, forming a single reaction boundary, which makes the system effectively one-dimensional. The motion of the reaction front with time is shown in Figure 3. This model is similar to that of Weiss et al.[30] for A + C → C where A is a one-dimensional continuous solute and C is a single trap. However, here both A and B are continuous (and A and B are like traps to each other). The results from the set of the reaction-diffusion equations for A and B, which are valid in the long-time limit, show that x_f (the position of the center of reaction front) scales with time as x_f $t^{1/2}$, while w (the width of the reaction front) scales as w $t^{1/6}$ and the production rate r_f (at $x = x_f$) scales as $t^{-2/3}$. We can also find that the global reaction rate R scales at $t^{-1/2}$, that is,

$$R = d\mathrm{C}/dt \quad t^{-1/2}.$$

This is similar to the results of Weiss et al.'s work for A + C → C (equation (22) for $d\mathrm{A}/dt$).

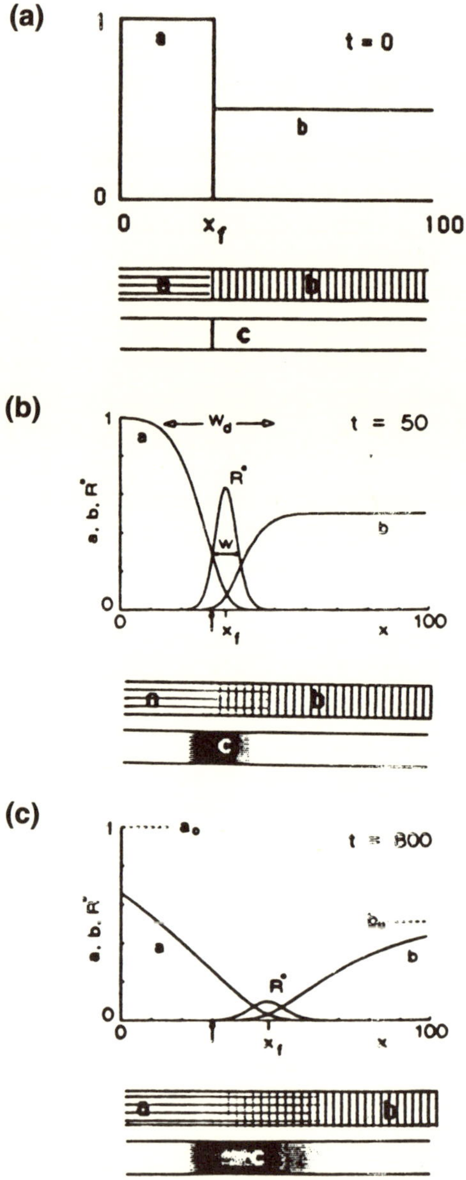

Figure 3. The motion of the reaction front with time. Length *(x)*, time *(t)*, the densities *(a,b)* of the reagents (A,B), and the magnified production rate ($R^* = 100R = 100\ kab$) of C are scaled to be all dimensionless. As an initial condition, we used: $a = 1, b = 0$ for $x < 30$, and $a = 0, b = 0.5$ for $x > 30$ (the $t = 0$ position of the center of the front, x_f is shown by an arrow). The top figures of (b) and (c) are from Ref. 41.

B. Steady Source Reactions

Steady source reactions are of prime importance in biology, ecology, geology, technology, and many other areas.[4] They give a wider range of phenomena than batch reactions due to three factors. 1) The richness of source-term structures (see below). 2) The inclusion of strange topologies (e.g., "dust" or "fractal dust") which would not support batch reactions for long. 3) Ordering phenomena that for batch reactions are only achievable at extremely long times, that is, at vanishing densities, but under steady source conditions may develop even at high densities.

1. *Source-Term Structures*

The most common source term is *uniformly random in both space and time* (e.g., when CO and O_2 are introduced into a catalytic chamber). One of the surprising results has been the fact that the mere presence of such a source, combined only with diffusion and reaction, may lead to highly self-ordered systems.[4,42] Another common source produces particles *correlated in space and/or time*. For instance, when H_2 particles land on a platinum surface they dissociate into *pairs* of atoms; or when electrons and holes are produced by photons hitting a photovoltaic cell, they are produced *geminately*. Alternatively, the source may be a hot beam of dissociated AB molecules, landing as pairs of A and B atoms on a surface, *correlated in time but not in space* (another example being ion-pairs created in a high energy radiation track). The particles may be able to react as they land on top of other particles (vertical reaction) or they may not, depending on the chemistry or physics of the situation. If they are not able to react they may be discarded or they may land in the immediate vicinity, depending again on the particular reaction. Sources with *vertical reaction* may produce particle distributions that differ drastically from those produced by sources with *no vertical reaction*. Also, the atoms on a surface may occasionally *desorb*, and the excitons produced in a solid have a natural decay time, independent of the annihilation or fusion reaction. Such a "monomolecular" (first order) decay process adds another *random* (negative) contribution to the source terms. All such possible source structures have significant effects on the outcome, in addition to the well-known source-term distinctions, such as uneven A and B production terms (alternatively viewed as a strong A source and a weaker B source). Furthermore, initial conditions may vary, such as an initially empty lattice (the usual case), versus a fully or partially occupied one. We also note that the source structure usually determines the *conservation* laws. For instance, a time-correlated, pairwise production of A and B particles (in an A + B → 0 reaction with no decay) will guarantee an *exact global* conservation of the equality $C_A = C_B$. On the other hand, a geminate production of A and B will guarantee both an *exact global and an exact local* conservation of $C_A = C_B$ (i.e., no segregation). However, a uniformly random source term gives only a *Statistical global* conservation of $C_A = C_B$, and, under certain circumstances (e.g., no desorption) this may lead to long-time realizations with $C_A >> C_B$ or vice versa.

An obvious question of interest is whether a given source term will produce a steady-state system. The answer may depend on a combination of source structure and medium dimensionality, or a combination of other conditions. There have been several recent surprises in this respect such as the demonstration of saturation (a mathematical catalyst poisoning),[23-26,43] which lends importance to future work aimed at clarifying this issue. If a steady-state is achieved, the next question is: Is the particle distribution uniformly random or ordered? The self-ordering process itself may either hinder or aid the generation of a steady-state. The self-ordering is, therefore, an interesting related question.

Certain source terms may generate a steady-state that mimics an equilibrium state. For instance, strictly geminate pair creation, with an annihilation reaction (A + A → 0 or A + B → 0), is expected to set up a steady-state that mimics equilibrium. We expect equilibrium states to be totally (randomly) disordered (by the second law of thermodynamics). Such source terms and steady-states are thus of special interest. However, the steady-states which are self-ordered (i.e., far from equilibrium) are of even more interest. It is these states that are responsible for the anomalous reaction rate laws. They may even lead to much speculated upon phenomena, such as biomorphogenesis,[23,24] which are in apparent (though not real) contradiction with the *second law*.

A + A Reactions

Analytical solutions exist so far only for a limited number of cases and are restricted to one dimension. For the A + A → 0 reaction,

Racz[44] has solved "one and a half" cases. For the geminate landing case the solution is complete: For the steady-state reaction the rate law is classical (second order) and the particle distribution is uniformly random. This is actually expected for all simple geminate landing cases, because, as pointed out above, such steady-states map onto equilibrium states,

$$A + A \leftrightarrow A_2 \tag{24}$$

and for equilibrium states we expect this classical behavior. On the other hand, for uniformly *random* landing, where a steady-state is also achieved, the rate law is *third order* in density. Furthermore, the particle distribution is *non-random* (see below). However, Racz[44] did not derive a functional form for such a distribution.

A recent paper by Doering and Ben-Avraham[33] gives a similar derivation for the one-dimensional A + A → A problem. Again a steady-state is found for the random (and geminate?) source terms. The rate law (for the random source) is again *third order* in density, and the particle distribution is non-random. In addition, the *interparticle distance distribution* was derived specifically (in terms of Airy functions). This paper demonstrates analytically how a non-random particle distribution develops from a random landing distribution.

Simulation-based results were obtained for all dimensions (or spectral dimensions, for fractals), both higher and lower than unity. Equations (12) and (13) actually give, for irreversible A + A reactions,

$$-dC_A/dt = kC_A^x - R, \tag{11$'$}$$

where R is the source term. As mentioned above, for these A + A reactions,

$$x = 1 + 2/d_s \qquad d_s < 2 \tag{25a}$$

$$x = 2 \qquad d_s > 2 \tag{25b}$$

while for the marginal (critical dimension $d = 2$), one indeed obtains $x = 2$ but a logarithmic correction appears:

$$-dC_A/dt = kC_A^2/\ln C_A - R \tag{26}$$

A more rigorous theoretical argument for the form of the last three equations has been derived recently,[26] also including particle correlation functions as well as higher-order terms for higher densities and modifications for source terms correlations. The logarithmic correction is typical of dynamical behavior at critical dimensions.[2]

A + B Reactions

This class of reactions is the most abundant, important, intriguing and challenging. The first demonstration that a completely segregated steady-state can be achieved with a *uniformly random source term*[42] was a real intellectual surprise. While the simulation demonstrations were performed on fractal lattices (Sierpinski gasket[45]) and critical percolation clusters,[46] it turns out[27,41] that this phenomenon also occurs in one- and two-dimensional media, and, theoretically, even in three-dimensional media (however, only for infinite systems[47]). In addition, it has been shown[42] that segregation occurs also for systems made of walkers (A) and "sitters" (B), though the distributions and the rate laws may differ. However, the most interesting aspect of steady-source reactions is the role of the source structure and statistics. The most subtle differences in the nature of the source may cause very dramatic differences in particle distributions and rate laws. An illustration of this is given in Table 2 (Table I of Clement et al.[25]). We see here how the segregation length Λ is affected by the source *conservation laws* (strict vs. statistical), source *correlation length* (δ) and nature (random vs. fixed), by the presence of vertical annihilation and by the presence of a "negative" source term, such as *desorption* or natural *decay* (with first order rate constant *K*). The reaction-diffusion parameters that enter the quantitative expression are the diffusion constant *(D)*, the source rate *(R)*, the particle size *(a)*, the lattice size *(L)*, and of course the dimensionality *(d)*.

A + C Reactions

It was already shown a decade ago (Agranovich and Galanin[16]) that for a *trapping reaction*, A + T → T, on a one-dimensional lattice with a steady, uniformly random source of A particles, the trapping rate is *linear* in A density but *quadratic* in *trap* density. This is true for both random and even (superlattice) trap distributions (though the rate coefficients differ):

$$\text{Rate} \quad C_A C_T^{\,2} \tag{27}$$

This result has also been derived via a somewhat different approach by Lopez-Peacock and Keiser.[48] This non-classical rate law is again due to a non-uniform A particle distribution.

Table 2. Scaling Behavior of Λ, Typical Size of Segregated Domains, as a Function of the Euclidean Dimension d and Various Source Conditions. Cases (a) and (b) are, Respectively, Random and Correlated Landing (δ is the separation between landing pairs) *with* Conservation of the Number of A's and B's on the System. For Cases (c), the Conservation Condition is Removed and a First-Order Decay Mechanism is Included (with a constant K). The System Size is L, the Size of a Particle is a, and the External Rate of Arrival of Particles is R.

Conservation Conditions	*Source Case*			$d = 1$	$d = 2$	$d = 3$
(a) Strict	random landing (random δ)		$K = 0$	$L/6$	$a \ln L/a$	$3a$
(b) Strict	correlated landing (fixed δ)		$K = 0$	δ	$a \ln\delta/a$	$2a$
(c) Statistical	Independent landing	Non-vertical annihilation	$K = 0$	saturation*	saturation	saturation
		Vertical annihilation	$K = 0$	ξ**	$a \ln\xi/a$	$3a$
		Symmetric desorption	$K \neq 0$	ξ**	$a \ln\xi/a$	$3a$

Notes: *Saturation means that only one of the species occupy all the system.

**For vertical annihilation: $\xi = \sqrt{D/a^d R}$

For symmetric desorption: $\xi = \sqrt{D/K}$

Based on preliminary simulations[49] and theoretical work,[31] the last equation can be generalized in the following way:

$$\text{Rate} \sim C_A C_T^{2/ds} \qquad d_s < 2 \tag{28a}$$

$$\text{Rate} \sim C_A C_T \qquad d_s > 2 \tag{28b}$$

where d_s is an effective spectral dimension and the marginal dimensionality is again two (at $d = 2$ there is probably again a logarithmic correction to equation (28b)).

III. EXPERIMENTS

Examples of chemical reactions, photochemistry, and photophysics have been selected in order to illustrate the peculiarities of diffusion-controlled reaction kinetics in low dimensions.

A. Photochemistry in Pores of Membranes

The photodimerization of anthracene in solution was probably the first well-studied diffusion-limited reaction.[50] The bimolecular reaction $A + A \rightarrow A_2$ (A = anthracene) followed the classical kinetic laws. A similar reaction provided the first study of fractal chemical kinetics:

$$N^* + N^* \rightarrow N_2^{**} \rightarrow N + N + h\nu \tag{29}$$

Here N^* is a naphthalene molecule excited to its first triplet state and N_2^{**} is the transitory dimer in the excited first singlet state.[5,6] The experiment is carried out in a solution embedded in various porous membranes (nylon, acetate, paper; see Figure 4). The naphthalene molecules diffuse inside the solvent which is constricted by the walls of the pores.

Figure 5 shows the pores of some of the channel-pore membranes used as "test tubes" for the reactor. These cylindrical tubes are 6000 nm long and have radii ranging from 1000 to 10 nm. Typical molecular diffusion lengths in these experiments are on the order of 100 nm. We thus expect to see a range of kinetic behaviors, from three-dimensional to one-dimensional, depending on the molecular diffusion parameters. Indeed,[5] the heterogeneity exponent h (see equation 9) ranges from $h = 0.07$ for 1000 nm pores to $h = 0.4$ for 25 nm pores. This is in good agreement with the theoretical limits, that is, $h = 0$ for three-dimensional media and $h = 1/2$ for one-dimensional media. More-

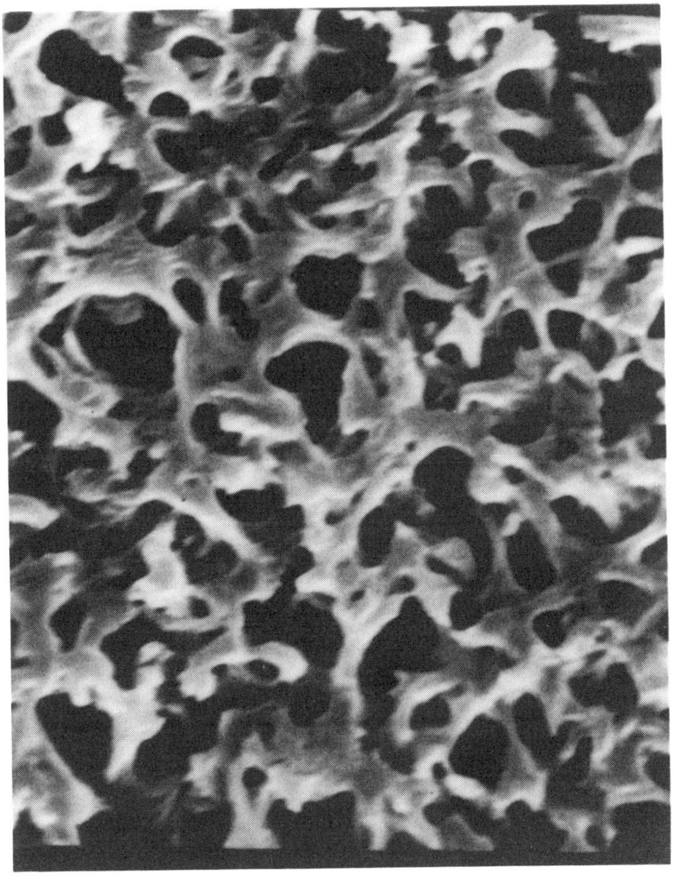

Figure 4. Porous nylon membrane (Gelman, Ann Arbor, MI, U.S.A.). Magnification about 10^4.

over, for an intermediate pore radius (100 nm) one observes a crossover in time, from $h = 0.06$ in the millisecond regime to $h = 0.48$ in the sub-second regime. We note that the uncertainties are between 0.05 and 0.1.

Earlier experiments,[6] carried out in ordinary porous membranes made of nylon (Figure 4), acetate, or paper give h values in the range of 0.3 to 0.4. This can be interpreted simply as some statistical averaging over various pore domains ranging between one-dimensional and three-dimensional topologies. Alternatively, one can ascribe to such membranes "effective" spectral dimensions d_s (see equation 10), with values between 1.2 and 1.6.

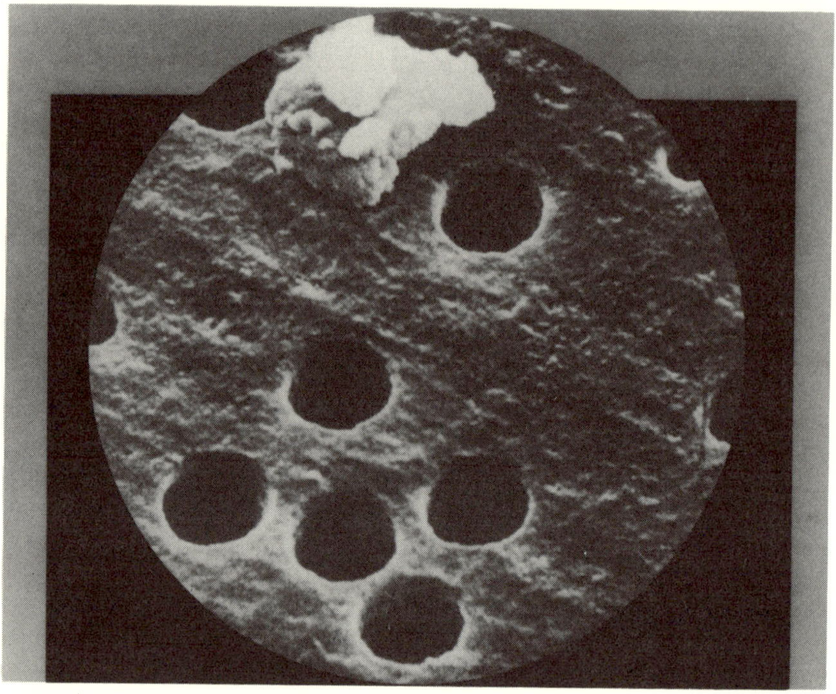

Figure 5. Polycarbonate channel-pore membranes (Nuclepore, Pleasanton, California). Magnification about 10^4. Channel length is 6 μm and the radius of the channel is 1 μm.

B. Photophysics in Polymer Blends

Excitation kinetics in molecular aggregates and materials is of biological interest (e.g., photosynthesis and neuron transmission[11,12]), morphological interest (e.g., studies of polymeric materials[7-9]), and potential technological interest (nano-devices[51]). Both A + A and A + C type reactions have been studied under tightly controlled conditions using an extremely dilute blend of P1VN (poly-1-vinyl-naphthalene) in PMMA (poly-methyl-methacrylate). The electronic excitation is confined to the naphthalene moieties of the P1VN. Triplet excitations are utilized because of their limited range of energy transfer (4 to 8 Å). In the absence of neighboring P1VN chains, the excitation transport is thus limited to a single chain and reflects its one-dimensional topology. We therefore limit ourselves to the very dilute blends (about

0.01% P1VN or less) where the P1VN chains are effectively isolated from each other.[9]

$A + A \rightarrow$ Product *Reaction (Homofusion)*

With short pulse laser excitations[9] the observed phenomenon is that of triplet-triplet fusion ("annihilation"):

$$T + T \rightarrow S \tag{30}$$

The singlet excitations (S) have a natural lifetime of about 100 ns and are monitored via the resultant fluorescence (UV), while the triplet excitations (T) have a natural lifetime of about 3 seconds and are monitored via their phosphorescence (green).

For the most dilute samples measured (e.g., 0.05%) the heterogeneity exponent was found to be about 0.4, thus approaching the theoretical expectation, $h = 1/2$, for one-dimensional confinement of the reaction kinetics. On the other hand, for pure P1VN material (100%) the experimental result is $h = 0.02 \pm 0.02$, in good agreement with the expectation for three-dimensional (classical) systems ($h = 0$).

$A + C \rightarrow C +$ Products *Reaction (Heterofusion)*

Under prolonged UV excitation of P1VN blends, excimers are produced,[7–9] both in the singlet and the triplet states. The excimers are essentially defect sites with trapped excitations. The excimer sites represent a locally transformed orientation of aromatic rings. This deformation is associated with a *reduction* in excitation energy (difference between excited state and ground state). The hopping excitations (excitons) are energetically trapped at such sites if the energy reduction exceeds the thermal energy. The mechanism of triplet-triplet annihilation can now be written, to a good approximation[7,8] as:

$$T_F + T_E \rightarrow T_E \tag{31}$$

where T_F is the free triplet excitation and T_E the trapped excimer excitation. The excimer sites are thus "catalytic" sites where the free excitations are "trapped" and consumed in an annihilative reaction.

From equations (8) to (10) we expect the heterogeneity exponent $h = 1/2$ for $d = 1$ and $h = 0$ for $d = 3$, in agreement with the exact analytical results[30] for binary reaction kinetics (equations 22 and 23).

Table 3. Time Exponents from Experiments

Reactor Size	C_{Cu}	C_{tetra}	α	β	δ
4×2 mm	10^{-3} M	5×10^{-5} M	0.48	0.16	0.53
	10^{-3} M	7.5×10^{-5} M	0.53, 0.48	0.15, 0.20	0.51, 0.54
	2×10^{-3} M	6×10^{-5} M	0.52	0.16	0.53
2×2 mm	2×10^{-3} M	6×10^{-5} M	0.60	0.22	0.52

Note: *Cu is Cu^{++} and tetra is disodium ethyl bis(5-tetrazolylazo) acetate trihydrate, where the solvent is 0.15% agarose/water. C_{Cu} and C_{tetra} are $t = 0$ concentrations. Here $x_f \sim t^{\alpha}$, $w \sim t^{\beta}$, and $R \sim t^{-\delta}$.

Measurements have been conducted for very dilute blends[9] (0.01% and 0.005%) and the average result is $h = 0.50 \pm 0.03$, as expected for truly one-dimensional reaction topologies.

C. Bimolecular Reaction in Capillary with Gel

The chemical reactions were performed in a rectangular or a square glass capillary. The binary complex reaction

$$(Cu^{++} + \text{disodium ethyl bis(5-tetrazolylazo)acetate trihydrate} \rightarrow \text{1:1 complex}) \tag{32}$$

was carried out in a gel inside the glass reactor, allowing efficient diffusion but little or no convection. At time 0, there is a sharp boundary between A and B, and that is where product C is being formed (see Figure 3). The absorbance of the product C was monitored at every fixed period by moving a detector (PMT) and a light source (green LED), which are fixed on a stepping motor, along the reaction front domain, while the reactor is fixed in time.

The results are summarized in Table 3 below. These results agree well with the theoretical expectations ($\alpha = 1/2$, $\beta = 1/6$, $\delta = 1/2$) given in Section BI5.

IV. CONCLUSIONS

The new qualitative and quantitative approach to chemical reaction kinetics is expected to have some major repercussions in chemistry, biology, earth sciences, and others where heterogeneous reactions are of importance. There are preliminary reports concerning enzymatic reactions in cells and biodegradation in the ground. Very recently we have demonstrated[52] the non-classical rate laws, employing industrial catalysts (platinum islands supported on alumina) and the most elementary chemical reaction on this surface: H + D → HD. On the more speculative level, reactive self-organization may have played a role in biomorphagenesis,[53,54] that is, the evolution of life. It may also play a role in the microfabrication of materials (we have already used some aspects of this in the nano-fabrication of a subwavelength light-source[51]). Both pleasant and unpleasant surprises may thus await us as we progress towards micro and nano-chemistry.

ACKNOWLEDGMENT

This work was supported by NSF grant DMR-911622 and the Donors of the Petroleum Research Fund, administered by the American Chemical Society.

REFERENCES

1. de Gennes, P. G. *J. Chem. Phys.* **1982,** *76,* 3316.
2. Havlin, S.; Ben-Avraham, D. *Adv. Phys.* **1987,** *36,* 695; Rudnick, J.; Gaspari, G. *Science* **1987,** *237,* 384.
3. Argyrakis, P.; Kopelman, R. *J. Chem. Phys.* **1980,** *72,* 3053; Klymko, P. W.; Kopelman, R. *J. Phys. Chem.* **1983,** *87,* 4565.
4. Kopelman, R. *Science* **1988,** *241,* 1620.
5. Prasad, J.; Kopelman, R. *Chem. Phys. Lett.* **1989,** *157,* 535.
6. Prasad, J.; Kopelman, R. *J. Phys. Chem.* **1987,** *91,* 265.
7. Li, C-S.; Kopelman, R. **1990,** *23,* 2223.
8. Li, C-S.; Kopelman, R. *J. Phys. Chem.* **1990,** *94,* 2135.
9. Shi, Z-Y.; Li, C.S.; Kopelman, R. In *Polymer Based Molecular Composites,* Mark, J. E.; Shaefer, D. W., Eds.; Proceedings of Materials Research Society: Pittsburgh, 1989.
10. Drake, M. Private communication.
11. Fauman, E. B.; Kopelman, R. *Comments on Molecular and Cellular Biophysics,* **1989,** *6,* 47.
12. Dissado, L. A. *Comments on Molecular and Cellular Biophysics,* **1987,** *4,* 143.
13. Hoshen, J.; Kopelman, R. *J. Chem. Phys.* **1976,** *65,* 2817.
14. Klymko, P.; Kopelman, R. *J. Phys. Chem.* **1982,** *86,* 3686.
15. Ovchinikov, A. A.; Zeldovich, Ya. B. *Chem. Phys.* **1978,** *28,* 215, and references cited therein; Kuzovkov, V. N.; and Kotomin, E. A. *J. Phys. C.* **1984,** *17,* 2283.
16. Agranovich, V. M.; Galanin, M. D. *Electronic Excitation Energy Transfer in Condensed Matter*; North Holland: Amsterdam, 1982.
17. Smoluchowski, M. V. *Z Phys. Chem.* **1917,** *92,* 129; Keizer, J. *J. Phys. Chem.* **1982,** *86,* 5052, and references cited therein.
18. Klafter, J.; Bluman, A.; Zumofen, G. *J. Stat. Phys.* **1984,** *36,* 561.
19. Kopelman, R. *J. Stat. Phys.* **1986,** *42,* 185.
20. Kang, K.; Redner, S. *Phys. Rev. Lett.* **1984,** *52,* 955.

21. Alexander, S.; Orbach, R. *J. Phys. (Paris) Lett.* **1982,** *43,* L625.
22. Blumen, A.; Klafter, J.; Zumofen, G. In *Optical Spectroscopy of Glasses*, Zschokke, I., Ed.; Reidel: Dordrecht, 1986, p. 199.
23. Lindenberg, K.; West, B. J.; Kopelman, R. In *Proceedings of Conference on Noise and Chaos in Nonlinear Dynamical Systems*; Catelin, S.; Moss, F., Eds.; Cambridge University Press: Cambridge, MA, 1989, p. 142.
24. Clement, E.; Sander, L. M.; Kopelman, R. *Phys. Rev. A,* **1989,** *39,* 6455.
25. Clement, E.; Sander, L. M.; Kopelman, R. *Phys. Rev. A,* **1989,** *39,* 6466.
26. Clement, E.; Sander, L. M.; Kopelman, R. *Phys. Rev. A,* **1989,** *39,* 6472.
27. Harmon, L. A.; Li, L.; Anacker, L. W.; Kopelman, R. *Chem. Phys. Lett.* **1989,** *163,* 463.
28. Parus, S. J.; Kopelman, R. *Phys. Rev. B* **1989,** *39,* 889.
29. Argyrakis, P.; Kopelman, R. *Phys. Rev. A* **1990,** *41,* 2114; Argyrakis, P.; Kopelman, R. *Phys. Rev. A* **1990,** *41,* 2121.
30. Weiss, G.; Kopelman, R.; Havlin, S. *Phys. Rev. A* **1989,** *39,* 466.
31. Clement, E.; Kopelman, R.; Sander, L. M. *Europhys. Lett.,* **1990,** *11,* 707.
32. Doering, C. R.; Ben-Avraham, D. *Phys. Rev. A* **1988,** *38,* 3035.
33. Ben-Avraham, D.; Doering, C. R. *Phys. Rev. Lett.* **1989,** *62,* 2563.
34. Torney, D. C.; McConnell, H. M. *J. Phys. Chem.* **1983,** *87,* 1441.
35. Anacker, L. W.; Parson, R. P.; Kopelman, R. *J. Phys. Chem.* **1985,** *89,* 4758.
36. Bramson, M.; Griffeath, D. Ann. Probab. **1980,** *8,* 183.
37. Toussaint, D.; Wilczek, F. *J. Chem. Phys.* **1983,** *78,* 2642.
38. Zumofen, G.; Blumen, A.; Klafter, J. *J. Chem. Phys.* **1985,** *83,* 3198.
39. Lindenberg, K.; West, B. J.; Kopelman, R. *Phys. Rev A* **1990,** *42,* 890.
40. Li, L. Ph.D. Thesis, University of Michigan, Ann Arbor, MI, 1989.
41. Galfi, L.; Racz, Z. *Phys. Rev. A* **1988,** *38,* 3151.
42. Anacker, L. W.; Kopelman, R. *Phys. Rev. Lett.* **1987,** *58,* 289; *J. Phys. Chem.* **1987,** *91,* 5555.
43. Lindenberg, K.; West, B. J.; Kopelman, R. *Phys. Rev. Lett.* **1988,** *60,* 1777.
44. Racz, Z. *Phys. Rev. Lett.* **1985,** *55,* 1707.
45. Mandelbrot, B. B. *The Fractal Geometry of Nature*, Freeman: San Francisco, 1983.
46. Newhouse, J. S.; Kopelman, R. *J. Chem. Phys.* **1986,** *85,* 6804; *J. Phys. Chem.* **1988,** *92,* 1538.
47. West, B. J., Kopelman, R.; Lindenberg, K. *J. Stat. Phys.* **1989,** *54,* 1429.
48. Peacock-Lopez, E.; Keizer, J. *J. Chem. Phys.* **1988,** *88,* 1997.
49. Anacker, L. W.; Kopelman, R. *Science at the John von Neumann National Supercomputer Center, 1989*; Consortium for Scientific Computing: Princeton, NJ, 199; p. 29.
50. Luther, R.; Weigert, F. *Z. Phys. Chem.* **1905,** *53,* 385.
51. Lieberman, K.; Harush, S.; Lewis, A.; Kopelman, R. *Science* **1990,** *247,* 59.
52. Holder, J. S.; Kopelman, R. In *Dynamics in Small Confining Systems*; Drake, J. M., Klafter, J., and Kopelman, R. Eds.; Materials Research Society: Pittsburgh, PA, 1990; pp. 217–220.
53. Turing, A. M. *Philos. Trans. R. Soc.* **1952,** London, Ser. B 237, 37.
54. Field, R. J. *Am. Sci.* **1985,** *142.*

STATE-RESOLVED SIMPLE BOND-FISSION REACTIONS: EXPERIMENT AND THEORY

Hanna Reisler and Curt Wittig

Advances in Chemical Kinetics and Dynamics, Volume 1, pages 139–185

ISBN: 1-55938-161-2

The rate constant k(E,J) which we have learned how to calculate in Part I does not represent a measureable property, for it is not feasible to observe the decay of an undisturbed isolated molecule of specified energy and angular momentum.

W. Forst, Theory of Unimolecular Reactions, 1973

I. INTRODUCTION

The unimolecular decomposition of polyatomic molecules is a central topic to much of chemistry and molecular physics, particularly the field of gas-phase chemical kinetics.[1-3] Its study has spanned many generations of scientists,[4] evolving to a powerful predictive tool, with mathematical simplicity and elegant physical basis. No wonder it has maintained a following for seven decades. The field has also benefited from the many scientific advances of the last decade, and it is now possible to carry out both experiments and calculations of unprecedented detail and sophistication. Consequently, the purpose of this review is to highlight recent results and directions, both in experiment and theory. This "highlighting" is meant to be illustrative rather than exhaustive, with many examples taken from work carried out in the authors' laboratories. Experimental systems will be limited to those in which single values of the energy in excess of reaction threshold, $E^{\dagger}$, are used, as opposed to samples where thermal equilibrium exists among the molecular *E, V, R, T* degrees of freedom. Where complementary methods, experimental and/or computational, have been employed, this also dictates preference. A thorough coverage of the literature will not be attempted.

Many modern technologies and experimental techniques have been applied to the study of unimolecular decomposition, with added precision accompanied by an array of increasingly fundamental questions. Isentropic nozzle expansions have found a niche in sample preparation, since they can be used to cool parent rotations to a few K;[5] vibrations can also be cooled efficiently, although not as well as rotations. This enables gaseous molecules to be excited under collision-free conditions to well-defined total energies and low rotational angular momentum states using a variety of photoexcitation methods. Alternately, double resonance methods, in which intermediate eigenstates have good angular momentum quantum numbers, can be used with room temperature samples to select very narrow ranges of angular momenta, even when the terminal levels are overlapped or within a continuum, without the restriction to low angular momenta that applies to expansion-cooled samples. So far, this "vibrationally mediated photochemistry" has been used mostly to study direct photodissociation processes,[6-8] although results on unimolecular reactions have now begun to appear.[9]

Methods of preparing excited molecules vary, but optical excitation with laser sources has clearly surpassed chemical activation as the method of choice for preparing reactants with a narrow range of energies. Internal conversion (IC) and/or intersystem crossing (ISC) can be efficient and straightforward to implement and have been used successfully by several groups.[10-40] Although systems inevitably retain some electronically excited-state character, this can be very small, with the wavefunction dominated by the reactive potential energy surface (PES). For example, if S_1 and S_0 are strongly coupled, $S_1 \leftarrow S_0$ absorption can be followed by unimolecular decomposition on a PES that is quite accurately called S_0. We will not discuss radiationless decay phenomena per se, as this has been done many times,[41] and ISC/IC are only vehicular to preparing energized molecules that subsequently decompose via one or more unimolecular decomposition mechanisms.

It is also possible to access vibrational levels above dissociation threshold directly, by pumping overtones and/or combination bands, a method pioneered by Crim and coworkers.[42] For example, pumping highly excited hydrogen stretching vibrations in small molecules with low dissociation thresholds, like H_2O_2 and HN_3, has proven quite successful, and allows initial vibrational excitation to be well characterized in terms of local modes, mixtures of normal modes, and so on.[42,43] With H_2O_2, initial hydrogen stretching excitation is converted internally to motion along the reaction coordinate, yielding two OH fragments which can be probed in great detail.[42] The experiments are a paradigm of the cleverly conceived idea carried to fruition using the best of modern spectroscopic tools. The case of HN_3 offers even further insight using a small and hopefully tractable system.[43] These experiments involve *ns*- and *ps*-resolution rate measurements, nascent state distributions, overtone pumping, and vibrationally mediated photodissociation, and are carried out with care and thoroughness. The lowest product channel, $NH(X^3\Sigma) + N_2$, is spin-forbidden, and enables singlet-triplet coupling to be examined carefully for a prototypical system. This has already spawned major theoretical activity, and an insightful explanation of the results, including symmetry-based propensities for the $NH(X^3\Sigma^-)$ spin-rotation *e,f* levels, has been provided by Alexander et al.[44] The spin-allowed product cnannel, $NH(a^1\Delta)$ + N_2, is only ~ 1.5 eV above the triplet channel, and is readily observed when the HN_3 internal energy exceeds the reaction threshold.[45]

High-sensitivity product detection using laser induced fluorescence (LIF) and multiple photon ionization (MPI) is now a commonplace laboratory tool in physical chemistry and can be applied with great ease to *some* small fragments (e.g., CN, OH, NH, NO, CH_2, etc.), but with difficulty or not at all to the majority of polyatomics. Product state distributions (PSDs) are obtained easily from routine spectroscopic measurements, except for M_J distributions, which can be obtained from somewhat more elaborate measurements of alignment. Correlated distributions (i.e., for a given internal state of one fragment, there is the corresponding distribution of internal states in the other fragment) can be obtained from state-specific center-of-mass (CM) kinetic energy distributions, for example using Doppler shift methods.[46] Vector properties such as anisotropic fragment recoil directions in the lab, M_J distributions, **v–J** correlations, **μ–J** correlations, **μ–v** correlations, and so on,[47,48] while invaluable for understanding direct photodissociation, are often washed-out on the long timescale of unimolecular decay by molecular rotation, loss of K as a good quantum number, CM kinetic energy *distributions*, and so on. Certain vector properties such as **v–J** correlation remain,[48] but in general, vector properties can easily be masked by the broad, state-specific CM kinetic energy distributions characteristic of unimolecular decomposition via loose transition states.

Time domain measurements of unimolecular reactions using ultrashort laser pulses probe the elementary rate process directly, and decomposition time constants have been measured for a number of prototypical molecules at single $E^\dagger$ values, with $E^\dagger$ tunable over a broad range of values. Pioneering experiments by Zewail and coworkers have been carried out with H_2O_2,[49] NCNO,[38,39] and CH_2CO,[40] and several results will be presented below. As with PSDs, the use of expansion-cooled samples, state-specific product detection, internal conversion, overtone pumping, and so on, gives information that can challenge the most rigorous of current theories.

In the material that follows, we first outline several theoretical models widely used to interpret data. These rely on statistical mechanics, but sometimes include empirical constraints, based on accepted physical principles, that improve their predictive capacity. There is a limit to how far one can or should go with the different phase space, adiabatic channel, variational RRKM, separate statistical ensembles, and other models. In our opinion, rates and PSDs can now be brought into acceptable agreement with the main features of the PES using such theories, and it is time to move to more fundamental approaches.

Next, experimental results are presented for several prototypical systems for which the data bases are particularly good, emphasizing those where both PSDs and rates have been obtained. We concentrate heavily on the unimolecular reaction of NCNO, for which the most complete sets of data are available. However, work on CH_2CO,[22,30,31] which is still in progress, promises to yield a comparable comprehensive data base to further test theoretical predictions. We also discuss other alipathic nitroso molecules, since they exhibit many common features in their spectroscopy and photochemistry which allow direct comparisons between them. By choosing a specific nitroso molecule, a desired feature of the dissociation can be highlighted. As mentioned above, the measurements have reached a high level of sophistication, providing precise values of experimental parameters and results. Hopefully, this will enable dynamical calculations to be carried out beyond the level of trajectories (which continue to yield amazingly good results for even very quantal systems[50–52]), probably with a mixed quantum-classical formalism, at least for the smaller systems. It is at this crossroads that we expect to see a marked evolution in our understanding of why physical systems have such an uncanny ability to become statistical at the least provocation. Finally, some directions for future work will be pinpointed.

II. THEORETICAL CONSIDERATIONS

In this section, several theoretical models will be discussed that bear on the kinds of unimolecular decomposition reactions being reviewed, that is, simple bond fission yielding two free radicals. These models are based on statistical mechanics, using the PESs only to provide inputs to the statistical models. Detailed dynamics studies (i.e., trajectory) will not be covered, nor will the optical excitation, IC and ISC processes that prepare reactant molecules. The PESs and associated dynamics of unimolecular reactions give rise to long reaction times that involve complicated nuclear motions. It was a true stroke of genius, or at least healthy pragmatism, that replaced such a dynamically complex system with a statistical approach to unimolecular reactions. Throughout the years, these transition state theories have had several forms, one of the most successful and widely used being RRKM theory for the calculation of microcanonical rate constants.[1,2] By doing statistics at transition states, reaction rates can be calculated that agree with experimental results over a wide range of conditions, such as reactant concentration, temperature, and total pressure.

As mentioned above, it has recently become common to study such reactions using expansion-cooled molecules that have been excited to specified, variable total energies, obtaining PSDs and product CM kinetic energy distributions, as well as microcanonical rate constants. In many cases, the latter have been obtained using direct, real-time measurements, which have now been extended to the *ps/fs* time domain.[38–40,49] In addition, several correlated distributions have been reported. Although both rates and PSDs are direct consequences of the PES and associated molecular dynamics, different features of the PES can govern different mechanistic aspects. Here, we are concerned solely with loose transition states for which there is no reverse barrier that might impose its biases on product excitations.

Experimental PSDs deriving from loose transition states may differ some in detail, but none display the pronounced signatures that characterize marked dynamical bias, for example, population inversions. Occupancies of non-degenerate states always decrease with energy, and PSDs can frequently be assigned temperatures. The experimental PSDs can be compared to statistical distributions, despite some model-dependent latitude in deciding exactly what should be called statistical. However, relative to simply noting temperature-like behavior in one or more degrees of freedom, comparison to a calculated statistical *distribution* is the best statistical measure of a PSD. Specifically, experiments have shown that different degrees of freedom sometimes display *different* temperatures, and from first principles it follows that even temperature itself is not a proper measure for a microcanonical system having a small number of degrees of freedom.

Thus, a number of theories have been developed over the years, based primarily on equal access to the molecular phase space, subject to the constraints of conservation of total energy and angular momentum.[53–57] These theories evolved from two major scientific communities involved in the study of unimolecular reactions, namely those working with ions and those working with neutrals, with parallel, essentially simultaneous developments. Nowadays, a standard example of these formalisms is the phase space theory of unimolecular reactions (PST) developed by Light and coworkers,[53-55] which can be used to calculate full distributions including correlations, in all degrees of freedom, subject to the above constraints. It is identical in structure to the statistical theories used to describe ion breakdown.[56,57]

A. Phase Space Theory

In PST, the transition state is taken as the loosest possible, with fragments assigned their asymptotic (i.e., fully separated) vibrational and rotational character. A low centrifugal barrier defines the transition state, usually using an attractive potential of the form R^{-n}, where n is near 6. Small changes in n around $n = 6$ are reflected weakly in the calculated product excitations. The resulting equations are simple enough in principle, although the calculations can require a decent-sized computer. A typical set of formulae is given below:

$$E \geq V_{max} \tag{1}$$

$$V(r) = V_{el}(r) + B(r)L(L + 1) + E_{\infty}(v,J) \tag{2}$$

$$L(L + 1) \leq 6\mu C_6^{1/3}(E_T/2)^{2/3} \tag{3}$$

$$\boldsymbol{J}_{total} = \boldsymbol{J}_1 + \boldsymbol{J}_2 + \boldsymbol{L} \tag{4}$$

where $E_{\infty}(v,J)$ is the internal energy of the fragments at infinite separation, V_{max} is the centrifugal barrier, $\boldsymbol{J}_1$ and $\boldsymbol{J}_2$ are the total angular momenta of the separated fragments and $\boldsymbol{L}$ is the corresponding orbital angular momentum of the pair with L its quantum number, $B(r)$ is the rotational constant for orbital motion, $V_{el}(r)$ is the one-dimensional interfragment electronic potential, μ is the reduced mass of the pair, and C_6 is the constant used to describe the attractive part of the potential, $V_{att}(r)$, with the term $V_{att}(r) = -C_6/r^6$. E_T is the relative translational energy of the separate fragments. The specific forms of the fragments dictate the angular momentum degrees of freedom. Calculations for combinations of diatoms and spherical tops are easy, symmetric tops are straightforward, and asymmetric rotors are not done because of the added complexity. Rates can also be calculated by computing the flux through the transition state, but are usually too large except at threshold—not surprising, given the choice of transition state.

Calculated PSDs using PST are considered a benchmark against which comparisons can be made. They were never intended to mimic reality for a large class of unimolecular reactions, although agreement has often been remarkable, especially for rotational degrees of freedom, as shown below. However, as more and increasingly refined PSD data became available, a propensity was noted in which product vibrations were often hotter than product rotations, relative to the PST predictions. While one can always state that it is simply due to the PES, this is not satisfactory. Can some general feature of unimolecular reactions be discerned that enables us to understand this observation in simple physical terms before dealing explicitly with the full PESs? Since product rotations are involved, one naturally looks to the parent degrees of freedom that *evolve* to product rotations, such as low frequency bends, torsions, and so on, and this is discussed below.

B. Statistical Adiabatic Channel Model

The statistical adiabatic channel model (SACM) of Troe and Quack,[58–63] which was initially developed to calculate rates, deals with the evolution of certain parent vibrations to product angular momentum states by constructing adiabatic channels that allow the transitional modes to evolve to fragment rotations and translations, and interpolating between the reactant and product ends of phase space using simple formulae. The conserved modes are those parent vibrations that remain as vibrations in the products, while the transitional modes are those parent vibrations that evolve to product R,T excitations. In this way, barriers arise for much of the product phase space that would be accessible in a PST calculation. The barriers are due to the partly vibrational character of the transitional modes; that is, zero-point and potential energies. Reaction rates are consequently lower, in better agreement with experiment than rates calculated using PST. Although the theory is no longer statistical (i.e., it has *adiabatic* channels), it is to be commended for taking a simple physical principle and using a minimum number of adjustable parameters to fit data. Clearly, more parameters mean a better fit, but this can quickly defeat the purpose of using statistical mechanics. The SACM equations are very similar to those of PST, as shown below:

$$V(r) = V_{el}(r) + \{E_a(r_e) - E_{a\infty}\} \exp[-\alpha(r - r_e)] + E_{a\infty} + B(r)P(P + 1) \quad (5)$$

$$P = L + (J_{total} - L) \exp[-\alpha(r - r_e)] \quad (6)$$

where the subscript a refers to a specific adiabatic channel, and α is the parameter used to interpolate between reactant and products. The attractive interfragment potential is usually Morse-like ($e^{-\beta r}$), so the whole interpolation is characterized by two parameters, α and β. Note that equations (5) and (6) revert to PST for the case $\alpha = \infty$.

Since the product rotational phase space is contracted relative to PST, it was initially thought that SACM could also reconcile the general observation that product rotational excitation is usually lower that vibrational excitation. However, this does not work, because the barriers along the reaction coordinate in SACM must ultimately pass their energy on to product rotational and translational degrees of freedom. In our initial publication describing the unimolecular reaction of CF_3CN,[64] we reported agreement between SACM calculations and the experimental results. However, we neglected the energy of the SACM barriers, which must evolve into additional R,T excitations, and this should be included when comparing with experimental results. Thus, product R,T excitations deriving from SACM would *not* be colder than vibrations when effects due to the exit channel barriers are added, contradicting the experimental observations. In fact, the exit channel barriers might even be expected to pass on dynamical bias. Moreover, in some of the experiments where initial conditions could be controlled very well, such as with specific total energy and low angular momentum, it was shown that the products are very cold indeed at threshold (e.g., $E_R(CN) = 0.4\ cm^{-1}$ for $NCNO^{\dagger}$ decomposition near threshold),[10,21] indicating no reverse barrier. Similar results were also reported for ketene.[31]

C. The Separate Statistical Ensembles Method

In our earliest work in which product rotations were shown to be colder than product vibrations, IR multiple photon dissociation (IRMPD) was used to drive the unimolecular reaction $CF_3CN \rightarrow CF_3 + CN$, with CN detected state-selectively up to $v = 4$.[64-66] These experiments prepared excited molecules under collision-free conditions with a range of energies in excess of reaction threshold, $E^{\dagger}$, and the CN fragment was detected state-selectively. PST could not reconcile the results, nor could SACM, for the reasons mentioned above. Similarly, the fact that a range of $E^{\dagger}$ values was produced could not be used to reconcile the results, since a bias favoring vibrational excitation is not expected for *any* $E^{\dagger}$ value. It occurred to us that product vibrational excitation, even at a single value of CF_3CN internal energy, derived from the 3N-6 parent vibrational degrees of freedom, while product R,T excitations were fundamentally different. The transitional vibrational modes *combine* with the three parent rotational degrees of freedom to make up the full set of product R,T degrees of freedom. Here, translation is in the parent CM system. Thus, if parent molecules are rotationally cold, this leads naturally to fragment rotational excitation being lower than fragment vibrational excitation. These early CF_3CN experiments led to the NCNO studies summarized below; the latter being an extremely well-characterized unimolecular reaction in which both the CN and NO fragments are detected in all of their low-lying quantum states, with translational energy resolution and correlations between fragment states, for a large range of energies above reaction threshold.

The premise behind the SSE method is to calculate PSDs subject to the physical constraint mentioned above, namely that product vibrational excitation derives only from 3N-6 degrees of freedom for a nonlinear parent, while product R,T excitations derive from a combination of the transitional modes with the parent rotational degrees of freedom, for *every* combination of product vibrational quantum numbers.[67] This is the essence of the method. Exactly how the vibrational and R,T calculations are done, and whether additional constraints are adopted, such as the reactant-product interpolation of SACM, is another matter. As the name implies, separate ensembles are required for calculating thc product vibrational distribution (using 3N-6 parent degrees of freedom) and product R,T excitations (using transitional modes + parent rotation, for each combination of product vibrational states). The method differs from PST in one important respect that deserves mention. PST calculates vibrational distributions from a phase space that is larger than in the corresponding SSE calculation, since it includes all product excitations, while SSE does not. The SSE method definitely uses one separate ensemble that contains just 3N-6 degrees of freedom to compute product vibrational excitations. In physical terms, the product vibrational distribution is established earlier along the reaction coordinate than are product R,T excitations, and therefore the vibrations "uncouple" before the R,T distributions are fully established. This is a sensible assumption since parent rotations are not expected to mix with the transitional modes before the transition state configuration is reached.

Once the product vibrational distribution is established, the computation for the *R,T* degrees of freedom is straightforward. For each combination of energetically allowed product vibrational quantum numbers, the energy in the transitional modes is obtained from energy conservation, and this is allocated amongst product degrees of freedom using some statistical algorithm, for example PST. The resulting PSD is stored with a weighting factor appropriate to the product vibrational distribution obtained in the first part, and the calculation is then repeated for all allowed values of product vibrational quantum numbers. The individual (weighted) distributions are summed to get the final distribution in whichever degree of freedom is under consideration.

Each of the ensembles in which product *R,T* excitations are calculated uses a larger phase space than for 3N-6 degrees of freedom (i.e., essentially 3N-3, but with vibrational populations determined separately). In physical terms, product *R,T* degrees of freedom have their parentage in both the transitional modes *and* overall parent rotation. If parent molecules are rotationally cold, product degrees of freedom will appear colder than product vibrations, because of this dilution. This second step of the SSE calculation is like PST with a constraint that product vibrational excitation derives only from parent vibrational excitation, not parent *V,R* excitation. An important advantage of the SSE method is its simplicity. It can be implemented with no more computational effort than PST, since calculating the vibrational distribution is straightforward. It is less computationally intensive than SACM with its many channels[58-63] or variational RRKM theory.

We have applied the SSE method to the case of NCNO, as shown below, with uncanny success, even though the 3N-6 vibrational part of the calculation was very crude—using a loose transition state assumption for the transitional modes and analytic formulas for their state densities. PST was used to allocate product *R,T* excitations, and excellent agreement was obtained for both CN and NO products, vibrations as well as rotations, and a wide range of energies.[16-18] The issues of contributions from both S_0 and T_1 and the multiple electronic surfaces arising from NO($X^2\Pi$) + CN($X^2\Sigma$) could be avoided in these PSD estimates, but must be dealt with in the rate calculations, as discussed by Marcus and coworkers.[68-71] Note that for energies where product vibrations are not energetically allowed, SSE is identical to PST. For energies where product vibrations *are* accessible, PST and

SSE will differ most in systems where parent rotational excitation is minimal, as with expansion-cooled samples.[67]

D. Variational RRKM Theory

Marcus and coworkers have extended RRKM theory to deal explicitly with the $V \rightarrow R,T$ transitional modes, calculating reaction rate constants and PSDs for NCNO and CH_2CO unimolecular decomposition reactions.[68–72] They used realistic PESs to locate transition states variationally, from which conserved vibrations were assumed to evolve adiabatically to products. Both reaction rates and product vibrational distributions were determined at the variational transition states. Product R,T excitations were then obtained using PST further along the reaction coordinate, that is, at transition states lying at the centrifugal barriers, as with the SSE method.

Comparisons with the rate constant measurements of Zewail and coworkers indicate good agreement for both NCNO and CH_2CO.[40,71,72] Product V,R,T excitations were also calculated for both NCNO and CH_2CO. For the case of NCNO, the calculations make use of portions of an ab initio PES to count vibrational levels and find transition states,[73] unifying the treatment of rates and PSDs. This is a very important point which has not been addressed adequately in the past. It should be possible to calculate both rates and PSDs for a single PES, but not necessarily with one kind of transition state for a single total energy. Variational transition states vary with energy, but in addition, the PST transition states used to calculate product R,T excitations are further out along the reaction coordinate than the variational one used to determine rates and product vibrational distributions. By using simple systems such as NCNO and CH_2CO as prototypes, it should be possible to test the calculations against measured quantities over a wide range of conditions. Hopefully, full PESs for these systems will become available in the near future.

As with SSE, the agreement between experimental and calculated PSDs is very good, as shown below. In calculating PSDs, the SSE and variational RRKM methods differ little if any in underlying premises. However, the thorough calculations of Marcus and coworkers *and* the very good agreement with measured rates and PSDs, underscore the correctness of the models. Although it had been appreciated for some time that rates and PSDs might use different parts of the PES, this approach provides the actual connection between PSD and rate calculations. The SACM model, with its adiabatic exit channel barriers for typical α/β ratios, would not predict the PST-like behavior of the NCNO and CH_2CO systems (see below). However, it might still do well at the rates, so a comparison of the two methods would be instructive.

E. Information Theoretic Approach

No discussion of the topic of statistical processes in chemical physics would be complete without mention of the many contributions of R. D. Levine and coworkers, particularly R. B. Bernstein.[74–77] The information theoretic approach starts with a statistical distribution and uses Lagrange multiplier methods to analyze distributions for signatures of physical constraints that are manifest as non-statistical PSDs. Their criterion for what is statistical is called a "prior distribution," which is obtained by assigning all product states their statistical weights, without conserving angular momentum. The latter sounds horrible, but only causes problems for small systems. Prior distributions also ignore the *flux* into product states, but this does not make too large a difference either in practice; PST *could* be used for the priors, but is infrequently done and is unnecessary for large molecules. Deviations from statistical behavior can be catalogued using "surprisal analyses," that is, finding the Lagrange multipliers needed to characterize the distribution. This can allow even quite complicated data sets to be parameterized using a small number of constants. The methodology benefits from the pioneering work of Claude Shannon in developing the field of information theory, and the interested reader can find a textbook-level presentation in Baierlein's *Atoms and Information Theory*.[78]

The advantage of this method over all others lies not in semantics about priors and constraints, but in the ease with which it can be implemented by experimentalists. Calculations can be set up and carried out in the same afternoon, which allows *first-cut* analyses to be carried out on experimental results immediately. This enables statistical analyses to be used quickly and routinely for a broad range of systems, with only minor caveats. For example, one should not be too quick to attribute small deviations from prior distributions to something unusual in the dynamics. The ease with which this method can be implemented is demonstrated below for the case of *t*-BuNO, whose 42 *V,R* degrees of freedom would prove needlessly tedious if explored using PST, SSE, or variational RRKM.

III. EXPERIMENTAL OBSERVATIONS AND COMPARISONS WITH THEORY

State-selective studies of simple bond-fission unimolecular reactions with well-characterized monoenergetic molecular ensembles have been carried out to date on only a few molecules. Most notable are the detailed studies of H_2O_2,[42,43] CH_2CO,[22,30,31,40] NCNO[10–21] and several aliphatic nitroso molecules (RNO; R = H, CF_3, CF_2Cl, $CFCl_2$, C_3F_7, and $C(CH_3)_3$).[21,27,79–85] In what follows, we concentrate on the photoinitiated unimolecular reactions of NO containing molecules, using examples mainly from our own work on NCNO and *t*-BuNO. Work on other molecules is described only when it confirms or contrasts specific aspects, and the reader is referred to the original papers for fuller descriptions.

Several types of experiments can be done to probe different aspects of unimolecular reaction mechanisms. These include (1) spectroscopic studies using LIF of expansion-cooled samples; (2) measurements of fluorescence lifetimes of selected absorption features; (3) photofragment yield spectra (i.e., variation of product yields as a function of photolysis wavelength); (4) product appearance rates, and (5) *E,V,R,T* distributions in products. These measurements allow investigations of the spectroscopy, couplings between electronic surfaces, IVR, the nature of the dissociative surface(s), and the dissociation mechanisms. Below, examples are given that relate to each of these aspects, and comparisons with theory are presented and discussed.

A. Spectroscopy and Photophysics

Aliphatic nitroso compounds share several common features in spectroscopy, electronic structure and photophysics, with radiationless decay processes allowing energized S_0 and/or T_1 molecules to be prepared under well-controlled experimental conditions. Figure 1 is a schematic of the three lowest potential curves of a hypothetical nitroso compound viewed along the R–NO reaction coordinate. The two molecules in this class for which parts of the potential surfaces have been accurately calculated are HNO[86] and NCNO,[73] but the basic shapes of the potentials are common to all. The $\tilde{X}^1A'$ ground state has no barrier in the absence of rotation and values of D_0 range from 13,860 $\pm$ 80 cm^{-1} for CF_3NO[34] to 17,085 $\pm$ 5 cm^{-1} for NCNO.[10] The S_1 state ($\tilde{A}^1A''$) often has its origin $\sim$ 14,000 cm^{-1} above S_0 (except for NCNO,

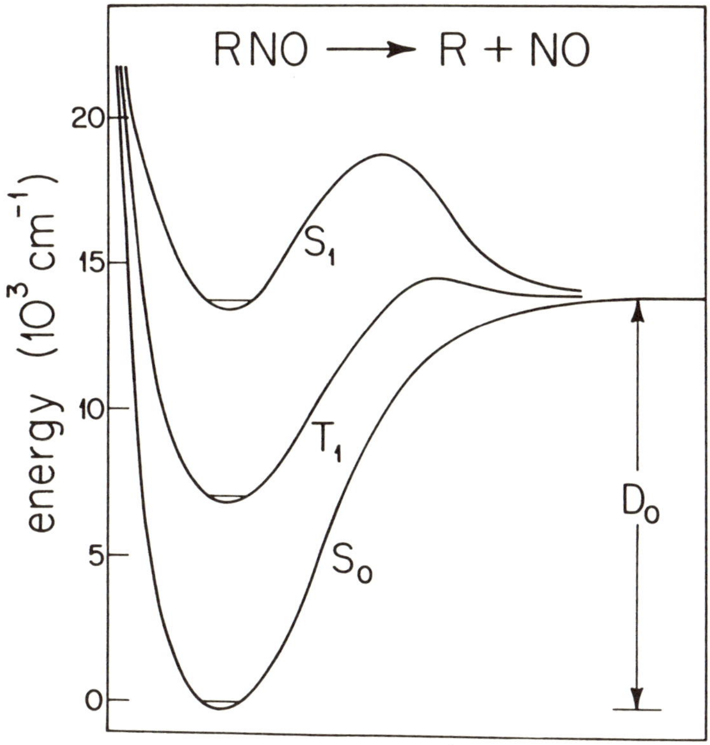

Figure 1. Schematic diagram of the three lowest potential surfaces of a hypothetical nitroso compound along the R–NO reaction coordinate.

whose S_1 origin lies at 11,339 cm^{-1}).[11,21] This state correlates to ground state products via a large barrier (e.g., ~ 7000 cm^{-1} for NCNO),[73] since the initial correlation to excited state NO undergoes an avoided crossing.[21] Little is known about the lowest triplet surface, and its origin and vibrational modes have been measured only for HNO.[88] Calculations for HNO place the T_1 state midway between S_0 and S_1, that is, ~ 7000 cm^{-1} above S_0,[79,89] in good agreement with the measured value of 6280 cm^{-1},[88] and suggest a small barrier in the dissociation channel on the T_1 surface.[89] For NCNO, the calculated barrier on T_1 is 2100 cm^{-1}.[73] The location and barrier heights for the triplet states of other nitroso molecules are unknown, and the participation of low-lying triplet surfaces in the dissociation will be discussed later.

The $\tilde{A}^1A'' \leftarrow \tilde{X}^1A'$ ($\pi^* \leftarrow n$) spectrum is discrete and can be rovibrationally resolved near the S_1 origin, provided the molecule is not sufficiently large to cause congestion and overlap. Such spectra were partially analyzed for HNO,[79,90] DNO,[91] NCNO,[11] CH_3NO,[92] CF_3NO,[36,37,82,92] CF_2ClNO[83] and *t*-BuNO.[27] In all cases but NCNO, LIF of the parent molecule was used. With NCNO, fluorescence is very weak, and two-photon absorption to a dissociative state was employed.[11] The first photon excites the molecule via the $S_1 \leftarrow S_0$ system, and a second photon carries it to a dissociative surface yielding CN($X^2\Sigma$), which is monitored by LIF. By scanning the excitation laser while monitoring CN($X^2\Sigma$), the absorption features of NCNO were obtained. While, strictly speaking, this measurement does not yield an absorption spectrum, it is very close, since the CN($X^2\Sigma$) state distribution changes slowly with $S_1 \leftarrow S_0$ excitation wavelength. In the jet-cooled spectrum of NCNO, the most Franck–Condon active modes near threshold are the NCO and NCN bends (ν_4 and ν_5).[11] A common feature of the aliphatic R-NO compounds is their eclipsed ground state and staggered S_1 state with respect to the dihedral (torsional) angle along the C-N bond.[27,37,93] As a consequence of this change in geometry upon $\pi^* \leftarrow n$ excitation, the torsional progression has a broad Franck–Condon profile.[27,36] Indeed, CF_3NO,[37,82] *t*-BuNO,[27] and CF_2ClNO[83] jet-cooled spectra show a convergent progression of peaks with rapidly increasing intensity to higher energy and a maximum near the top of the torsional potential barrier. Most of the dissociation studies reported to date involve excitation of members of the torsional progression. A second dominant progression in RNO molecules is the CNO bend.

As mentioned above, vibrationally excited RNO molecules in S_0 or T_1 are prepared via S_1 excitation followed by radiationless transitions. This is a powerful technique, since excitation can be efficient, samples can be cooled using nozzle expansions, and a wide and often nearly continuous range of excitation energies is accessible. The couplings between the optically excited S_1 state and S_0 and/or T_1 are important and may lead to dissociation from one or both of these surfaces. In radiationless transitions, molecules are divided into "small" and "large" categories, depending on their density of states and coupling matrix elements.[41] The nitroso molecules studied so far range from "small" (e.g., HNO, NCNO) to "large" (e.g., CF_3NO, *t*-BuNO, C_3F_7NO).

It is well known that in nitroso compounds the $\pi^* \leftarrow n$ transition is orbitally forbidden, and therefore the $\tilde{A}^1A''$ radiative lifetime is quite long (15–30 μs).[20,94] The fluorescence lifetimes, however, vary greatly from molecule to molecule, and depend on the competing nonradiative processes. In HNO, levels below D_0 have fluorescence lifetimes comparable to the radiative lifetimes despite the numerous perturbations evident in the spectra.[90(c),95] Above D_0, levels coupled to S_0 or T_1 dissociate very rapidly, and the fluorescence quantum yield drops to zero as expected.[79,90(c)] HNO definitely belongs to the "small molecule limit."

A particularly interesting case involves NCNO, which, like the other nitroso compounds, has a structured absorption spectrum.[11] However, unlike the other RNO molecules, its dissociation threshold is higher than the S_1 origin, thereby allowing the study of non-radiative couplings of the originally excited S_1 state over a wide range. The spectroscopic studies of Noble et al. showed that numerous perturbations exist between S_1 and S_0 even near the S_1 band origin, with bending and out-of-plane motions being the promoter modes.[11] Qian et al. measured the fluorescence lifetimes, τ_f, of NCNO below D_0 and found lower limits in the range 20–150 μs, limited by the physical removal of excited molecules from the detection region (the nozzle expansion uses an Ar carrier), and intra-beam relaxation processes.[20] The fluorescence lifetimes are longer than the radiative lifetime τ_{rad} (15 μs) in the entire region between the S_1 origin and D_0, without evidence of state specificity. We note, however, that the lack of observed state specificity may be a result of the transport-limited nature of the experimental arrangement. Evidence of state specificity is obtained by observing the linewidths of the transitions in the absorption spectrum.[11] Often, there are striking differences in the linewidths of adjacent lines, and transitions involving bending modes appear to be much broader than those with quanta of NO and CN stretch. The latter are narrow even above D_0.

Since $\tau_f > \tau_{rad}$, NCNO's behavior below D_0 is characteristic of the small molecule limit, although we note that a short dephasing component cannot be ruled out. S_0 state densities are typically 50 and 250 per cm^{-1} near the S_1 origin and D_0, respectively.[18] Thus, the number of states that can couple to a single optically excited state ϕ_s is low, and the molecular eigenstates, Ψ_n, are given by $\Psi_n = a_s^n \phi_s + \Sigma b_1^n \phi_1$, where ϕ_s is the optically excited state and ϕ_1 is a state to which ϕ_s couples. Below D_0, there are no possibilities for irreversible radiationless decay, and ϕ_s is "diluted" by the coupled states, yielding $\tau_f > \tau_{rad}$. With 5 ns laser excitation, the width of each Ψ_n level, γ_n, is governed by the fluorescence lifetime of the mixed states leading to $\gamma_n \sim 5 \times 10^{-5}$ cm^{-1} (3 MHz). Thus, below D_0 and with the densities of states

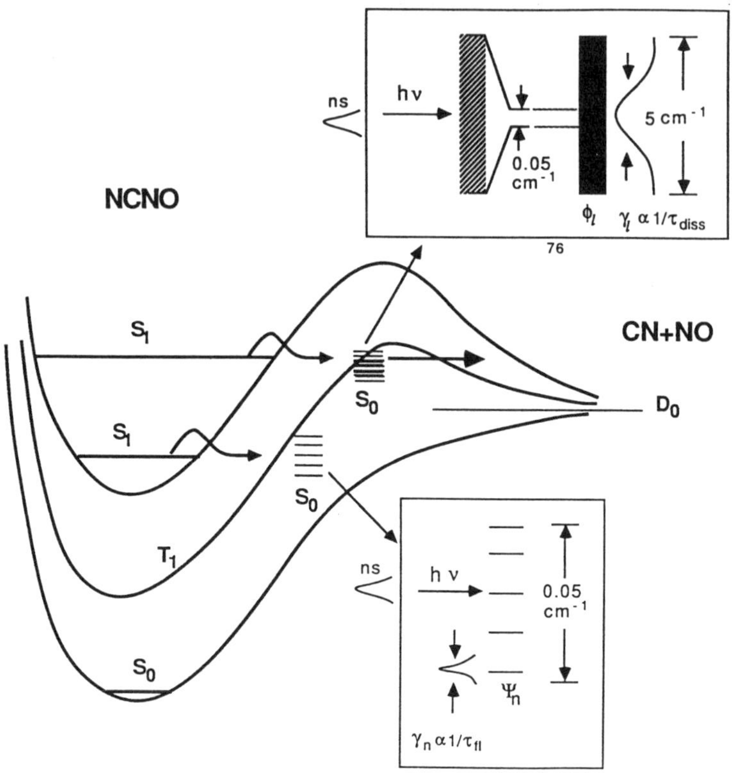

Figure 2. Schematic description of nonradiative processes and decay widths relevant to S_1-S_0 couplings above and below D_0. The laser pulse duration is 5 ns, corresponding to a coherent width of 10^{-3} cm^{-1}. The bottom inset describes the situation for $E^{\dagger} < D_0$. The Ψ_n are molecular eigenstates whose decay widths γ_n are $< 5 \times 10^{-5}$ cm^{-1}. We assume $\rho_1 = 100$ per cm^{-1}. The upper inset describes the situation for $E >> D_0$, where γ_1 is given by the decay width of the S_0 levels ($\gamma_1 = 5$ cm^{-1} for $\tau_{diss} = 1$ ps). Because of the broad decay width, many levels in S_0 overlap and are excited coherently. We assume $\rho_1 = 300$ per cm^{-1} (from ref. 18).

given above, coherent excitation of a single molecular eigenstate can be achieved, as required by the small molecule limit.[18] This situation is shown schematically in the lower inset in Figure 2.

The situation changes dramatically for excitation energies greater than D_0. Because of fast dissociation, γ_1 is now determined by the unimolecular decay rate, and when $\tau_{diss} = 1$ ps, $\gamma_l = 5$ cm^{-1} (see top inset in Figure 2). Under these conditions, many levels can be coherently excited, and this situation has been termed "weak mixing by efficient damping" in radiationless transition theory.[96] Hence, due to the large decay width above D_0, the statistical limit of irreversible decay may actually be achieved, converting NCNO into a "large molecule." In this case, the decay rate can be estimated using Fermi's Golden rule:

$$\Gamma_{IC} = 2\pi |V_{sl}|^2 \rho_{eff} \qquad (7)$$

where Γ_{IC} is the decay rate of the optically accessible state and V_{sl} is the coupling matrix element. As a consequence, at high $E^{\dagger}$ the radiationless transition rate may become comparable or smaller than the dissociation rates, thus influencing the appearance times of the products.[18] Measurements of the rates of unimolecular reactions initiated by radiationless transitions are therefore limited only to those excess energies where the dissociation rate is the limiting step. The near-threshold dissociation rates measured by Khundkar et al. (Section III.C.) appear to be below this limit, and yield the true unimolecular rates.[39] We note that the product state distributions are not affected by changes in the nonradiative rates, and should remain statistical.[34]

In NCNO, the S_0 vibrational density of states, $\rho(S_0)$, is only ~ 250 per cm^{-1} near D_0, while for CF_3NO and *t*-BuNO, $\rho(S_0) \sim 10^6$ and 10^{12}, respectively.[15,21,27] Thus, these molecules definitely belong to the large molecule limit, where practically irreversible radiationless decay is observed, rendering $\tau_f << \tau_{rad}$. Indeed, in the case of jet-cooled CF_3NO, the measured fluorescence decay times near the electronic origin are < 200 ns, decreasing rapidly with excitation energy,[34] and the fluorescence quantum yield (Φ_f) is $< 10^{-2}$.[37] Spears and Hoffland give a detailed account of the couplings between S_1 and S_0 in CF_3NO and identify the R-NO torsion as a promoting mode.[37] Here, state-specific effects are apparent, and different vibrational modes in S_1 exhibit different fluorescence lifetimes. In *t*-BuNO, as expected, the fluorescence lifetimes are even shorter, 30–40 ns near the band origin.[27] At higher energies above threshold (600–1000 cm^{-1}), Φ_f decreases abruptly, probably due to efficient mixings within S_1 which tend to promote the couplings with the bath states.[27] In *n*- and *i*-C_3F_7NO only extremely weak fluorescence is observed throughout the entire S_1 system, in accord with the very large densities of states in these molecules.[85]

A more careful examination of the fluorescence decay by Pfab and coworkers has shown that many absorption features in CF_3NO, CF_2ClNO, and $CFCl_2NO$ exhibit bi-exponential decay.[82-84] These bi-exponential decay profiles are interpreted as indicating coupling to levels of the triplet surface, even for levels close to the dissociation threshold. It appears that ISC to T_1 plays an important role in the radiationless decay of aliphatic nitroso compounds.

B. Unimolecular Reactions: Product State Distributions

Near-Threshold Behavior

Measurements of photofragment yield spectra and energy distributions near dissociation threshold are very important in identifying barriers to dissociation and testing unimolecular reaction models. By using nearly monoenergetic ensembles of expansion-cooled reactant molecules, and tuning the photolysis laser in the near-threshold region while monitoring selected rovibrational levels of the fragments, differences between theories of unimolecular reactions can become apparent, and the dissociation mechanism can be unraveled.

NCNO predissociates to CN and NO upon electronic excitation when $h\nu$ exceeds D_0 (17,085 $\pm$ 5 cm^{-1}).[10] We find that at reaction threshold, the products contain essentially *no* internal excitation.[10] CN is observed only in $v'' = 0$ and $> 95\%$ of the products are in $N'' = 0$; thus the average internal energy of CN is < 0.4 cm^{-1}. This means that even the zero point energies of the transitional bending modes are converted into energy used in the breaking of the NC-NO bond. Such complete energy transformation can occur only when parent vibrations evolve into product motions via a loose transition state in the absence of a barrier. Careful sub-Doppler resolution measurements of the CN($N'' = 0$) line profiles near dissociation threshold are adequately fit by a Gaussian lineshape, and the two main sources contributing to the observed linewidth are the laser bandwidth and the transverse velocity components of the parent molecules in the molecular beam.[10,18] Within the experimental uncertainty, no translational energy from the dissociation needs to be added to the simulated CN Doppler profiles. NO rotational distributions near threshold are just as cold. Thus, at dissociation threshold, CN and NO are born with almost no internal or relative kinetic energy, and therefore an exit channel barrier is ruled out.[10,18]

The appearance thresholds for higher rotational levels of CN coin

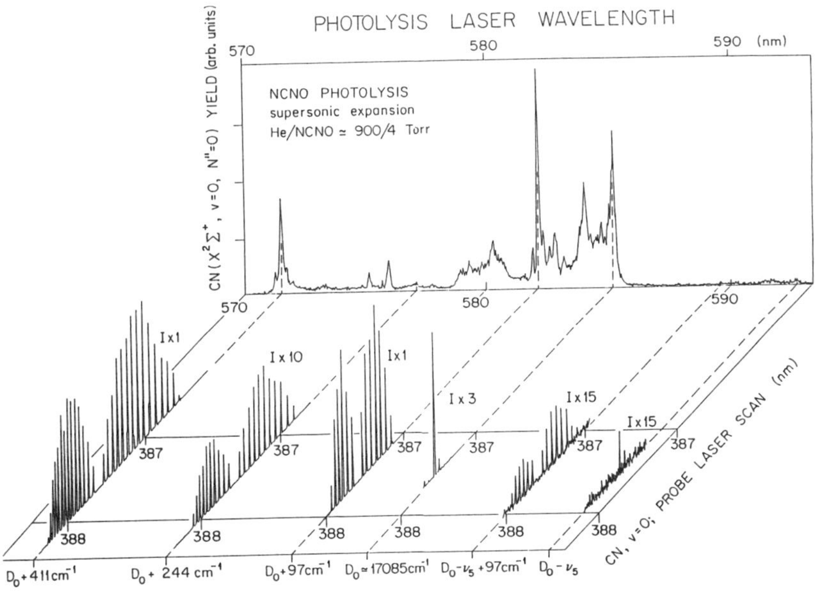

Figure 3. CN LIF spectra at different photolysis wavelengths, following the photodissociation of expansion-cooled NCNO. The upper panel shows the photofragment yield spectrum obtained by monitoring $CN(X^2\Sigma, v'' = 0, N'' = 0)$ by LIF via the $B^2\Sigma \leftarrow X^2\Sigma$ system. The individual CN LIF spectra at specific photolysis wavelengths are displayed perpendicular to the horizontal axes. The two CN LIF spectra on the right derive from excitations of ν_5 "hot bands" (from ref. 16).

cide with the corresponding thermochemical values, and the CN and NO distributions are fit very well by PST. With increasing $E^\dagger$ the CN distributions evolve smoothly and monotonically, and no state specific effects are observed. Examples of near threshold $CN(v'' = 0)$ rotational distributions are shown in Figure 3.[10,16]

Recent experiments by Chen et al. on the photofragmentation of ketene near the threshold for forming singlet methylene also show conclusively that the fit to PST is unique, with no evidence of exit channel barriers.[31] As with NCNO, radiationless decay prepares excited molecules on S_0 and/or T_1, with S_0 character predominating in the reaction leading to singlet products. These authors measured photofragment yield spectra of selected rovibrational levels of singlet CH_2 and observed structure in the threshold region arising from the opening of rotational levels of the CO fragment. The spectra could be fit very well with PST, but not with SACM using $\alpha/\beta = 0.5$, that is, the parameter ratio recommended by Quack, Troe, and coworkers for simple bond-fission reactions.[60–63] As mentioned above, SACM predicts an effective barrier for each dissociation channel, which arises because parent transitional modes correlate *adiabatically* with product R,T excitations. In PST, parent excitations evolve *nonadiabatically* into product excitations, so barriers do not appear in the exit channel of a loose transition state reaction. For the near-threshold behavior of NCNO and ketene, the measurements provide quite strong evidence that the channel barriers predicted by SACM do not exist.

Nascent Product Excitations

Nascent product excitations in state-resolved unimolecular reactions are extremely important, since they provide the data bases needed for rigorous tests of the theories, and can help in identifying the nature of the dissociative PES and the existence of barriers to dissociation. Below, we use examples from our own work to demonstrate the mechanistic information inherent in product state distributions (NCNO), and the participation of S_0 and/or T_1 (*t*-BuNO).

CN and NO E,V,R,T Distributions from NCNO† Decomposition. NCNO was chosen as a prototypical molecule for testing theories of unimolecular reactions because it behaves in many ways like other nitroso molecules, having an $\tilde{A}^1\,A'' \leftarrow \tilde{X}^1\,A'$ spectrum in the near IR and visible which is at least partly predissociative.[11,24] However, it has the distinct advantage that both CN and NO fragments can be studied quantitatively by LIF using wavelengths not significantly absorbed by NCNO. As with all nitroso species, the geometry changes accompanying $\pi^* \leftarrow n$ electronic excitation ensure that transitions to a large energy region of the upper state are Franck–Condon allowed. As a result, it proved possible to study product E,V,R,T excitations in fine detail with $E^\dagger$ in the range 0–5000 cm^{-1}. Even in such a small molecule, because of low frequency bending vibrations, it was necessary to use jet-cooled samples to prevent effects due to hot-bands, which could dominate near reaction threshold.[11,13,24] The monoenergetic data on NCNO represent the most extensive and detailed set of product distributions following a unimolecular reaction reported to date.[10,16–18]

When product vibrations are energetically inaccessible, the rotation

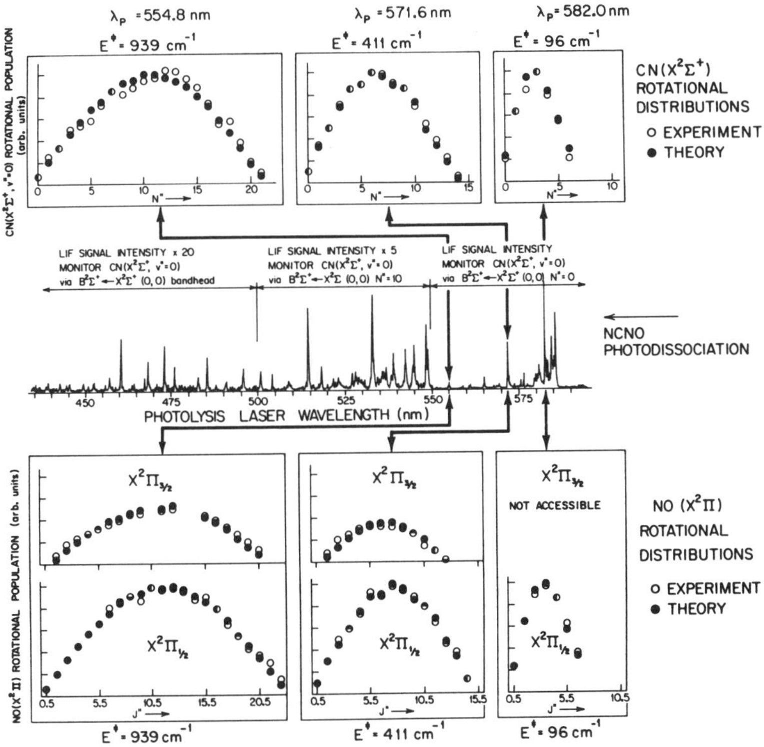

Figure 4. $CN(X^2\Sigma, v'' = 0)$ and $NO(X^2\Pi, v'' = 0)$ rotational populations for several $E^\dagger$ below the $NO(X^2\Pi, v'' = 1)$ threshold. The photofragment yield spectrum in the center is obtained by monitoring CN while changing the photolysis wavelength. The peaks correspond to peaks in the absorption spectrum. The upper entries show calculated and experimental CN distributions for $E^\dagger$ = 939, 411, and 96 cm^{-1}, and the lower entries show the corresponding NO distributions. For NO, the spin-orbit states are shown separately and the relative $^2\Pi_{3/2}$ and $^2\Pi_{1/2}$ populations are to scale for each $E^\dagger$. The calculated $^2\Pi_{3/2}$ population is normalized to the experimental value (from ref. 17).

al distributions of both CN and NO are fit quite accurately using PST.[16,17] Figure 4 shows the agreement between theory and experiment for $0 < E^\dagger < 940$ cm^{-1}.[17] The photodissociation spectrum for NCNO in this region is also shown, with an abrupt onset at 585.3 nm (17,085 cm^{-1}). The spectrum is very dense at these wavelengths, with presumably extensive couplings to S_0 and/or T_1, and possibly some intramolecular vibrational redistribution within S_1, since its origin is at 11,339 cm^{-1}.[11]

As $E^{\dagger}$ increases, product vibrations become accessible (the $v'' = 1$ energies of NO and CN are 1876 and 2043 cm^{-1}, respectively), and for $E^{\dagger} > 1900$ cm^{-1}, deviations from PST can be seen; that is, PST overestimates rotational excitation and underestimates vibrational excitation compared to the experimental observations.[16,17] Thus, as product vibrations become accessible, it is no longer possible to predict V,R,T distributions using PST alone, as was done for $E^{\dagger} < E[\mathrm{NO}(v'' = 1)]$, and the SSE method has to be used.[16,17,67] First, the product vibrational distribution is calculated using a loose transition state assumption to equilibrate 3N-6 parent degrees of freedom. This provides CN(v'') and NO(v'') distributions for a given $E^{\dagger}$. Next, for each set of product vibrations [e.g. CN($v'' = 0$) and NO($v'' = 1$)], the available energy is distributed among the product R,T degrees of freedom using PST. Each of these ensembles, in which the product vibrations are fully specified, mixes degrees of freedom whose parentage is NCNO rotation with those parent transitional vibrations being converted to product R,T motions. For the case of NCNO, 6V and 3R parent degrees of freedom are being converted to 2V, 4R, and 3T product degrees of freedom, and product vibrations appear "hotter" than product R,T excitations, as discussed in Section II.C.

The CN($X^2\Sigma, v'' = 0$) rotational distributions at $E^{\dagger}$ values where NO($X^2\Pi, v'' = 1$) is accessible match the SSE calculations more closely than the PST calculations alone.[16] This agreement persists at all $E^{\dagger}$ for which data are available, and is shown in reduced form in Figure 5, where the experimental and calculated (PST, SSE) values of $\langle E_{rot} \rangle$ vs. $E^{\dagger}$ are shown for CN $v'' = 0$ and 1. One admittedly loses information in reducing complete distributions to a single value of $\langle E_{rot} \rangle$, but this is the only way to display in compact form data obtained over a large range of $E^{\dagger}$. Clearly, the calculations match the experimental results very well.

For NO($X^2\Pi$), the agreement between experimental and calculated quantities is comparable to the case of CN.[17] This is shown in Figure 4 for low $E^{\dagger}$, and in Figure 6 for the case of $E^{\dagger} = 2348$ cm^{-1} and NO($X^2\Pi, v'' = 1$). The measured vibrational distributions are compared with PST and SSE calculations and SSE matches the results better than PST, as shown in Table 1. Thus, for the case of the unimolecular reaction of NCNO†, which has been studied in detail for $0 < E^{\dagger} < 5000$ cm^{-1}, there is no hint of non-statistical behavior in the nascent CN and

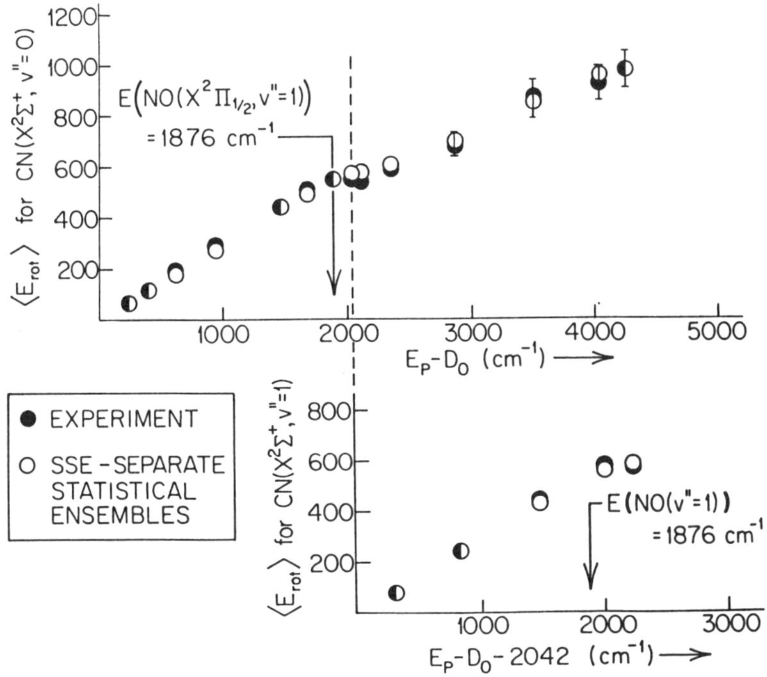

Figure 5. Experimental and calculated (SSE) values of $<E_{rot}>$ for CN($X^2\Sigma$,v'') vs. $E_{h\nu}$-$D_0(v'')$. The upper and lower entries are for $v'' = 0$ and 1, respectively. The SSE calculations are in excellent agreement with the experimental results at all excess energies. Vertical arrows at 1876 cm^{-1} mark the energy threshold for the NO($X^2\Pi_{1/2}$,$v'' = 1$) channel (from ref. 16).

NO *V*,*R* distributions, which can be described well by the SSE model. In contrast to product *V*,*R* excitations, is NO spin-orbit excitation ($^2\Pi_{1/2}$ and $^2\Pi_{3/2}$ states) which is "cold" relative to the *V*,*R* degrees of freedom.[17] Similar cold NO spin-orbit distributions were found for several other nitroso compounds,[32–35,83,84] although a quantitative explanation of such populations is yet to emerge.

After the SSE model calculations were reported, Klippenstein et al.[71] have used the variational RRKM theory of Wardlaw and Marcus,[69] described in Section II.D., to calculate both the unimolecular reaction rates and the product state distributions. In Figure 7, we present both the experimental and the calculated CN rotational distributions at $E^\dagger$ = 2348 cm^{-1}. Clearly, the variational RRKM theory gives a better fit

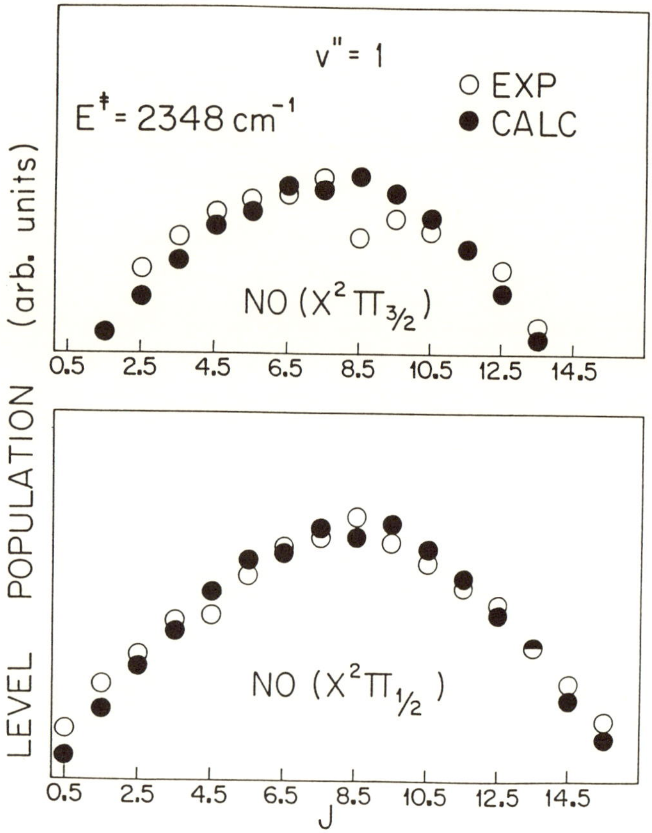

Figure 6. NO(v'' = 1) $^2\Pi_{1/2}$ and $^2\Pi_{3/2}$ experimental and calculated (SSE) rotational populations for NCNO dissociation at $E^\dagger$ = 2348 cm^{-1}, obtained in the dissociation of NCNO. The spin-orbit ratio is fixed at the experimental value for ease of comparison of theory and experiment (from ref. 17).

than PST, and is just as successful as SSE in describing the rotational distributions when NO vibrations are accessible. The vibrational distributions are equally well described, as can be seen in Table 1.

The model described by Klippenstein et al. depends on the NCNO PES.[71] Thus, an accurate ab initio PES for the transition state region will improve the accuracy of the calculations. In the absence of such a PES, approximate surfaces are constructed by using spectroscopic, Lennard-Jones and Varshni potential parameters for NCNO. The same model, with the same potential parameters, is used with equal success

Table 1. Relative Product Vibrational Populations for Different Energies in Excess of the NCNO Reaction Threshold

E† *(cm⁻¹)*	*Experiment**	PST**	SSE**	Var. RRKM***
	$[CN(X^2\Sigma, v'' = 1)/\Sigma_{v''}[CN(X^2\Sigma, v'')]$			
2348	0.07 ± 0.02	0.034	0.08	0.08
2875	0.16 ± 0.02	0.11	0.17	0.14
3514	0.20 ± 0.03	0.17	0.22	0.17
4050	0.24 ± 0.03	0.20	0.24	0.23
4269	0.27 ± 0.04	0.21	0.26	0.22
	$[NO(X^2\Pi, v'' = 1)/\Sigma_{v''}[NO(X^2\Pi, v'')]$			
2348	0.12 ± 0.03	0.07	0.14	0.16

Notes: *From ref. 17.
**From ref. 17, values shown are for $J_0 = 5$; with $J_0 = 0$ or 10, there is a small change.
***From ref. 71.

to predict the unimolecular reaction rates of NCNO, as described in Section III.C.[71]

Recently, Qian et al. carried out experiments that probed correlations between the internal energies of the two photofragments (i.e., for a specific $CN(X^2\Sigma, v'', N'')$ state, what are the corresponding NO states).[18] This was achieved by measuring the translational energy of selected rovibrational levels of CN, and deriving the associated NO internal energies using energy and angular momentum conservation. From energy conservation:

$$E_T = E^{\dagger} - [E_V(\mathrm{NO}) + E_R(\mathrm{NO}) + E_E(\mathrm{NO}) + E_V(\mathrm{CN}) + E_R(\mathrm{CN})] \quad (8)$$

and from angular momentum conservation:

$$\boldsymbol{J_0} = \boldsymbol{J_{CN}} + \boldsymbol{J_{NO}} + \boldsymbol{L} \quad (9)$$

where $\boldsymbol{J_0}$ is the total angular momentum, $\boldsymbol{J_{CN}}$ and $\boldsymbol{J_{NO}}$ are rotational angular momenta, and $\boldsymbol{L}$ is the orbital angular momentum (which is measured as relative translation). For each state of CN there is a distribution of NO states, and consequently a distribution of translational energies. In the statistical theories, these distributions are subject only to the constraints of the model. However, if there are dynamical biases associated with the correlations between the quantum state distributions of the fragments, this can cause deviations from a statistical prediction. Although both CN and NO internal state distributions fit statistical predictions very well,[16,17] such *global* product state

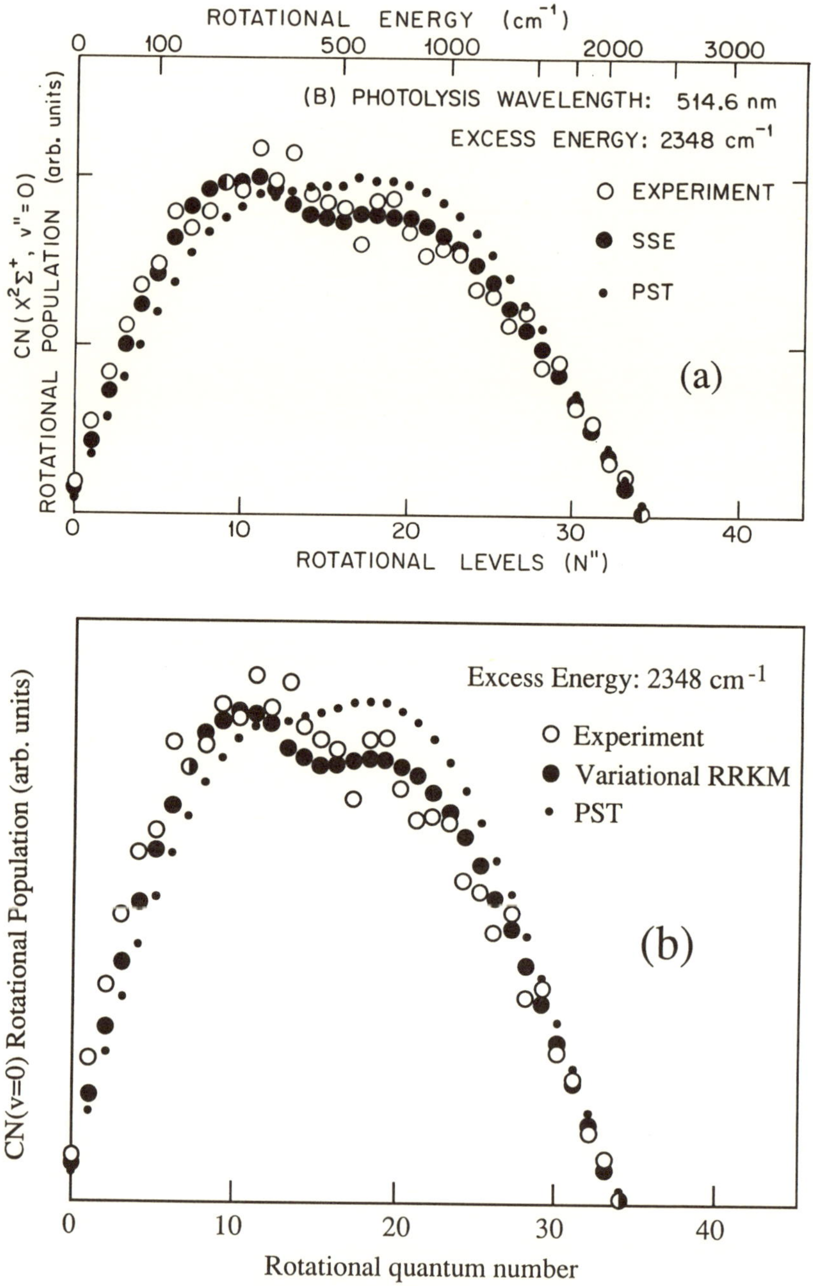

Figure 7. Plots of CN($v'' = 0$) rotational populations for NCNO dissociation at $E^{\dagger} = 2348$ cm^{-1}, comparing experimental results and PST calculations to: (a) SSE calculations (from ref. 67); (b) variational RRKM calculations (from ref. 71).

distributions may obscure biases that depend on the shape of the PES along the reaction coordinate. These are better discerned by looking at *correlated* distributions. For example, the distribution of NO internal states associated with a specific CN internal state may be non-statistical, even though global CN and NO distributions are quite statistical.

In order to check for such correlation, Qian et al. recorded LIF spectra of CN fragments with sub-Doppler resolution.[18] Since the recoil velocity distribution is expected to be isotropic, it is possible to compare the experimental lineshapes with the calculated profiles assuming a statistical distribution amongst the NO internal states. The $R(7)$ and $P(10)$ transitions of the CN $B^2\Sigma \leftarrow X^2\Sigma$ system were chosen for these measurements because the F_1 and F_2 spin-rotation branches were not completely resolved, and therefore the valley between the two peaks could serve as a sensitive measure of the fit.

In Figure 8, we compare the experimental and simulated CN(N'' = 10) Doppler profiles for $E^\dagger$ values of 1670 and 2348 cm^{-1}. In the latter, NO(v'' = 1) is already populated, and thus PST and SSE predict different line profiles. Note that although the Doppler profile at $E^\dagger$ = 1670 cm^{-1} appears broader than at 2348 cm^{-1}, the calculated average kinetic energy associated with it is slightly smaller (443 vs. 480 cm^{-1}). Clearly, the experimental results at 2348 cm^{-1} are better fit with SSE than PST, and the *shapes* of the computed profiles are sensitive to differences in the translational energy *distributions*. All the measured line profiles at $411 < E^\dagger < 2348$ cm^{-1} are well fit with the same PST or SSE model that is used to simulate the global CN and NO state distributions.[18] The model has no adjustable parameters, and the results confirm that the SSE model works also for the correlated distributions. We find less R,T excitation and more vibrational excitation than predicted by PST at excess energies where product vibrations can be excited. From simple energy conservation, it follows that the Doppler profiles for high rotational states of CN are correlated with low NO(N'') states, indicating that the orbital angular momentum of the separating fragments is directed oppositely to the CN rotational angular momentum, since the NO angular momentum is low. In summary, the correlated distributions show excellent agreement with SSE, with no hint of barriers, and confirm our previous conclusions which were based on the global CN and NO distributions.

Before closing, we turn to another aspect that relates to the mechanism of NCNO dissociation, namely, the nature of the transition state.

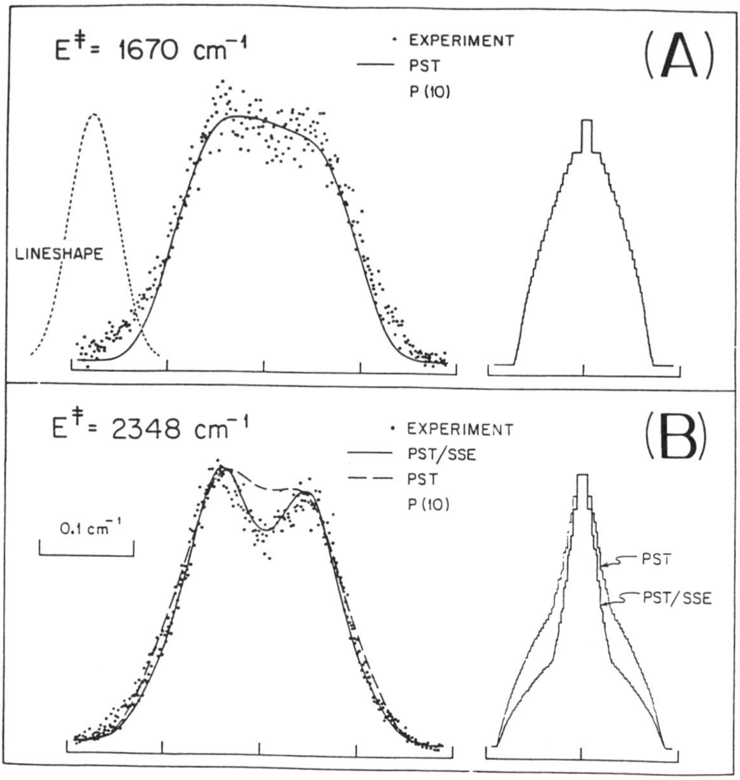

Figure 8. Doppler profiles for the CN $P(10)$ line obtained at (A)$E^{\dagger}$ = 1670 cm^{-1} and (B) $E^{\dagger}$ = 2348 cm^{-1}. On the right hand side, Doppler profiles calculated using the PST and SSE models at infinitely narrow laser linewidth are shown. These are convoluted with the experimental laser linewidths (shown on the left hand side of part A) to yield the simulated spectra which are compared with the experimental data (dots). Notice that SSE gives a better fit to the data at $E^{\dagger}$ = 2348 cm^{-1} than does PST. Also, the profiles of the two spin-rotation components at $E^{\dagger}$ = 2348 cm^{-1} appear narrower than those at 1670 cm^{-1} despite the fact that the average translational energy associated with the former is higher (see Table 1 and the text; from ref. 18).

This can be partially probed by determining the relative populations of the NO($X^2\Pi$) Λ-doublet components. In the Hund's case (b) limit, the $\Pi(A')$ Λ-doublet component has the lobe of the singly occupied $p\pi$ orbital in the plane of rotation, while the $\Pi(A'')$ component corresponds to the $p\pi$ orbital lying perpendicular to this plane.[97] Since NO belongs to an intermediate a/b Hund's case, the above geometrical interpretation is valid only for high-J NO states, and therefore mechanistic implications inferred from Λ-doublet propensities apply only to the high-J subset of product molecules. In the NO $A^2\Sigma \leftarrow X^2\Pi$ transition, Q-branch lines probe the Π(A″) Λ-doublet component, while P- and R-branch lines probe the Π(A′) Λ-doublet component.[97] The Λ-doublet ratios were measured at two excess energies, 2348 and 2875 cm^{-1}, and for $J'' > 30.5$, there was a small preference for the Π(A′) component: Π(A′)/Π(A″) = 1.3 ± 0.2.[18]

Since dissociation occurs from S_0, which is of A′ electronic symmetry, and CN($X^2\Sigma^+$) is also A′, we expect NO($X^2\Pi$) to be A′ for planar dissociation dynamics. In this case, the singly occupied NO $p\pi$ orbital will be preferentially oriented in the plane of rotation. The NCNO ground state is planar, and the small preference for the Π(A′) Λ-doublet component suggests that this is preserved to some extent during dissociation, at least for high NO rotational levels. In our measurements, these high J''_{NO} values, for which the simple geometric interpretation of the Λ-doublets applies, are close to the highest J''_{NO} values allowed by energy conservation, and are therefore associated with CN fragments having low J''_{CN}. The small subset of NO fragments with J''_{NO} near the thermochemical limit is thus correlated with CN fragments that are hardly rotating. The NCNO transition state associated with this subset may be viewed as pseudo-triatomic, with almost no torque exerted on the CN. Thus, the propensity for planar dissociation for these molecules is not surprising.

Dissociations on S_0 and/or T_1. The detailed and extensive set of product state distributions and their correlations obtained for the unimolecular reaction of NCNO provided the experimental data bases required for testing statistical models, and making the necessary modifications (e.g., SSE, variational RRKM). We emphasize that the required corrections are not large, and are probably important only for small molecules such as NCNO, H_2O_2 or ketene. For larger molecules (e.g., CF_3NO, t-BuNO)[27,34] the V,R distributions in the absence of exit-channel barriers are fairly well fit even by a simple prior calculation, in which only energy conservation is taken into account. In these larger nitroso compounds, the observation of deviations from statistical behavior at higher $E^\dagger$ usually signify the opening of a second, competing dissociation channel, such as T_1, that has a barrier to dissociation. Dissociation on T_1 can be identified even in the absence of a barrier by product appearance times, which are shorter on T_1 owing to its lower density of states, and these measurements are described in more detail in the next section.

The identity of the dissociative surface is not easily determined in unimolecular reactions induced by laser initiated radiationless transitions, because IC and ISC can lead to excitation of S_0 and T_1, respectively, while T_1-S_0 coupling can lead to further population of S_0. In the case of *t*-BuNO, it proved possible to distinguish clearly between dissociation on S_0 and T_1. *t*-BuNO is structurally similar to the other nitroso compounds, but has 39 vibrational degrees of freedom. Consequently, S_0 and T_1 have high densities of states, and this results in fast (i.e., ns time-scale) radiationless decay of S_1 levels.[27] On the other hand, RRKM theory predicts that unimolecular reaction is considerably slower (e.g., several μs for dissociation on S_0 at 500 cm^{-1} above D_0), and this permits temporal separation of the radiationless decay and the S_0 unimolecular reaction rates. The predissociation of jet-cooled *t*-BuNO following laser excitation in the $\tilde{A}^1A'' \leftarrow \tilde{X}^1A'$ system has been studied in both the frequency and time domains, and it has been possible to separately measure radiationless transition rates and unimolecular reaction rates in real time.[27] Five types of experiments were carried out: (1) LIF spectra of the *t*-BuNO parent, (2) fluorescence lifetimes of *t*-BuNO, (3) NO yield spectra (i.e., photodissociation spectra), (4) NO appearance times and (5) LIF spectra of nascent NO.

The *t*-BuNO fluorescence lifetimes (and hence the radiationless transition times) are short, < 45 ns. The NO appearance times, however, show two distinct regimes. At $E^† < 650$ cm^{-1} ($D_0 = 13{,}940 \pm 30$ cm^{-1}; S_1 origin $= 13{,}911$ cm^{-1}) lifetimes > 3.5 μs are observed, whereas at higher $E^†$ a fast risetime component (< 15 ns, the time resolution of the experiment) becomes apparent and soon dominates. Over a narrow $E^†$ region near the threshold for the appearance of the fast component, a double exponential appearance curve is observed, but the relative amplitude of the slow component decreases rapidly with excess energy. The "fast" and "slow" NO contributions can be seen clearly in the photofragment yield spectra displayed in Figure 9. Photofragment yield spectra were recorded by fixing the probe laser wavelength at the P_{11} bandhead of NO and detecting NO via one-photon LIF.[27] Figure 9 shows spectra obtained with pump/probe delays of ~ 60 ns and 3.2 μs, compared to the LIF spectrum of *t*-BuNO in the same excitation region. The spectra show clearly a group of peaks at wavelengths > 685 nm that have long appearance times. Although the "fast" and "slow" appearance rates could easily be observed, accu-

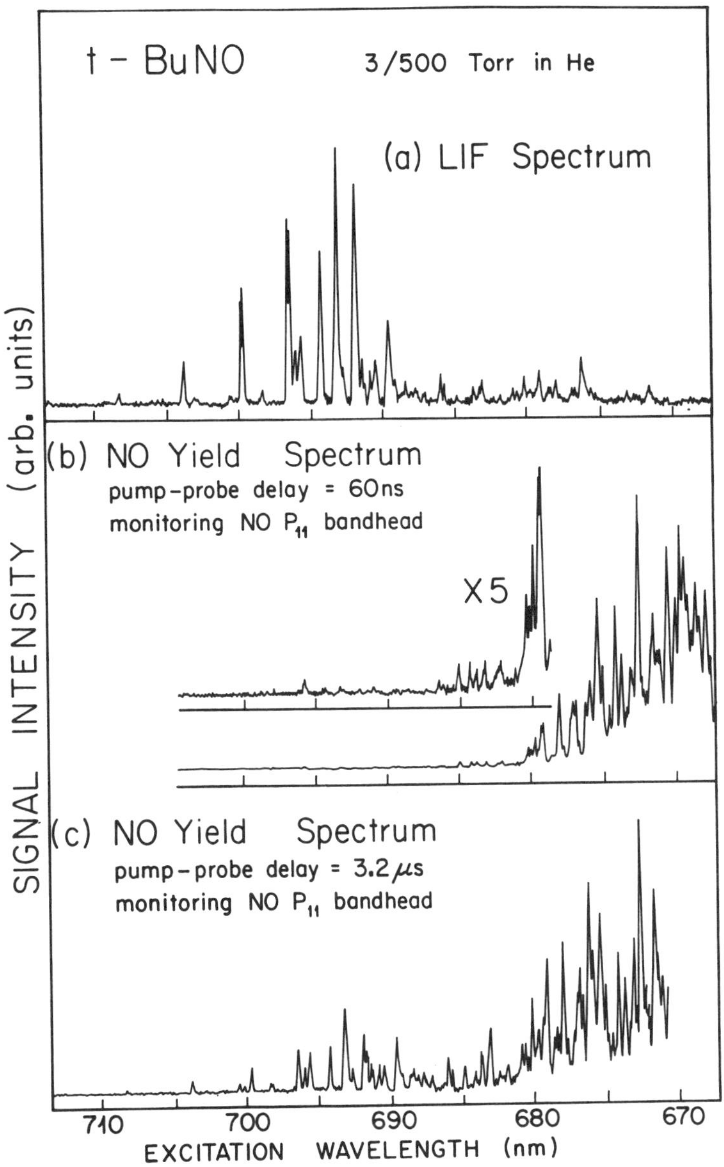

Figure 9. *t*-BuNO LIF and NO photofragment yield spectra as a function of the photolysis laser wavelength. (a) is *t*-BuNO LIF; (b) and (c) are the NO photodissociation yield spectra recorded with the probe laser set to the P_{11} bandhead of NO at 226.3 nm, with laser delays of 60 ns and 3.2 μs, respectively. All spectra are corrected for laser energy (from ref. 27).

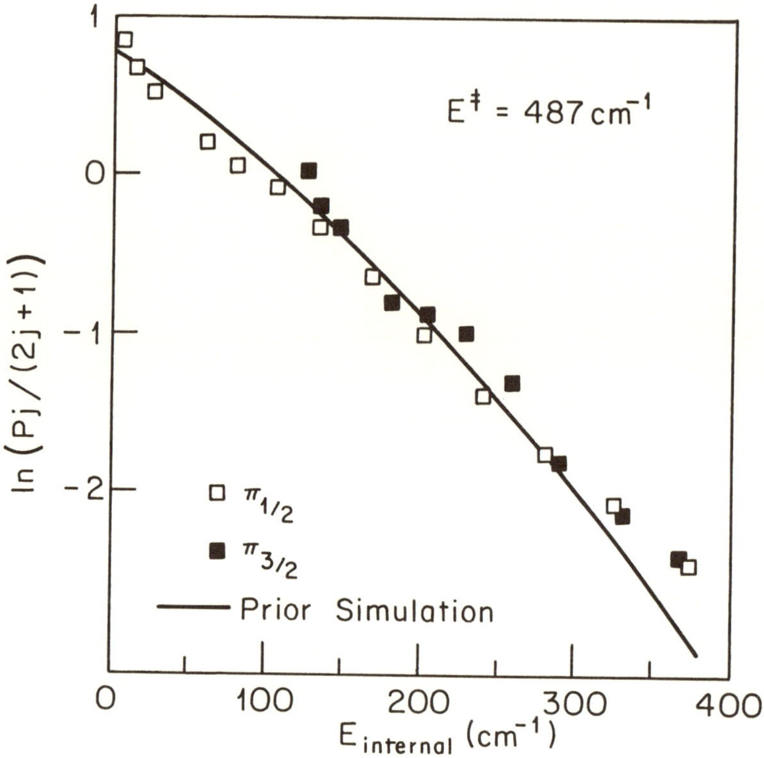

Figure 10. Nascent NO rotational state distributions ($\Pi_{1/2}$ and $\Pi_{3/2}$) following *t*-BuNO photolysis at 696.67 nm compared to a prior calculation using $E^{\dagger} = 487$ cm^{-1}. Note that the curves for the two spin states of NO coincide (from ref. 27).

rate dissociation rates could not be obtained. Measurements of the slow rates were limited by removal of the molecules in the supersonic jet from the region of the probe laser. Thus, only lower limits on the dissociation lifetimes were derived. The measurements of the fast rates at $E^{\dagger} > 650$ cm^{-1} were limited by the time resolution of the experiment (~ 15 ns).

The experimental results suggest that two separate processes are responsible for NO formation. At $E^{\dagger} < 650$ cm^{-1}, dissociation proceeds only on S_0. However, when $E^{\dagger} > 650$ cm^{-1}, the barrier to dissociation via T_1 is exceeded and direct dissociation via this channel competes with dissociation on S_0. This mechanism is further supported by NO($X^2\Pi$) rotational energy distributions. At $E^{\dagger} < 650$ cm^{-1}, dissociation on S_0 leads to nascent NO distributions which can be fit using statistical partitioning of $E^{\dagger}$ with a loose transition state and no exit channel barrier (Figure 10). The spin-orbit population ratio, (i.e., $[^2\Pi_{3/2}]/[^2\Pi_{1/2}]$), is statistical as well. Conversely, above the small ($\sim$ 650 cm^{-1}) barrier to dissociation on T_1, *non-statistical* behavior is observed, as shown in Figure 11. The NO rotational distributions are colder than statistical, and most notably, the spin-orbit population ratio is much smaller than statistical. The results indicate that the two processes, $T_1 \rightarrow S_0$ and $T_1 \rightarrow$ *products* occur at comparable rates near the top of the T_1 barrier. This causes the bi-exponential NO appearance rates, since some molecules dissociate on T_1 very quickly (< 20 ns), whereas others remain trapped on S_0 for up to several μs. Variations in

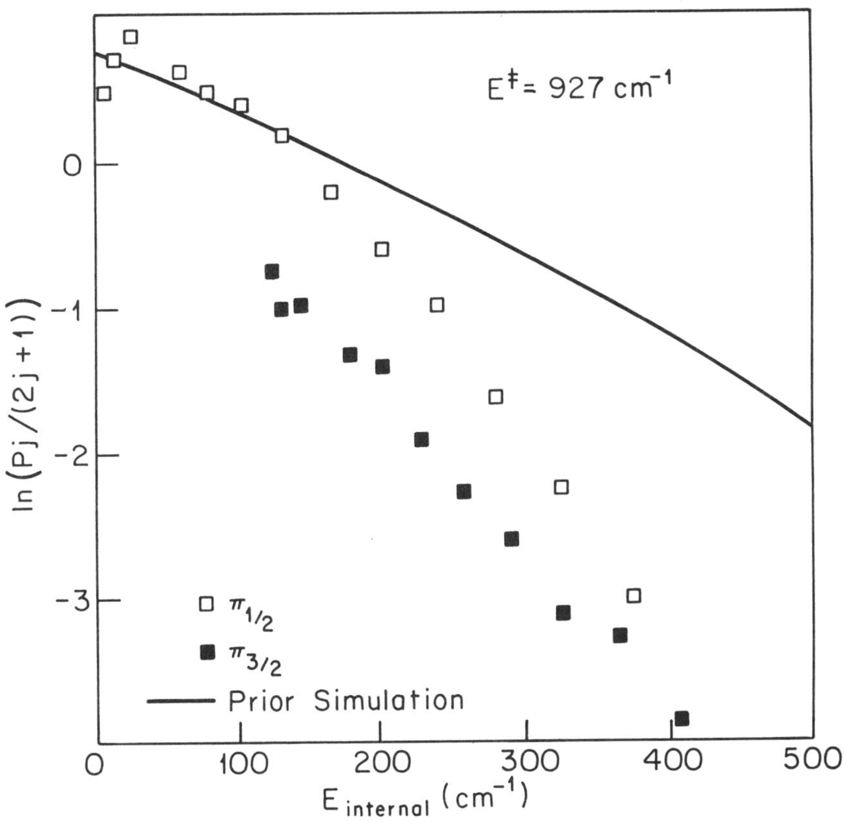

Figure 11. Nascent NO rotational state distributions ($\Pi_{1/2}$ and $\Pi_{3/2}$) following *t*-BuNO photolysis at 673.09 nm compared to a prior calculation using $E^{\dagger}$ = 927 cm^{-1}. Note that the NO spin states are anomalously populated (i.e., $[\Pi_{3/2}] << [\Pi_{1/2}]$) (from ref. 27).

the relative amplitudes of the fast and slow components probably reflect variations in the $S_1 \rightarrow T_1$ and/or $T_1 \rightarrow S_0$ rates. At the top of the T_1 barrier, NO is internally cold due to the smaller excess energy at the barrier, compared with dissociation on S_0. However, because of the barrier, the products apparently gain kinetic energy in the exit channel. The recoil energy appears principally in the form of translation (plus some rotation), and will be partitioned between the fragments according to conservation of momentum, not densities of states. Thus, *t*-BuNO offers a unique opportunity for studying a simple bond fission reaction above a barrier.

Well above the T_1 barrier, most of the *t*-BuNO molecules that have undergone ISC into T_1 will dissociate on T_1 instead of undergoing ISC to S_0, since $T_1 \rightarrow$ *products* becomes faster than $T_1 \rightarrow S_0$. RRKM calculations yield T_1 dissociation times < 10 ns when the T_1 origin is ~ 5000 cm^{-1} above S_0.

In conclusion, *t*-BuNO is typical of a large aliphatic nitroso compound whose predissociation proceeds on S_0 and T_1. It presents not merely another test case for statistical theories, but also a case for which the radiationless transition and unimolecular reaction rates can be separated in time. Indeed, it has proved possible to observe ISC and/or IC, and competitive dissociation on two surfaces (S_0 and T_1), the latter having a small barrier to dissociation. A similar type of barrier, leading to nonstatistical PSDs was proposed by Bitto et al. to explain the unimolecular reaction of ketene on the triplet surface.[22] We emphasize, however, that a barrier on T_1 is not required in nitroso molecules, and the occurrence and size of the barriers differ from one molecule to another. The barriers arise from avoided crossings and depend on the interactions with other surfaces correlating diabatically with excited product states. Ab initio calculations exist for HNO and NCNO; the former has a small barrier on T_1,[86] the latter a rather substantial one (2100 cm^{-1}).[73] Furthermore, the existence of a barrier does not necessarily lead to nonstatistical PSDs.

The participation of T_1, with or without a barrier, is probably rather common in RNO compounds, especially at high $E^{\dagger}$. Based on NO appearance times, we concluded that dissociation on T_1 is rather important in the unimolecular reactions of *n*- and *i*- C_3F_7NO.[85] In this case, the NO rotational distributions were statistical, and the participation of T_1 was inferred from the NO appearance times, as described in the next section, and from comparisons between the NO photofragment yield spectra and the absorption spectra.[85] The deviations from statistical behavior at high $E^{\dagger}$ in the unimolecular reaction of CF_3NO may also be caused by the opening of the T_1 dissociation channel.[34,82] In this case, appearance times are dominated by the radiationless transition rates, and thus cannot help in distinguishing between dissociation on S_0 and T_1. Participation of T_1, possibly without a barrier, was also implicated in the unimolecular reaction of CCl_2FNO.[84] Differences between the absorption or LIF spectra and the photofragment yield spectra indicated that state-specific channels may correlate with dissociation on T_1.[84]

C. Unimolecular Reactions: Rate Measurements

The determination of state-selective unimolecular rates can be an exceedingly difficult task. Unimolecular reaction rates of large molecules can be very slow, especially near threshold, and thus physical transport of excited molecules and relaxation put an upper limit on the fragment appearance times that can be measured. In cases involving laser-initiated radiationless transitions, dissociation on several surfaces leading to the same fragment may produce additional complications. For small molecules, for example, NCNO, CH_2CO, CF_3NO, the unimolecular reaction rates are fast, thereby alleviating problems associated with physical transport and relaxation. However, as discussed in Section III.A., product appearance rates do not necessarily reflect the unimolecular reaction rates, but may be influenced, or even dominated, by nonradiative decay rates.

The situation is simpler in limiting cases. As discussed above, in *t*-BuNO the unimolecular reaction rates on S_0 near threshold are much smaller than the radiationless transition rates and can be separated temporally.[27] Therefore, the product appearance time, even though often only an upper limit, can serve to distinguish between dissociation on S_0 and T_1. Sometimes, the unimolecular reaction times on S_0 are so long that products are not observed at all on the time scale of the experiment (a few μs), and only products from dissociation on T_1 can be monitored. A good case in point is the dissociation of *n*- and *i*-C_3F_7NO following excitation to S_1.[85] In these two compounds, the appearance times of NO are short (< 1 μs), but the dissociation *quantum* yield is << 1. It is concluded that only dissociation on T_1 is detected, but this is the *minor* channel, and most of the molecules dissociate on S_0 on a time scale much longer than the observation time. An illustrative example involves the dissociation of *n*-C_3F_7NO.[85] No

NO signal is observed up to 1 μs delay at wavelengths shorter than the S_1 origin and D_0 (730 nm). Signals appear only at photolysis wavelengths below 650 nm, and rise towards shorter wavelengths in a part of the spectrum where the n-C_3F_7NO absorption cross sections are actually *dropping*. This led us to conclude that the *observed* NO in the photodissociation of C_3F_7NO derives from T_1, and the $S_1 \rightarrow S_0$ and $S_1 \rightarrow T_1$ radiationless decay channels are competing, with branching ratios that vary with energy. In these particular molecules, the NO rotational distributions at $E^\dagger < 2000\ cm^{-1}$ could be well fit by the prior model, and thus there was no need to invoke the existence of a barrier to dissociation on T_1. Differences in densities of states and coupling matrix elements may govern the branching ratios in this case.

At the other extreme stands the unimolecular reaction of CF_3NO, the first nitroso compound for which the NO fragment was directly monitored.[32] Here, the NO appearance times are equal to the CF_3NO fluorescence times in both 300 K and jet-cooled samples.[32–35] Near threshold, the appearance times are much longer than those calculated using RRKM theory for reaction on S_0. Thus, the lifetimes reflect the radiationless transition rates rather than the unimolecular reaction rates. In fact, the measured rates have been used to study the nonradiative couplings in CF_3NO.[37,82] Spears and Hoffland identified the torsions as the promoting modes for the nonradiative couplings, and found better agreement between radiationless transition theory and experiment if IC to S_0 was involved, at least near threshold.[37] Dyet et al. analyzed the bi-exponential fluorescence decay curves and concluded that ISC is important at well.[82] The NO rotational energy distributions obtained with excitation energies near D_0 (13,880 ± 80 cm^{-1}) were consistent with a statistical distribution of energy (prior) in the products.[34] Thus, although the dissociation rate is much larger than the radiationless transition rate, all memory of the initial S_1 excitation is erased, and energy randomization on the dissociative surface is complete prior to dissociation. As more highly excited vibronic states of S_1 are selected, the product energy distributions begin to show signs of nonstatistical behavior, although the deviations are small. These deviations may reflect a greater involvement of T_1 (see the section on t-BuNO).

The major payoff in measuring appearance times comes from the opportunity to obtain monoenergetic unimolecular reaction rates, k_{uni}, and compare them with statistical theories. The pioneering experiments of Khundkar et al. yielded a set of rates for the unimolecular reaction of NCNO near threshold.[38,39] As discussed in Section III.A., NCNO is a "small/intermediate molecule" with regard to its radiationless transitions. The measured NO appearance rates above threshold rise fast, and reach 10^{11} s^{-1} at $E^{\dagger}$ 600 cm^{-1}.[39] With such fast rates, care should be exercised in deciding whether the rates reflect the true unimolecular reaction rates, or are influenced by the preceding radiationless transitions. In the case of NCNO, the observed rates increased by three orders of magnitude on going from near threshold to $E^{\dagger} = 700$ cm^{-1}. Such a fast increase is typical of unimolecular reaction rates, but rather improbable for changes in the radiationless transition rate with energy. Thus, near threshold the measured rates reflect predominantly the unimolecular reaction rates.[39]

When the rates measured by Khundkar et al. are extrapolated to higher $E^{\dagger}$, $\tau_{uni} \leq 1$ ps is obtained. These lifetimes are shorter than the parent rotational times, and therefore some anisotropy in the fragment spatial distributions is expected. Qian et al. looked for such anisotropy in the Doppler profiles of CN obtained in the photodissociation of NCNO at $E^{\dagger} = 2348$ cm^{-1}, but within their experimental uncertainty, no anisotropy was discerned.[18] This may indicate that the radiationless transition rate starts to be rate limiting. However, the lack of anisotropy can also be attributed to one or more of the following processes: (1) unimolecular reaction rates slower than those predicted by statistical theories; (2) nonplanar motions during the dissociation, and (3) reduced initial alignment for the expansion-cooled NCNO.[18]

The rates measured by Khundkar et al.[38,39] could not be simulated by the standard, packaged RRKM theory, with any reasonable parameters of the transition state, and they were substantially lower than the PST predictions, when the electronic degeneracy was treated simply as a constant over the energy range $0 < E^{\dagger} < 700$ cm^{-1}, with contributions also included from S_1 and T_1.[39] Klippenstein et al. recently computed the rates by using variational RRKM theory and obtained good fits with the experimental results at $0 < E^{\dagger} < 700$ cm^{-1}, as shown in Figure 12.[71] A similarly good fit was obtained using PST and assuming reaction only on S_0.[71] At higher $E^{\dagger}$, the differences between the rates calculated by PST and variational RRKM are expected to increase, with PST giving rates higher than the experimentally observed ones. In variational RRKM, the transition state lies at the position of minimum flux along the reaction coordinate, and moves inward with increasing $E^{\dagger}$.[68–72] In PST, the transition states are assumed to lie at the (quite large) fragment separations associated with centrifugal barriers.[53–57] Thus,

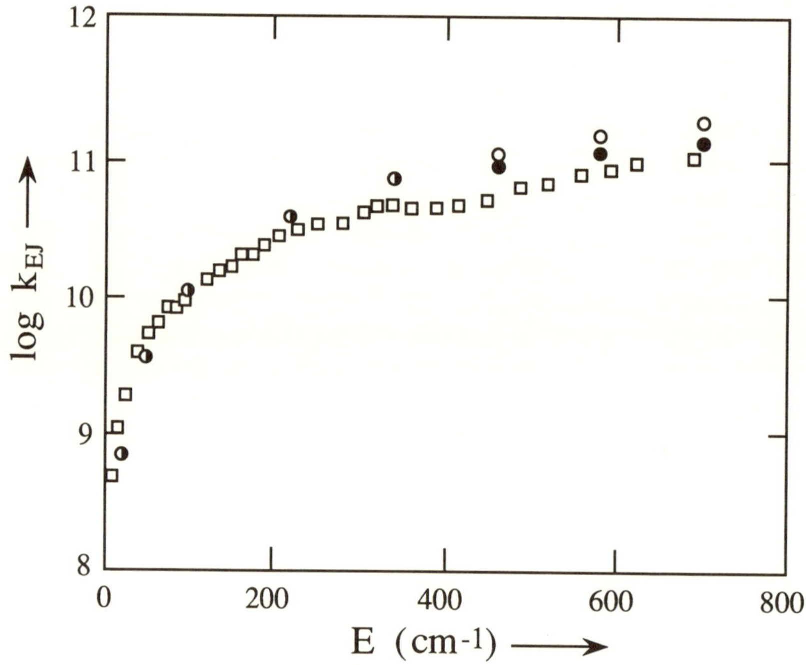

Figure 12. Plots of the experimentally determined unimolecular rates for NCNO dissociation (boxes) compared with the rates calculated using variational RRKM theory. The open and filled circles denote calculations with different parameters assumed for the NCNO PES (from ref. 71).

rates predicted by the variational RRKM theory are predicted to be progressively smaller than those calculated by PST with increasing $E^{\dagger}$.

In conclusion, the unimolecular reaction rates in NCNO can be reasonably well described within the framework of statistical theories. Very recently, the variational RRKM model was also used successfully to fit unimolecular reaction rates in ketene at $0 < E^{\dagger} < 5000$ cm^{-1}.[40] At these high $E^{\dagger}$, the differences between the experimental rates and those predicted by PST are substantial.

IV. CONCLUDING REMARKS

The work described in this review amply demonstrates the evolution in our understanding of unimolecular reactions, and the level of precision and detail with which these processes can now be interrogated. The fascination in studying unimolecular reactions derives partly from the generality of mechanisms involving statistical partitioning of energy. Thus, results obtained for small molecular systems that are readily amenable to detailed experimental studies, as well as rigorous theoretical treatments, can often be extrapolated to large molecules. As a consequence, the theoretical models can become truly predictive for a large number of systems. The results obtained with NCNO and ketene (yielding singlet methylene), two molecules exhibiting simple bond-fission reactions without a barrier, underscore this point; their unimolecular reactions can be described using very similar models. However, the use of small molecules as prototypical systems for statistical unimolecular processes carries a caveat: complete IVR is not guaranteed, and deviations from the statistical models may indicate incomplete energy randomization in the parent vibrational degrees of freedom prior to dissociation. H_2O_2 may fall into this category, since its OH stretch is not easily coupled to the other vibrational degrees of freedom. More work on small molecules is definitely needed to find out the limitations of the statistical descriptions.

A great advance in our ability to obtain a comprehensive picture has been achieved with the recent measurements of monoenergetic unimolecular rates, in addition to PSDs. This has enabled Marcus and coworkers to formulate a comprehensive, unified statistical theory that encompasses both rates and quantum state distributions. A single statistical theory can now describe all the experimental observations, at least for NCNO and ketene. Several simpler theoretical treatments have been quite successful in addressing various aspects of the dissociation, but more experiments are needed to test their validity and scope. Also, the influence of parent rotational excitation on the unimolecular rates and PSDs can now be assessed by using double resonance two-frequency excitation schemes.

Another result that emerged from studies of small molecules is the role of the dissociative PES in governing the locations of the transition states for rates and PSDs. It was previously believed that simple bond-fission reactions proceed with no exit-channel barriers, and thus are insensitive to the shape of the PES. This is often true when product distributions are observed, but the recent work of Marcus and coworkers has shown that the rates, in particular, are sensitive to the location of the transition state, and an overall agreement with all the experimental results requires knowledge of the PES. Previous discrepancies between SACM, PST, and the variational RRKM theory may have derived from the (explicit or implicit) use of PESs of different shapes.

When measuring PSDs of simple bond-fission reactions proceeding without a barrier, the detailed shape of the PES is not very important, and thus PSDs are not sensitive probes of the potential along the reaction coordinate. In contrast, when the simple bond-fission reaction proceeds over a barrier (e.g., as a result of avoided crossings or conical intersections), PSDs have proven to be rather sensitive to exit-channel barriers, exhibiting deviations from statistical behavior even for relatively small barriers (e.g., 600 cm^{-1} in *t*-BuNO). However, little theoretical and experimental work exists that addresses rates, PSDs and their relation to the shape of the PES, and more examples such as *t*-BuNO, HN_3, and ketene (yielding triplet methylene) are needed. Such processes are expected to be particularly important when activation is achieved via electronic excitation followed by radiationless transitions, and the dissociation can proceed on surfaces other than the ground electronic PES (e.g., T_1).

Almost two decades ago, Wendell Forst, a giant in the field of unimolecular reactions, made the statement given at the beginning of this review.[1] Since then, the rapid advances in this field have clearly exceeded even his wildest expectations. While lessening our appetite for bold prediction, they have nevertheless fired our enthusiasm, and we anticipate great discoveries ahead.

Note added in proof: Since this review was written, several exciting developments have occurred in our understanding of unimolecular reactions proceeding without barriers, much in-line with the optimistic expectations stated so boldly in the concluding paragraph of this review. Lack of space prevents us from doing justice to them, nor can we put them in perspective. We will therefore state only a few, with some references, and the interested reader is encouraged to peruse the original papers for further details. Our understanding of IVR in the parent molecule and the effect of parent rotational excitation have been greatly enhanced by recent meticulous and detailed work by Rizzo, Crim, and their coworkers on such molecules as H_2O_2 and HNO_3.[98–100] This work provides clues to the importance of incomplete IVR as the cause of deviations from statistical behavior. Very recently, stepwise increases in the unimolecular reaction rates, $k(E)$, of NO_2 and ketene were observed for the first time[101,102] and may signify the opening of vibrational channels at the transition state. This will undoubtedly lead to more theoretical work on this issue.[103] In addition, statistical fluctuations in product state distributions were observed in the unimolecular reaction of NO_2 and interpreted as arising from interference of many overlapping resonances,[104] a phenomenon known in nuclear physics as Ericson fluctuations. Fluctuations in the *rates* of the unimolecular reaction of formaldehyde, a reaction proceeding over a large barrier, were observed previously and interpreted theoretically on statistical grounds.[105]

ACKNOWLEDGMENTS

We wish to thank our former colleagues, J. Pfab, I. Nadler, M. Noble, and C. X. W. Qian, whose enthusiasm, skill, dedication and insights made this series of studies such a rewarding endeavor. We also thank R. A. Marcus and C. B. Moore for sending us preprints and figures.

Research supported by the National Science Foundation and U.S. Air Force Office of Scientific Research and by the U.S. Army Research Office under the auspices of the Center for the Study of Fast Transient Processes.

REFERENCES

1. Forst, W. *Theory of Unimolecular Reactions*; Academic: New York, 1973.
2. Robinson, P. J.; Holbrook, K. A. *Unimolecular Reactions*; Wiley: New York, 1972.
3. Benson, S. W. *Thermochemical Kinetics*, 2nd ed.; Wiley: New York, 1976.
4. Benson, S. W. *Ann. Rev. Phys. Chem.* **1988,** *39,* 1.
5. Scoles, G. *Atomic and Molecular Beam Methods*; Oxford: New York, 1988.
6. (a) Likar, M. D.; Baggott, J. E.; Crim, F. F. *J. Chem. Phys.* **1989,** *90,* 6266; (b) Sinha, A.; Vander Wal, R. L.; Crim, F. F. *J. Chem. Phys.* **1989,** *91,* 2929; (c) Vander Wal, R. L.; Crim, F. F. *J. Phys. Chem.* **1989,** *93,* 5331.
7. (a) Ticich, T. M.; Likar, M. D.; Dübal, H. R.; Butler, L. J.; Crim, F. F. *J. Chem. Phys.* **1987,** *87,* 5820; (b) Sinha, A.; Vander Wal, R. L.; Butler, L. J.; Crim, F. F. *J. Phys. Chem.* **1987,** *91,* 4645; (c) Likar, M. D., Baggot, J. E.; Sinha, A.; Ticich, T. M.; Vander Wal, R. L.; and Crim, F. F. *J. Chem. Soc. Faraday Trans.* **1988,** *84,* 1483; (d) Likar, M. D.; Sinha, A.; Ticich, T. M.; Vander Wal, R. L.; Crim, F. F. *Ber. Bunsenges. Phys. Chem.* **1988,** *92,* 289.
8. (a) Andresen, P.; Schinke, R. In *Molecular Photodissociation Dynamics*; Baggott, J., Ashfold, M. N. R., Eds.; Royal Society of Chemistry, 1987; (b) Andresen, P.; Beushausen, V., Häusler, D.; Lülf, H. W.; Rothe, E. W. *J. Chem. Phys.* **1985,** *83,* 1429; (c) Schinke, R.; Engel, V.; Andresen, P.; Häusler, D.; Balint-Kurti, G. G. *Phys. Rev. Lett.* **1985,** *55,* 1180.
9. Luo, X.; Rieger, P. T.; Perry, D. S.; and Rizzo, T. R. *J. Chem. Phys.* **1988,** *89,* 4448.
10. Nadler, I.; Reisler, H.; Noble, M.; Wittig, C. *Chem. Phys. Lett.* **1984,** *108,* 115.

11. Noble, M.; Nadler, I.; Reisler, H.; Wittig, C. *J. Chem. Phys.* **1984,** *81,* 4333.
12. Reisler, H.; Radhakrishnan, G.; Sumida, D.; Pfab, J.; Chou, J. S.; Nadler, I.; Wittig, C. *Israel J. Chem.* **1984,** *24,* 259.
13. Nadler, I.; Pfab, J.; Reisler, H.; Wittig, C. *J. Chem. Phys.* **1984,** *81,* 653.
14. Nadler, I.; Pfab, J.; Radhakrishnan, G.; Reisler, H.; Wittig, C. *J. Chem. Phys.* **1983,** *79,* 2088.
15. Pfab, J.; Nadler, I.; Radhakrishnan, G.; Reisler, H.; Wittig, C. *J. Chem. Soc. Farad. Disc.* **1983,** *75,* 279.
16. Nadler, I.; Noble, M.; Reisler, H.; Wittig, C. *J. Chem. Phys.* **1985,** *82,* 2608.
17. Qian, C. X. W.; Noble, M.; Nadler, I.; Reisler, H.; Wittig, C. *J. Chem. Phys.* **1985,** *83,* 5573.
18. Qian, C. X. W.; Ogai, A.; Reisler, H; Wittig, C. *J. Chem. Phys.* **1989,** *90,* 209.
19. Reisler, H.; Wittig, C. *Ann. Rev. Phys. Chem.* **1986,** *37,* 307.
20. Qian, C. X. W.; Reisler, H.; Wittig, C. *Chem. Phys. Lett.* **1987,** *139,* 175.
21. Reisler, H.; Noble, M.; Wittig, C. In *Molecular Photodissociation Dynamics*; Bassot, J.; Ashford, M. N. R., Eds.; Royal Society of Chemistry, 1987.
22. Bitto, H.; Chen, I-C.; Moore, C. B. *J. Chem. Phys.* **1986,** *85,* 5101.
23. Buelow, S.; Noble, M.; Radhakrishnan, G.; Reisler, H.; Wittig, C.; Hancock, G. *J. Phys. Chem.* **1986,** *90,* 1015.
24. Pfab, J. *Chem. Phys. Lett.* **1983,** *99,* 465.
25. Pfab, J.; Häger, J.; Krieger, J. *J. Chem. Phys.* **1983,** *78,* 266.
26. Reisler, H.; Pessine, F. B. T.; Haas, Y.; Wittig, C. *J. Chem. Phys.* **1983,** *78,* 3785.
27. Noble, M.; Qian, C. X. W.; Reisler, H.; Wittig, C. *J. Chem. Phys.* **1986,** *85,* 5763.
28. Noble, M.; Qian, C. X. W.; Reisler, H.; Wittig, C. *J. Chem. Phys.* **1986,** *84,* 3573.
29. Polik, W. F.; Moore, C. B.; Miller, W. H. *J. Chem. Phys.* **1988,** *89,* 3584.
30. Nesbitt, D. J.; Petek, H. P.; Foltz, M. F.; Filseth, S. V.; Bamford, D. J.; Moore, C. B. *J. Chem. Phys.* **1985,** *83,* 223.
31. (a) Chen, I.-C.; Green, J.; Moore, C. B. *J. Chem. Phys.* **1988,** *89,* 314; (b) Chen, I.-C.; Moore, C. B. *J. Phys. Chem.* **1990,** *94,* 263; (c) Chen, I.-C.; Moore, C. B. *J. Phys. Chem.* **1990,** *94,* 269; (d) Green, W. H., Jr.; Chen, I.-C.; Moore, C. B. *Ber. Bunsenges. Phys. Chem.* **1988,** *84,* 389.
32. Asscher, M.; Haas, Y.; Roellig, M. P.; Houston, P. L. *J. Chem. Phys.* **1980,** *72,* 768.
33. Roellig, M. P.; Houston, P. L.; Asscher, M.; Haas, Y. *J. Chem. Phys.* **1980,** *73,* 5081.
34. Bower, R. D.; Jones, R. W.; Houston, P. L. *J. Chem. Phys.* **1983,** *79,* 2799.
35. Jones, R. W.; Bower, R. D.; Houston, P. L. *J. Chem. Phys.* **1982,** *76,* 3339.
36. DeKoven, B. M.; Fung, K. H.; Levy, D. H.; Hoffland, L. D.; Spears, K. G. *J. Chem. Phys.* **1981,** *74,* 4755.
37. Spears, K. G.; Hoffland, L. D. *J. Chem. Phys.* **1981,** *74,* 4765.
38. Knee, J. L.; Khundkar, L. R.; Zewail, A. H. *J. Phys. Chem.* **1985,** *89,* 4659.
39. Khundkar, L. R.; Knee, J. L.; Zewail, A. H. *J. Chem. Phys.* **1987,** *87,* 77.

40. Potter, E. D.; Gruebele, M.; Khundkar, L. R.; Zewail, A. H. *Chem. Phys. Lett.* **1989,** *164,* 463.
41. Jortner, J.; Levine, R. D. In *Photoselective Chemistry*, Part 1; Jortner, J.; Levine, R. D.; Rice, S. A., Eds.; *Adv. Chem. Phys.* **1981,** *50,* 1.
42. (a) Rizzo, T. R.; Hayden, C. C.; Crim, F. F. *J. Chem. Phys.* **1984,** *81,* 4501; (b) Ticich, T. M.; Rizzo, T. R.; Dübal, H. R.; Crim, F. F. *J. Chem. Phys.* **1986,** *84,* 1508; (c) Crim, F. F. *Annu. Rev. Phys. Chem.* **1984,** *35,* 657.
43. (a) Foy, B. R.; Casassa, M. P.; Stephenson, J. C.; King, D. S. *J. Chem. Phys.* **1989,** *90,* 7037; (b) Foy, B. R.; Casassa, M. P.; Stephenson, J. C.; King, D. S. *J. Chem. Phys.* **1988,** *89,* 608.
44. Alexander, M. H.; Werner, H. J.; Dagdigian, P. J.; *J. Chem. Phys.* **1988,** *89,* 1388.
45. Casassa, M. P. Private communication, 1990.
46. Nadler, I.; Mahgarefteh, D.; Reisler, H.; Wittig, C. *J. Chem. Phys.* **1985,** *82,* 3885.
47. Zare, R. N. *Angular Momentum*; Wiley: New York, 1988.
48. (a) Houston, P. L. *J. Phys. Chem.* **1987,** *91,* 5338 and references cited therein; (b) Simons, J. P. *J. Phys. Chem.* **1987,** *91,* 5378.
49. Scherer, N. F.; Zewail, A. H. *J. Chem. Phys.* **1987,** *87,* 97.
50. Schatz, G. C.; Fitzcharles, M. S. In *Selectivity in Chemical Reactions*; Whitehead, J. C., Ed., Kluwer: Dordrecht, Netherlands, 1988.
51. (a) Mayne, H. R.; Toennies, J. P. *J. Chem. Phys.* **1981,** *75,* 1794; (b) Götting, R.; Mayne, H. R.; Toennies, J. P. *J. Chem. Phys.* **1986,** *85,* 6396; (c) Mayne, H. R. *J. Chem. Phys.* **1980,** *73,* 217; (d) Barg, G.-D.; Mayne, H. R.; Toennies, J. P. *J. Chem. Phys.* **1981,** *74,* 1017.
52. Blais, N. C.; Truhlar, D. G. *J. Chem. Phys.* **1988,** *88,* 5457; **1985,** *83,* 2201.
53. Pechukas, P.; Light, J. C.; *J. Chem. Phys.* **1965,** *42,* 3281.
54. Pechukas, P.; Rankin, C.; Light, J. C. *J. Chem. Phys.* **1966,** *44,* 794.
55. Light, J. C. *Farad. Disc. Chem. Soc.* **1967,** *44,* 14.
56. Klotz, C. E. *J. Phys. Chem.* **1971,** *75,* 1526.
57. Klotz, C. E. *Z. Naturforsch.* **1972,** *27a,* 553.
58. Quack, M.; Troe, J. *Ber. Bunsenges. Physik. Chem.* **1974,** *78,* 240.
59. Quack, M.; Troe, J. *Ber. Bunsenges. Physik. Chem.* **1975,** *79,* 170.
60. Quack, M.; Troe, J. *Int. Rev. Phys. Chem.* **1981,** *1,* 97.
61. Quack, M.; Troe, J. *Ber. Bunsenges. Physik. Chem.* **1977,** *81,* 329.
62. Troe, J. *J. Chem. Phys.* **1981,** *75,* 226.
63. Troe, J. *J. Chem. Phys.* **1983,** *79,* 6017.
64. Beresford, J. R.; Hancock, G.; MacRobert, A. J.; Catanzarite, J.; Radhakrishnan, G.; Reisler, H.; Wittig, C. *Faraday Disc. Chem. Soc.* **1983,** *75,* 221.
65. Reisler, H.; Kong, F.; Renlund, A. M.; Wittig, C. *J. Chem. Phys.* **1982,** *76,* 997.
66. Reisler, H.; Kong, F.; Wittig, C.; Stone, J.; Thiele, E.; Goodman, M. F. *J. Chem. Phys.* **1982,** *77,* 328.
67. Wittig, C.; Nadler, I.; Reisler, H.; Noble, M.; Catanzarite, J.; Radhakrishnan, G. *J. Chem. Phys.* **1985,** *83,* 5581.

68. Marcus, R. A. *Chem. Phys. Lett.* **1988,** *144,* 208.
69. Wardlaw, D. M.; Marcus, R. A. *Chem. Phys. Lett.* **1984,** *110,* 230; *J. Chem. Phys.* **1985,** *83,* 3462; *J. Phys. Chem.* **1986,** *90,* 5383; *Adv. Chem. Phys.* **1988,** *70,* 231.
70. Klippenstein, S. J.; Marcus, R. A. *J. Phys. Chem.* **1988,** *92,* 3105.
71. Klippenstein, S. J.; Khundkar, L. R.; Zewail, A. H.; Marcus, R. A. *J. Chem. Phys.* **1988,** *89,* 4761.
72. Marcus, R. A. *Phil. Trans. Roy. Soc., Series A,* **1990,** in press.
73. Bai, Y. Y.; Segal, G. A. *Chem. Phys. Lett.* **1988,** *151,* 31.
74. Levine, R. D.; Kinsey, J. J. In *Atom-Molecule Collision Theory—A Guide for the Experimentalist*, Bernstein, R. B., Ed.; Plenum: New York, 1979; ch. 22.
75. Levine, R. D.; Bernstein, R. B. *Molecular Reaction Dynamics and Chemical Reactivity*, Oxford: New York, 1987.
76. Procaccia, I.; Levine, R. D. *J. Chem. Phys.* **1976,** *64,* 808.
77. Zamir, E.; Levine, R. D. *Chem. Phys.* **1980,** *52,* 253.
78. Baierlein, R. *Atoms and Information Theory*, Freeman: San Francisco, 1971.
79. Dixon, R. N.; Jones, K. B.; Noble, M.; Carter, S. *Mol. Phys.* **1980,** *42,* 455.
80. Dixon, R. N.; Noble, M. *Springer Series in Chemical Physics* **1979** *6,* 81.
81. Ernsting, N. P. *J. Chem. Phys.* **1984,** *80,* 3042.
82. Dyet, J. A.; McCoustra, M. R. S.; Pfab, J. *J. Chem. Soc. Faraday Trans. II* **1988,** *84,* 463.
83. (a) McCoustra, M. R. S., Dyet, J. A.; Pfab, J. *Chem. Phys. Lett.* **1987,** *136,* 231; (b) McCoustra, M. R. S.; Dyet, J. A.; Pfab, J. *Laser Chemistry* **1988,** *9,* 289.
84. Dyet, J. A.; McCoustra, M. R. S.; Pfab, J. Unpublished manuscript.
85. Kolodney, E.; Powers, P. S.; Iwata, L.; Reisler, H.; Wittig, C. *J. Chem. Phys.* In press.
86. (a) Salotto, A. W.; Burnelle, L. *Chem. Phys. Lett.* **1969,** *3,* 80; (b) Freedman, P. A. *Chem. Phys. Lett.* **1976,** *44,* 605.
87. Marudarajan, V.; Segal, G. A. *Chem. Phys. Lett.* **1986,** *128,* 1.
88. Ellis, H. B., Jr.; Ellison, E. G., *J. Chem. Phys.* **1983,** *78,* 6541.
89. (a) Wu, A. A.; Peyerimhoff, S. D.; Buenker, R. J. *Chem. Phys. Lett.* **1975,** *35,* 316; (b) Nomura, O.; Iwata, S. *Chem. Phys. Lett.* **1979,** *66,* 523.
90. (a) Dalby, F. W. *Can. J. Phys.* **1958,** *36,* 1336; (b) Bancroft, J. L.; Hollas, J. M.; Ramsay, D. A. *Can J. Phys.* **1962,** *40,* 322; (c) Dixon, R. N.; Noble, M.; Taylor, C. A.; Delhoume, M. *Farad. Disc. Chem. Soc.* **1981,** *71,* 125; (d) Dixon, R. N. *Philos. Trans. R. Soc., London Ser. A* **1982,** *307,* 603.
91. (a) Dixon, R. N.; Rosser, C. A. *J. Mol. Spectros.* **1985,** *110,* 262; (b) Petersen, J. C. *J. Mol. Spectros.* **1985,** *110,* 277.
92. Ernsting, N. P.; Pfab, J.; Römelt, J. *J. Chem. Soc. Farad. Trans. II* **1978,** *74,* 2286.
93. (a) Gordon, R. D.; Dass, S. C.; Robins, J. R.; Shurvell, H. F.; Whitlock, D. R. F. *Can. J. Chem.* **1976,** *54,* 2658; (b) Gordon, R. D. *Int. Rev. Phys. Chem.* **1986,** *5,* 121.
94. Calvert, J. G.; Pitts, J. N. *Photochemistry*, Wiley: New York, 1967.
95. Obi, K.; Matsumi, Y.; Takeda, Y.; Mayama, S.; Watanabe, H.; Tsuchiya, S. *Chem. Phys. Lett.* **1984,** *95,* 520.

96. Tramer, A.; Voltz, R. In *Excited States*; Lim, E. C., Ed.; Academic: New York, 1979; Vol. 4.
97. (a) Andresen, P.; Rothe, E. W. *J. Chem. Phys.* **1985,** *82,* 3634; (b) Alexander, M. H.; Dagdigian, P. J. *J. Chem. Phys.* **1984,** *80,* 4863; (c) Alexander, M H. et al. *J. Chem. Phys.* **1988,** *89,* 1749.
98. Fleming, P. R.; Li, M.; Rizzo, T. R. *J. Chem. Phys.* **1991,** 95, 865.
99. Fleming, P. R.; Li, M.; Rizzo, T. R. *J. Chem. Phys.* **1991,** 95, 1461.
100. Vander Wal, R. L.; Scott, J. L.; Crim, F. F. *J. Chem. Phys.* **1991,** 94, 1859.
101. Brucker,`G. A.; Ionov, S. I.; Chen, I; Wittig, C. *Chem. Phys. Lett.* **1992,** 194, 301.
102. Lovejoy, E. R.; Kim, S. K.; Moore, C. B. *Science* **1992,** 256, 1541.
103. Marcus, R. A. *Science* **1992,** 256, 1523.
104. Robie, D. C.; Hunter, M.; Bates, J. L.; Reisler, H. *Chem. Phys. Lett.* **1992,** 193, 413.
105. Miller, W. H., Hernandez, R.; Moore, C. B.; Polik, W. F. *J. Chem. Phys.* **1990,** 93, 5657.

APPLICATIONS OF CHEMICAL KINETICS

David M. Golden and Jeffrey A. Manion

Advances in Chemical Kinetics and Dynamics, Volume 1, pages 187–276

ISBN: 1-55938-161-2

I. INTRODUCTION

In this chapter we will describe some of the research in chemical kinetics at SRI International. The structure of research at an institution such as SRI means that we will describe endeavors that are often undertaken by overlapping groups of scientists ranging from postdoctoral associates to senior colleagues past and present. It will be apparent from the authorship of cited publications that contributions to them by various members of the Chemical Kinetics Department vary. In any case the research described herein is the result of a team effort. It is also true that research at SRI is likely to be associated with an applied problem. Our experience is that pursuit of solutions to these applied problems often leads to fundamental understanding that would not have been uncovered by pure exploratory research.

A. Applications of Chemical Kinetics

Quantitative understanding of the chemical mechanism involved is the key to control and prediction in several important practical areas. We, working at a contract research institute, pay particular attention to areas where our scientific interests overlap with these practical needs.

Atmospheric Chemistry

The need to understand atmospheric chemistry in relation to problems in the troposphere ranging from urban photochemical smog to acid rain, to the problem of understanding the "clean" troposphere as a source for stratospheric species, as well as problems in the stratosphere relating to anthropogenic ozone destruction in the global stratosphere and at the poles in spring, has occupied a good deal of our attention.

Combustion

If energy output were the only consideration in the design of combustors, whether for power production or transportation, chemical kinetics would probably be a minor consideration, since the oxidation of hydrocarbon fuels is controlled for the most part by the single chain branching reaction:

$$H + O_2 \rightarrow OH + O$$

However, an important consideration in almost all combustion operations is the formation and emission of pollutants. These may be the oxides of nitrogen, sulfur dioxide, and/or products of incomplete combustion. The trade-off in design parameters required to accommodate these disparate needs requires a reliable quantitative model for the chemistry. At SRI we have addressed the problems of chemical kinetics of combustion in both experimental and applied theoretical ways.

High Energy Materials

The chemical kinetics involved in problems of propellant burning are in some sense a subset of combustion-related kinetics. The chemicals that are involved are more complex than hydrocarbon fuels and often carry their own oxidizing capabilities (monopropellants). Studies of actual propellant molecules and appropriate model analogs have led to studies of the thermal decomposition of nitroalkanes, nitroaromatics, hydrazines, and nitramines. As always seems to be the case, studies of problems that would appear to be of interest only in regard to a particular application, often lead to the unveiling of some unique chemistry. Recently we have shown[1,2] that nitramines and nitroaromatics undergo unimolecular isomerization in which a nitro group transforms into a nitrite linkage that is short lived with respect to RO-NO cleavage.

Chemical Processing

It has been particularly gratifying to learn that data and methods that we have helped to produce or refine have been applied directly to the solution of commercial problems. We are aware of applications of our methods toward the control and modification of chemical processes by several chemical or petrochemical companies. Processes that may yield more than one product may be "tuned" to the most desirable products with the help of quantitative validated chemical mechanisms. In some cases we have, under proper non-disclosure agreements, helped with the implementation of our results to specific problems.

Electronics Manufacturing

The production of semiconductor-based electronic devices is a complex chemical process. By its nature this obviously includes gas-surface chemistry, whether involving etching or through deposition of species created in situ (CVD). We have embarked on a body of experimental studies that we envisage as pointing the way toward the ability to codify and extrapolate gas-surface data in a manner analogous to "Thermochemical Kinetics" for gaseous species.

B. Theme of Chemical Kinetics Research at SRI International

Operating with the view that understanding of complex chemical systems requires a quantitative mechanism (model), we have extended and applied "Thermochemical Kinetics" and performed experiments to enhance the gas phase data base for such application. We also have been performing experiments that we envision as leading to the appropriate data base for extending the general concepts of "Thermochemical Kinetics" to heterogeneous systems.

The name "Thermochemical Kinetics" comes from the title of a monograph[3] by Sidney W. Benson, that was written based on ideas largely developed during his tenure at SRI between 1963 and 1976. In essence, the methods described in that book can be divided into two areas, thermochemistry and thermal chemical kinetics. Under thermochemistry, both additivity rules for extending the empirical data base and statistical thermodynamic calculations of entropy and heat capacity are included. The discussions of thermal rate constants for gas phase processes is couched in the language of conventional canonical transition state theory (TST) with excursions into microcanonical theory to understand unimolecular reactions and the related problem of reactions that proceed via bound intermediates.

The use of a generalized formalism, such as TST, allows the overview of reaction rates and mechanisms that avoids the trap of thinking of a given reaction as a universe unto itself. Rather, experiments can be judged by how they conform with the body of knowledge. (It is, of course, always important to recognize when an experimental result is enlarging the scope of that knowledge.)

II. THERMOCHEMISTRY

Thermochemistry underlies the methods that we stress for understanding chemical kinetics. In any chemical model the reactions must be entered in both the forward and reverse directions. Since the rate constants are related through the equilibrium constant, a data base of thermochemical values ensures consistency in this process and, of course, obviates the need for half of the rate constants in the model. The widely used CHEMKIN code[4] computes reverse rate constants automatically from the forward rate constants and the equilibrium constant computed from such a thermochemical data base.

Conventional canonical transition state theory considers the rate constant in terms of differences in thermochemical properties of the "transition state" and the reactant(s). It behooves us to have a good understanding of the thermochemical values for stable species so that we may extrapolate to transient stable species such as free radicals, as well as to values for constructs, such as transition states.

A. Additivity Rules

Our objective is to obtain values for ΔH°_T, ΔS°_T, and $\Delta C^\circ_{p,T}$ for chemical reactions. The methods employed are fundamentally empirical. They are based on extrapolation and codification of measured data. Thermodynamic data may be obtained by several methods and are found in several tabulations.[5-9]

Given the chemical reaction,

$$aA + bB + \ldots \rightleftarrows pP + qQ + \ldots \quad (1)$$

a change in some thermodynamic property Φ is given by,

$$\Delta\Phi = p\Phi_P + q\Phi_Q + \ldots - a\Phi_A - b\Phi_B - \ldots \quad (2)$$

Atom Additivity

A simple disproportionation reaction of diatomic molecules may be written:

$$A_2 + B_2 \rightleftarrows 2\ AB \tag{3}$$

Any property for which $\Delta\Phi = 0$ obeys the law of atomic additivity. The property Φ is in reality an atomic property. It is very unlikely that thermochemical quantities would obey such a simple law, but the uncertainty limits might be adequate. To rephrase, $\Delta\Phi = 0 \pm \delta$ signifies a property for which the additivity relationship is obeyed within limits of $\pm\ \delta$. If atom additivity were adequate, a periodic table with one entry for each atom would allow the prediction of the property Φ for any molecule. This would be marvelously simple and would require an extremely small data base. Properties of any two hydrocarbons would be sufficient to predict all of the other hydrocarbons.

In reality, atom additivity is totally unsatisfactory for the estimation of enthalpies and only of marginal value in estimation of entropy and heat capacity.

Bond Additivity

A next level in the hierarchy that can be used for estimation of thermochemical properties is envisaged in terms of the reaction:

$$A-X-A + B-X-B \rightleftarrows 2\ A-X-B \tag{4}$$

If for this reaction, $\Delta\Phi = 0 \pm \delta$, the law of bond additivity is obeyed. Properties are associated with the chemical bonds, which are the same on both sides of the equation. For many relatively simple molecules containing C, H, O, N, S, F, Cl, Br, I, bond additivity is acceptably accurate for entropy and heat capacity ($\delta \approx \pm$ 1–2 cal mole^{-1} K^{-1}) for many species. Bond additivity requires a much greater data base than the atom additivity contributions. As an example, from methane and ethane properties, we obtain the C-H and C-C bond contributions and may predict properties of any alkane.

Group Additivity

The next step in the additivity hierarchy is defined through the reaction

$$A-XY-A + B-XY-B \rightleftarrows A-XY-B + B-XY-A \qquad (5)$$

If for this case $\Delta\Phi = 0 \pm \delta$ the law of group additivity is obeyed and properties are associated with an atom and its ligands. This is usually a very useful level of approximation. The data base required is relatively large, but still accessible. Sometimes combinations of bond and group additivity are necessarily employed.

The notation for groups is generally transparent. Usually a central atom is surrounded by its nearest neighbors. Occasionally, certain ligands are treated as pseudo atoms, and some peculiarities of notation have become embedded (i.e., instead of $C_d(C_d)(H)_2$, just $C_d(H)_2$). The value for any group $C(H)_3(X)$ is the same as $C(H)_3(C)$. Some examples are:

$CH_3CH_3 \Rightarrow 2C(C)(H)_3$
$CH_3CH_2CH_3 \Rightarrow 2C(C)(H)_3 + C(C)_2(H)_2$
$CH_3OH \Rightarrow C(O)(H)_3 + O(C)(H)$
$H_2C = CH_2 \Rightarrow 2C_d(H)_2$
$CH_3COCH_3 \Rightarrow 2C(H)_3(CO) + CO(C)_2$
$CH_3CH_2NO_2 \Rightarrow C(H)_3C + C(C)(H)_2(NO_2)$
$p-CH_3-C_6H_4-OH \Rightarrow C(H)_3C_B + 4C_BH + C_B-O + O-(C_B)(H)$

In order to describe all alkanes using group additivity, the four groups, $C(C)(H)_3$, $C(C)_2(H)_2$, $C(C)_3(H)$, and $C(C)_4$ would be required.

We find it profitable to tabulate group contributions to $\Delta H^\circ_{f,298}$, S°_{298}, and $C^\circ_{p,T}$ for various temperatures. We can calculate ΔH° and ΔS° for chemical reactions at temperatures different than 298 K, via:

$$\Delta H^\circ_T = \Delta H^\circ_{298} + \int_{298}^{T} \Delta C^\circ_p (T)dT \qquad (6)$$

We assume

$$\Delta C^\circ_p (T) = [\Delta C^\circ_p \{T\} - \Delta C^\circ_p \{298\}]/2 = \overline{\Delta C_p} \qquad (7)$$

$$\Delta H^\circ_T = \Delta H^\circ_{298} + \overline{\Delta C_p}(T-298) \qquad (8)$$

$$\Delta S^\circ_T = \Delta S^\circ_{298} + \int_{298}^{T} \Delta C^\circ_p (T)d \ln T = \Delta S^\circ_{298} + \overline{\Delta C_p} \ln \frac{T}{298} \qquad (9)$$

It is also necessary to point out that all the additivity contributions for entropy are for *intrinsic* entropy rather than for real symmetry-corrected entropy. This is understood in terms of $S^\circ = S^\circ_{int} - R \ln \sigma/n$, where σ is the rotational symmetry number for the molecule and n is equal to the number of optical isomers. The external rotational symmetry correction takes into account that if all rotational permutations were weighted equally, then indistinguishable configurations would be counted more than once. Thus, while heteronuclear diatomic molecules have $\sigma_{ext} = 1$, homonuclear diatomics have $\sigma_{ext} = 2$, since a simple 180° rotation, which permutes the atoms, yields an indistinguishable configuration. For polyatomic species, the same general dictum holds. The symmetry number, σ_{ext}, is the total number of ways that the molecule can be permuted to give an identical configuration by a simple rotation of the entire molecule. Examples: CH_4(12); benzene (12), SF_6 (24), and so on.

$$\sigma_{external} = \frac{\text{Rotational symmetry of fixed molecule}}{\text{Number of identical rotamers}}$$

Linear alkanes:	$\sigma_{ext} = 2$
NO_2	$\sigma_{ext} = 2$
CH_4	$\sigma_{ext} = 12$
Benzene	$\sigma_{ext} = 12$

The group contributions also need to be corrected for internal rotational symmetry. Thus, CH_3 groups have a three-fold rotational symmetry which has the effect of multiplying the external symmetry by 3 for every methyl group. Linear alkanes which have an external $\sigma = 2$ also have two-end CH_3s, and therefore, an overall symmetry number of $3 \times 3 \times 2 = 18$. Neopentane (C_5H_{12}) has an external symmetry of 12 (same as CH_4), and the four methyl groups contribute 3^4 for $\sigma = 12 \times 3^4 = 972$.

$$\sigma_{internal} = \text{Symmetry associated with internal rotors}$$

$$CH_3 \ \sigma_{int} = 3$$

$$\sigma \rightarrow \text{overall symmetry number} = \sigma_{external} \times \sigma_{internal}$$

$$\text{Linear alkanes: } \sigma = \sigma_{ext} \cdot \sigma_{int} = 2 \times 3^2 = 18$$

Finally, for molecules exhibiting optical activity, there is an entropy of mixing correction term of R 1n n, where n is the number of optical isomers. The overall correction to group additivity entropies is therefore given by:

$$S^o_T = S^o_{T,intrinsic} - R \ 1n \ \sigma/n \tag{10}$$

Group additivity in its purest form serves us well, but still falls short in several ways. We could attempt to correct this by going up the hierarchy one more step to an additivity that includes next-nearest-neighbor interactions, but the data base required would be so large as to be impractical. Thus we adapt and compromise by giving up the purity of the group additivity definition and using additive correction factors. These corrections are sometimes only to the heat of formation, but sometimes entropy and heat capacity need corrections as well. Corrections are tabulated for *cis* interactions, *gauche* interactions, and for cyclic compounds.

Tables of Additivity parameters may be found in "Thermochemical Kinetics."[3]

B. Structural Estimates of Entropy and Heat Capacity

Statistical Thermodynamics

If molecular properties, such as molecular weight, bond lengths, bond angles, and molecular frequencies are known, the entropy and heat capacity may be calculated using the rigid rotor harmonic oscillator (RRHO) model. In most cases the ideal gas approximation is valid. Appropriate formulas are presented in the following pages.

This method, more time consuming than group additivity, is also generally more accurate if the molecular parameters are known. To the extent that some inputs must be estimated (e.g., vibrational frequencies, rotational barriers, etc.), it becomes in effect a more detailed additivity scheme. A major advantage, however, is that the same techniques may be applied to transition states, hence allowing the estimation of A-factors, modeling of multi-channel systems, and so on. This will be discussed more fully in Section III.

Since we always know the molecular weight, we may compute

translational contributions to S and C_p accurately. With some reasonable guesses as to bond lengths and angles, we can compute I_m^3, the product of inertia. (Simple computer codes are available.) The S and C_p errors associated with small changes in I_m^3 are small.

Quantitative examples include the fact that changing the C-C distance in n-butane from 1.54 Å to 2.54 Å, increases the value of I_m^3 by a factor of three. This increases the rotational entropy by 1.1 cal mole^{-1} K^{-1}.

The biggest error that we will make in entropy (and C_p) calculations is in the vibrational contribution. Each molecule has 3N-6 vibrations, and larger molecules with low-lying frequencies (often corresponding to torsions or the bending motions of heavy groups) can be a problem. In the case of torsions, these modes undergo a temperature dependent shift from a torsional motion to a free rotation, with a resultant change in the contributions to S and C_p. The temperature regime in which this occurs is dependent on the shape and height of the rotational barrier. Although much progress has been made in this area,[10,11] uncertainties remain for many molecules. It is important to understand that even in these latter cases, the size of the entropy error will not be overwhelming, and the advantage of being able to compute limits is enormous. Of course, we could measure entropy calorimetrically if an accurate value is of importance, but we will also wish to extend these methods to species such as free radicals and to transition states, where experimental values are not accessible.

DEFINITIONS

Q = molar partition function

q = molecular partition function

$Q = \frac{1}{N_0!} q^{N_0}$; N_o = Avogadro's Number

Molar Thermodynamic Quantities

Helmholtz free energy: $A = -kT \ln Q$ (11)

Gibbs free energy: $G = A + PV$ (12)

Internal energy: $E = kT \left(\frac{\partial \ln Q}{\partial \ln T}\right)_v = -k \left(\frac{\partial \ln Q}{\partial (1/T)}\right)_v$ (13)

Enthalpy: $$H = E + PV \tag{14}$$

Entropy: $$S = \frac{E - A}{T} \tag{15}$$

$$= k \ln Q + k\left(\frac{\partial \ln Q}{\partial \ln T}\right)_v \tag{16}$$

Heat capacity at constant volume $$C_v = \left(\frac{\partial E}{\partial T}\right)_v = k\left(\frac{\partial \ln Q}{\partial \ln T}\right)_v + k\left(\frac{\partial^2 \ln Q}{\partial (\ln T)^2}\right)_v \tag{17a}$$

$$= \frac{k}{T^2}\left(\frac{\partial^2 \ln Q}{\partial (1/T)^2}\right)_v \tag{17b}$$

Ideal gases $$C_p = \left(\frac{\partial H}{\partial T}\right)_p = \left(\frac{\partial E}{\partial T}\right)_p + P\left(\frac{\partial V}{\partial T}\right)_p = C_v + R \tag{18}$$

Evaluation of Molar Partition Functions and Molar Thermodynamic Quantities in Terms of Molecular Parameters

For the translational contribution to thermochemical properties for one mole of ideal gas at a standard state of 1 atm:

$$Q^{\circ}_{\text{tran}} = \frac{(RT)^{N_o}}{N_o!}\left(\frac{2\,\pi\, mkT}{h^2}\right)^{3N_o/2} \tag{19}$$

$$E^{\circ}_{\text{trans}} = \frac{3}{2} N_o kT = \frac{3}{2} RT \tag{20}$$

$$C^{\circ}_{v(\text{tran})} = \frac{3}{2} R \;;\; C^{\circ}_{p(\text{tran})} = \frac{5}{2} R \tag{21}$$

$$S^{\circ}_{(\text{tran})} = 37.0 + \frac{3}{2} R \ln\left(\frac{M}{40}\right) + \frac{5}{2} R \ln\left(\frac{T}{298}\right) + R \ln (n) \tag{22}$$

where N_o = Avogadro's number 6.02×10^{23} molecules.

In the last equation M is the molecular weight in amu and n is the number of optical isomers. (This latter is really an entropy of mixing.)

Assuming the molecule to be a rigid rotor, we have the following for rotational degrees of freedom:

1. Linear molecule (2 degrees of freedom):

$$Q_{rot-2D} = \left(\frac{8\ \pi^2 IkT}{\sigma_e h^2} \right)^{N_o} \tag{23}$$

$$E_{rot-2D)} = RT \tag{24}$$

$$C_{v(rot-2D)} = R \tag{25}$$

$$S_{rot-2D} = 6.9 + R\ \ln\ (I/\sigma_e) + R\ \ln\ (T/298) \tag{26}$$

where σ_e is the external symmetry number of the molecule.

2. Nonlinear molecule (3 degrees of freedom):

$$Q_{rot-3D} = \left[\frac{\pi^{1/2}}{\sigma_e} \left(\frac{8\pi^2 I_m kT}{h^2} \right)^{(3/2)} \right]^{N_o} \tag{27}$$

$$E_{rot-3D} = \frac{3}{2} RT \tag{28}$$

$$C_{v(rot-3D)} = \frac{3}{2} \tag{29}$$

$$S_{rot-3D} = 11.5 + \frac{R}{2} \ln \left(\frac{I_m^{\ 3}}{\sigma_e^2} \right) + \frac{3}{2} \mathrm{R} \ln \left(\frac{\mathrm{T}}{298} \right) \tag{30}$$

3. Internal rotations (1 degree of freedom):
One-dimensional rotor:

$$Q_{rot-1D} = \left[\frac{\pi^{1/2}}{\sigma_i} \left(\frac{8\pi^2 I_r kT}{h^2} \right)^{1/2} \right]^{N_o} \tag{31}$$

$$E_{rot-1D} = \frac{1}{2} RT \tag{32}$$

$$C_{v(\text{rot}-1D)} = \frac{1}{2} R \tag{33}$$

$$S_{\text{rot-}1D} = 4.6 + R \ln\left(\frac{I_r^{1/2}}{\sigma_i}\right) + \frac{R}{2} \ln\left(\frac{T}{298}\right) \tag{34}$$

where σ_i is the internal rotational symmetry number.

For each vibrational degree of freedom, making the harmonic oscillator approximation, we have:

$$Q_{\text{vib}} = (1 - e^{-h\nu/kT})^{-N_o} \tag{35}$$

with the appropriate thermodynamic contributions to be evaluated from the foregoing equations. υ is the frequency of the vibration and, in general, the contributions of a single vibrational degree of freedom to S, C_p, and E tend to be small. For large molecules, these contributions may become sizeable. When $h\upsilon << kT$, we find:

$$Q_{\text{vib}} \underset{\substack{\text{Classical}\\\text{limit}}}{\longrightarrow} \left(\frac{kT}{h\nu}\right)^{N_o} \tag{36}$$

$$E_{\text{vib}} \to RT \tag{37}$$

$$C_{\text{vib}} \to R \tag{38}$$

$$S_{\text{vib}} \to R \ln\left(\frac{kT}{h\nu}\right) + R \tag{39}$$

In the opposite extreme, when $kT << h\upsilon$, then $Q_{\text{vib}} \to 1$ and all of the thermodynamic functions for vibration approach zero.

Only the lower frequencies contribute much to the entropy at 300 K, and small errors in assignment will still cause only small errors. The difference between a 200 cm^{-1} and 100 cm^{-1} frequency is only 1.2 cal mol^{-1} K^{-1} at 300 K.

The assignment of vibrational frequencies is an exercise in describing internal coordinates. The vibrational assignment ''locates'' each atom in space.

Consider the trans form of n-butane, C_4H_{10} (number of vibrations = $3N-6$ = 36):

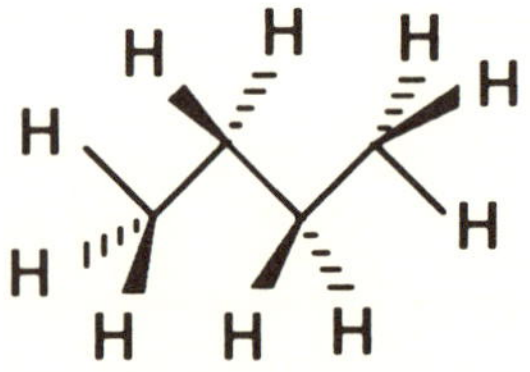

Structure 1.

We can break the assignment up into parts as follows:

13-Stretches:		
10 C-H		3100 cm^{-1}
3 C-C		1000 cm^{-1}
23-Bends		
Three due to Heavy Atoms:		
2 C-C-C		420 cm^{-1}
1 Et-Et	torsion	
Eight due to CH_2 groups:		
2 H-C-H	scissors	1450 cm^{-1}
2 CH_2	wag	1150 cm^{-1}
2 CH_2	twist	1150 cm^{-1}
2 CH_2	rock	700 cm^{-1}
Twelve due to CH_3 groups:		
6 H-C-H	scissors	1450 cm^{-1}
4 CH_3	rocks, wag, twist	1150 cm^{-1}
2 CH_3	torsion	

The assignment of frequencies is from "Thermochemical Kinetics."[3] The torsions are somewhat more complex. The easiest concept is to consider the groups (in the above case, CH_3 and C_2H_5) to be rotating against a very heavy mass. Thus we may compute a moment of inertia for this rotor and thus a partition function and appropriate thermochemical quantities corresponding to a free rotor. If the rotation is hindered, it behaves like a torsion at low temperatures $V >> RT$ and like a rotation at higher temperatures. In the intermediate regime, corrections to the free rotor values can be applied based on the barrier height, temperature, and reduced moment of the rotor.[3]

In the n-butane example, we see that each CH_3 rotor would contribute 3.6 cal mol^{-1} K^{-1} to the entropy if there were no barrier, while with a barrier of 3 kcal mol^{-1}, this is reduced by 1.4 cal mol^{-1} K^{-1} to a total of 2.2 cal mol^{-1} K^{-1} each. This is the same as 200 cm^{-1} torsion. The ethyl rotor must be treated differently since it is not rotating against a very heavy group, but only another ethyl. Thus, I_r is $I_{Et}/2$, and a free rotor would contribute 5.3 cal mol^{-1} K^{-1}. A 3 kcal mol^{-1} barrier reduces this to 3.9 cal mol^{-1} K^{-1}, the same as an 85–90 cm^{-1} torsion.

Cases with unsymmetrical rotors of similar mass are more complicated due to the unsymmetrical barrier and difficulties associated with calculation of I_r, but reasonable approximation methods exist and have been discussed elsewhere.[3,12]

The above methods may be applied to radicals if data on the vibrational frequencies is available. See references 13–15, for example. Some data on vibrational energy levels of transients has been conveniently compiled.[16–18]

There is one final point regarding the symmetry number and the calculation of thermodynamic parameters using the RRHO model. This concerns molecules which undergo inversion and therefore exhibit a double minimum vibrational potential. In such molecules, the classic example being NH_3, but also including non-planar alkyl and substituted alkyl radicals, and some ring compounds, assignment of the symmetry number can be confusing. For a XY_3 species, for example, one would generally assign the symmetry numbers of 6 and 3 to the planar and pyramidal forms, respectively. In the latter case, however, if the pyramid were rapidly inverting, its effective symmetry would be 6 and the question arises as to when the assignment of higher symmetry is justified.

This problem actually represents a breakdown of the RRHO model, rather than an ambiguity. The symmetry number, as generally discussed, represents a correction to the rotational partition function as a result of the Pauli principle: it takes into account that, due to nuclear spin, not all rotational states are available to molecules with identical nuclei (see any discussion of ortho, para hydrogen; e.g. McQuarrie[19]). This symmetry number refers to a spatially fixed, rigid molecule and, as such, vibrations are not considered to have an effect on the configuration of the molecule. If the molecule is allowed to invert, however, we are removing this restriction. The molecule is then permitted to have both "up" and "down" forms and must be assigned to a group in which all the operations which involve permutation of the positions and spins of the identical nuclei are considered. In the XY_3 case, this group must contain all the elements of the C_{3v} group plus those which interconvert the two C_{3v} configurations. This larger group is isomorphous with D_{3h}, and therefore has a symmetry of 6. An alternative way of saying this is that the external rotational symmetry is 3 but there is an additional internal symmetry of 2 associated with the inversion vibration. If the above view is taken, however, the inversion mode must be described as a double minimum vibration whose partition function is not correctly given by a single minimum harmonic oscillator.

Various possibilities with regard to the height of the potential barrier may be considered and are diagrammed in Figure 1.

In the limit of an infinite barrier (potential A), the energy levels are nearly harmonic, but they are also degenerate due to inversion doubling (the levels correspond to the symmetric and antisymmetric combinations of the wavefunctions of the "up" and "down" forms[20]). The partition function is therefore $Q_v = 2[1 - \exp(-h\nu/kT)]^{-1}$, rather than $Q_v = [1 - \exp(-h\nu/kT)]^{-1}$ as for a single minimum potential. This adds a term $R \ln 2$ to the molar vibrational entropy (see equation 16). Note that in this case exactly the same overall entropy would be obtained if the molecule were treated as non-inverting, and thus having C_{3v} symmetry ($\sigma = 3$), but also the non-degeneracy of a single minimum vibrational potential.

As the barrier to inversion is lowered, the increased interaction of the two wave functions causes the degeneracy of the levels to be split, with the degree of splitting increasing rapidly at higher vibrational levels as the top of the barrier is approached. In this case (potential B) the energy levels are no longer harmonic in the region near the top of the barrier. Even here the use of the RRHO approximation (i.e., C_{3v} symmetry and a single minimum harmonic oscillator with a wavenumber corresponding to, say, the $0 \rightarrow 1$ transition) is expected to be only modestly incorrect, with maximum deviation in the temperature regime where the vibrational levels near the top of the barrier are becoming populated. For greater accuracy, a quartic potential can be used for explicit calculation of the energy levels (and partition function therefrom) if enough information is available.[21] Care must be taken, however, that the summation over the vibrational levels is not truncated too early, particularly if values at high temperatures are desired. A further difficulty, of course, is that data or facile estimation methods do not exist for the energy levels of many species.

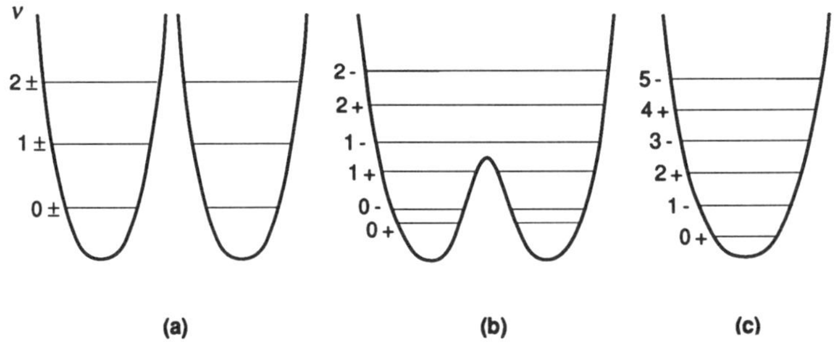

Figure 1. Potential functions and energy levels of a double minimum vibrator with a) infinite, b) intermediate, and c) negligible barrier to inversion.

As the inversion barrier is lowered further, the splitting continues to increase until, when the barrier is negligible (case C), the potential is simply that of a planar molecule, having obvious D_{3h} symmetry (σ = 6), and a partition function that is well described by the usual single minimum harmonic oscillator treatment. Notice that the frequency of the merged oscillator (case C, D_{3h}) is about one-half that of the separate oscillators (case A, C_{3v}) and that the classical partition function is the same for A and C.

To summarize, in the calculation of the thermodynamic functions for an inverting molecule, the external rotational symmetry is always that given by the spatially fixed, rigid molecule, but each inversion has an associated symmetry term of 2 that must be included if its partition function is calculated from a double minimum potential. A pyramidal XY_3 molecule thus has $\sigma = 3 \times 2 = 6$, or, in general $\sigma = \sigma_{ext} \times 2^m$, m = number of double minimum vibrations. This view is fundamentally more correct. However, since the usual RRHO model treats all modes as having single minimum harmonic oscillator potentials, in the RRHO approximation a pyramidal XY_3 molecule has $\sigma = 3$. In the limits of 0 and ∞ barriers the two treatments give identical results. In intermediate cases, the RRHO model should give fairly good results in most instances.

The major difficulty in calculation of the properties of unstudied or transient species remains the correct assignment of the structure (planar or pyramidal) and the estimation of the low-frequency bending vibrations. In the absence of detailed calculations, such cases must be assessed by analogy. For an excellent review of the structure of alkyl radicals, see references 15 and 22–24 for some other recent examples.

Molecular Model Compounds

The structural considerations make it apparent that model compounds offer another estimation method.

The n-butane molecule is similar in many ways to $CH_3OCH_2CH_3$, or CH_3NH-CH_2CH_3, or $CH_3CH_2CH_2OH$. The changes are simple to make and due almost entirely to symmetry and H-motions. The most difficult estimate would be changing rotational barriers, and we have seen that this is a small effect.

Many of the preceding comments are prologue to the estimation of free radical and transition state properties. Many radicals can be thought of as being derived from molecules by removal of an atom. That is, if we wish to have the properties of the radical R$^\cdot$, we may ask how we expect it to differ from the molecule R-H! In Table 1, it can be seen that C_2H_5 radical is "derived" from C_2H_6 with the only important changes being due to symmetry 18 → 6 for a gain of R ln 3 = 2.2 e.u. (e.u. = entropy units = cal mol^{-1} K^{-1}), an increase in the entropy due to the freeing[25] of the internal rotation, and the additional entropy due to the electronic state degeneracy for free radicals, accounting for a gain of R ln 2 = 1.4 e.u. (Of course, free radicals, such as allyl or benzyl which have a special delocalized character, will have to be treated a little differently.)

Table 1. Estimates of S^o_{300} (C_2H_5) and $C_{p,300}$ by Differences from C_2H_6

	ΔS^o	ΔC^o_p
Translation	Negligible (−0.1)	None
Rotation*	Negligible (−0.27)	None
Vibration		
One C-H stretch (3100 cm^{-1})	0	0
Two H-C-H bends (1450 cm^{-1})	0	−0.2
Hindered rotation: reduced moment	−0.25	
Barrier change from 3 kcal → 0 kcal	+1.4	−0.86
Symmetry and spin	+3.6	None
Totals	+4.4	−1.06

*If the C-C bond is 1.45 Å rather than 1.54 Å, there is an additional correction of about −0.3 due to lower moment. Ab initio calculations, see Reference 22, suggest a distance of 1.50 Å.

The final results, $S^0_{300}(C_2H_5) = 59.3$, $C_{p300} = 11.7$ cal mol^{-1} K^{-1}, compare well with the statistical mechanical results $S^0_{300} = 59.2$, $C_{p300} = 12.3$,[14] which are based on the matrix *IR* spectrum.[26–28]

It is also apparent from the foregoing that the largest effect to be expected in forming a cycle species from an open chain molecule will be the change of torsional motions into ring motions. For our purposes, it is sufficient to notice that the net entropy loss per rotor will usually be of the order 3–4 e.u.

C. Enthalpy Estimates

In general, we are restricted in estimating enthalpies of stable species to those for which we may apply additivity rules. (Some quantum mechanical calculational schemes do give good results as well.) However, certain enthalpy characteristics are transferable and useful in estimations.

Bond Dissociation Energies (BDE)

The enthalpy of the bond breaking reaction:

$$\text{A–B} \rightarrow \text{A}\cdot + \text{B}\cdot \tag{40}$$

is associated with the ''bond dissociation energy,'' or BDE. The value of ΔH for this reaction at 298K is the BDE.

Part of the chemist's ability to codify reactions comes from the fact that BDEs are largely transferable. Thus, in alkanes, all primary C-H bonds are 100 kcal mol^{-1},[14] all secondary are 97.5,[29,30] and all tertiary, 94 kcal mol^{-1}.[31] Note that we may codify these by listing radical heats of formation and/or BDEs themselves.

When the bond homolysis in question involves electronic rearrangement in the radical, as in the reaction:

$$H_2C{=}CH-CH_3 \rightarrow H_2C \mathrel{\underline{\cdots}} CH \mathrel{\underline{\cdots}} CH_2 + H \tag{41}$$

we find that the delocalization causes a stabilization, which is a transferable quantity. The above BDE in propylene is 87 kcal mol^{-1},[32,33] and the stabilization due to ''allylic delocalization'' (3 electrons in an orbital made from 3 *p*-orbitals) is about 13 kcal mol^{-1}. Many experimental values of BDEs have been determined over the past twenty years and many prototypical values exist.[33–35]

Pi-bond Energies

A specific type of bond dissociation energy can be defined by hypothetical reactions of the type:

$$H_2C = CH_2 \rightarrow H_2\dot{C} - \dot{C}H_2 \qquad (42)$$

The enthalpy change here is called D_π, the pi-bond energy. These transferable quantities make thinking about addition reactions particularly simple. The enthalpy change for the reaction:

$$H + C_2H_4 \rightarrow C_2H_5\cdot \qquad (43)$$

is the difference between the primary C-H bond being formed (100 kcal mol^{-1}) and the pi-bond energy being destroyed (64 kcal mol^{-1}). The reaction is 36 kcal mol^{-1} exothermic.

We find that C=O pi-bonds are much stronger than C=C, and that C=N pi-bonds are about the same as C=C. These facts all are useful ways to classify knowledge that we wish to bring to bear when trying to understand chemical mechanisms.

Strain Energy

We see quite often that the incremental energy associated with ring compounds in which the bond angles are bigger or smaller than in the normal open chain compounds is a rough constant. This is often called strain energy. The strain energy in cyclopropane is the difference between the group additivity enthalpy increment of 3 C-$(H)_2(C)_2$ groups ($3 \times -4.93 = -14.8$ kcal mol) and ΔH_f(cyclo-C_3H_6) = 12.7 kcal mol^{-1}, or about 27.5 kcal/mol. This number seems to be modestly constant for many three-membered rings. Similarly, we find strain energies of 26.1, 6.3, and 0.2 kcal mol^{-1} for four-, five-, and six-membered cycloalkanes. Such generalizations can be very useful when considering the energetics of transition states involving rings.

III. ESTIMATING AND COMPUTING RATE CONSTANTS

Since one of our ultimate goals is to develop quantitative models which are applicable over a wide range of temperatures and pressures—a range often beyond that for which there are data—it is frequently necessary to fit and extrapolate from the rate data at hand. Even for bimolecular reactions, conventional Arrhenius expressions are generally valid for only a few hundred degrees, and the degree of curvature must be assessed. In other reactions, additional complications may arise, including pressure dependence and multi-channel chemically activated pathways.

In the following sections we discuss some methods that have been developed which address these difficulties and allow accurate computation and numerical representation of rate constants based on experimental data. In the absence of data, these same techniques can be used to estimate rates of unstudied reactions. This can be a great help in determining reaction mechanisms or focusing on the most important reactions in a complex system.

A. Bimolecular Metathesis Reactions

Rate Constant Formats from Transition State Theory

Bimolecular metathesis reactions are cast into the simplest transition state theory (TST) representation. Recognizing that the TST expression for the rate constants is given by:

$$k(T) = \frac{kT}{h} \exp\,(\Delta S_T^{\ddagger}/R)\exp\,(-\Delta H_T^{\ddagger}/RT) = \frac{kT}{h}\frac{Q^{\ddagger}(T)}{Q(T)}\exp\,(-\Delta H_o^{\ddagger}/RT) \tag{44}$$

we use expressions of the form:

$$k(T) = AT^{B}\exp(-C/T) \tag{45}$$

where the parameters A, B, and C may be related to molecular properties of the reactants and the transition state.

$$A = [k/h\ (300) <\Delta C_p^{\ddagger}/R>]\ \exp\,[(\Delta S_{300}^{o\ddagger} - <\Delta C_p^{\ddagger}>)/R] \tag{46a}$$

$$B = (<\Delta C_p^{\ddagger}> + R)/R \tag{46b}$$

$$C = [\Delta H_{300}^{\ddagger} - <\Delta C_p^{\ddagger}\ (300)\ ]/R \tag{46c}$$

Where:

$$<\Delta C_p^{\ddagger}> = (\Delta C_{p,T}^{\ddagger} + \Delta C_{p,300}^{\ddagger})/2 \tag{47}$$

$\Delta S_{300}^{o\ddagger}$ is the entropy change from reactants to transition state at 300 K, referred to a standard state of one atmosphere.

k, h, and R have their usual meanings as Boltzmann's constant,

Planck's constant and the gas constant (in the same units as ΔS and ΔC_p), respectively. If equations (46a), (46b), and (46c) are used in equation (45) the rate constant will have units of $atm^{-1}\ s^{-1}$.

Modeling and Extrapolating H Abstraction Reactions

Several authors have used various schemes for estimating transition state properties. We use methods closely related to those described by Benson[3] and reiterated recently by Cohen et al.[36,37] In a recent paper[38] we have illustrated the method for a series of H-atom metatheses. The structure of the transition state is chosen by keeping the X--H--R angle at 180 degrees and fixing the X--H bond and the R--H bond distances to be approximately 0.3 Å greater than their values in the species X-H and R-H. The frequencies are chosen from the list of normal and partial bond frequencies in "Thermochemical Kinetics."[3] Where this list is incomplete, we have added some in a consistent manner. Internal rotations are treated as described therein, with barriers being chosen as in normal molecules except for those involving rotation around the partial bonds of the transition state where barriers of zero were always assigned.

The important generality is that no single reaction is treated as a universe unto itself, rather a self-consistent approach is used.

The six reactions chosen (O, OH, and CH_3 abstracting H from ethane and neopentane) are meant to be illustrative of the simple TST method as we use it. It seems clear that this consistent way of extrapolating rate data for metathesis reactions will suffice for most modeling purposes. The enthalpy change was obtained by forcing the TST calculation to reproduce the measured rate constant at a chosen temperature. (The effects of varying input parameters on the calculated rate constants have been discussed in detail.[36]) Tables 2 and 3 give the molecular and transition state properties used in the calculations, while Table 4 lists our suggestions for the three-parameter fits to the temperature dependence of these six rate constants. The value of $\Delta H^{\ddagger}_{o}$ that results from our procedure is also listed in Table 4. The parameters given in Table 4 for OH + C_2H_6 yield rate constants which are in excellent agreement with the recent experimental values.[39]

It is of particular interest that the above self-consistent approach works equally well for the reaction of methy radicals with ethane as it

Table 2. Properties of O, OH, CH_3, C_2H_6, and C_5H_{12} Used in TST Calculations

O-Atom		*OH-Radical*	
16,5,3	A	17,2,2	A
3,158.265,1,266.977	B	1,0,1,0.87,0.87	C
		3735	D
		2,139.7	B
CH_3-Radical		*C_2H_6*	
15,2,1	A	30,1,1	A
6,0,6,1.816,1.816,3.308	C	17,1,6,31.84,31.84,4	C
3002,3184,3184,1383,1383,580 –	D	2975,2975,2975,2975,2975,2975	D
		1190,1470,1470,1470,1470,1380	
		1380,1190,995,820,820	
C_5H_{12} (neopentane)		3	E
72,1,1	A	1.5	F
41,4,12,113.321,113.321,113.321	C	0,0,3000,0,0,0	G
2964,2964,2963,2963,2963,2959	D		
2959,2959,2883,2882,2882,2882			
1470,1470,1470,1459,1459,1459			
1458,1458,1383,1383,1383,1368			
1266,1266,1266,1035,1035,981			
981,981,900,900,900,724			
403,403,403,355,355			
3,3,3,3	E		
3,3,3,3	F		
0,0,4100,0,0,0	G		
0,0,4100,0,0,0			
0,0,4100,0,0,0			
0,0,4100,0,0,0			

Notes: A Mass/AMU, electronic degeneracy, number of electronic states.
B Degeneracies and energies (cm^{-1}) of higher electronic states.
C Number of vibrations, number of internal rotations, external symmetry number, moments of inertia (AMU-Å^2).
D Frequencies (cm^{-1}).
E Symmetry numbers for internal rotations.
F Moments of inertia for internal rotations (AMU-A^2).
G Barriers for internal rotation (cal) [V_1, V_2, V_3, V_4, V_5, V_6].

does for all the others. Currently promulgated values for this process[40,47] use T^4 dependence for the rate constant, supposedly to accommodate H-atom tunnelling, although this type of correction does not appear in the same compilations for other similar reactions considered. Tunnelling may very well play a role in these reactions, but it appears to be subsumed by the protocol that we have followed.[36]

Table 3. Properties of the Transition States Used in TST Calculations

O—H—C_2H_5	
46,6,1	A
19,1,1,104.1,92.5,18.09	C
3100,3100,3100,3100,3100,2200	D
1450,1450,1450,1450,1150,1150	
1150,1150,1000,700	
500,250,250	
3	E
3.0	F
0,0,3000,0,0,0	G

HO—H—C_5H_{11}	
89,2,1	
45,5,1,272.756,272.050,114.03	A
3700,3100,3100,3100,3100,3100	C
3100,3100,3100,3100,3100,3100	D
2200,1450,1450,1450,1450,1450	
1450,1450,1450,1450,1450,1150	
1150,1150,1150,1150,1150,1150	
1150,1000,1000,1000,1000,1000	
700,500,420,420,420,420	
420,250,250	
3,3,3,3,1	E
3,3,3,50,1.0	F
0,0,4100,0,0,0	G
0,0,4100,0,0,0	
0,0,4100,0,0,0	
0,0,3000,0,0,0	
0,0,0,0,0,0	

O—H—C_5H_{11}	
88,6,1	A
43,4,1,265.334,265.203,113.119	C
3100,3100,3100,3100,3100,3100	D
3100,3100,3100,3100,3100,2200	
1450,1450,1450,1450,1450,1450	
1450,1450,1450,1450,1150,1150	
1150,1150,1150,1150,1150,1150	
1000,1000,1000,1000,700,500	
420,420,420,420,420,250,250	
3,3,3,3	E
3,3,3,50	F
0,0,4100,0,0,0	G
0,0,4100,0,0,0	
0,0,4100,0,0,0	
0,0,3000,0,0,0	

CH_3—H—C_5H_{11}	
87,2,1	A
51,5,1,269.117,269.107,116.499	C
3100,3100,3100,3100,3100,3100	D
3100,3100,3100,3100,3100,3100	
3100,3100,2200,1450,1450,1450	
1450,1450,1450,1450,1450,1450	
1450,1450,1450,1450,1150,1150	
1150,1150,1150,1150,1150,1150	
1150,1150,1000,1000,1000,1000	
700,500,420,420,420,420	
420,250,250	
3,3,3,3,3	E
3,3,3,50,3.0	F
0,0,4100,0,0,0	G
0,0,4100,0,0,0	
0,0,4100,0,0,0	
0,0,3000,0,0,0	
0,0,0,0,0,0	

CH3—H—C_2H_5	
45,2,1	A
27,2,1,106.643,95.002,21.281	C
3100,3100,3100,3100,3100,3100	D
3100,3100,2200.1450,1450,1450	
1450,1450,1450,1450,1150,1150	
1150,1150,1150,1150,1000,700	
500,250,250	
3,3	E
3.0,2.75	F
0,0,0,0,0,0	G
0,0,3000,0,0,0	

HO—H—C_2H_5	
2,47,2,1	A
21,2,1,124.544,111.128,19.81	C
3100,3100,3100,3100,3100,3100	D
2200,1450,1450,1450,1450,1150	
1150,1150,1150,1000,1000,700	
500,250,250	
3,1	E
3.0,.69	F
0,0,3000,0,0,0	G
0,0,0,0,0,0	

Notes: See Table 2.

Table 4. Parameters for Modeling Some Metathesis Reactions

Reaction	$A/cm^3\ mol^{-1}s^{-1}$	*B*	*C/K*	$\Delta H_0^{\ddagger}/cal\ mol^{-1}$
$O + C_2H_6$	8.975×10^7	1.924	2864	6310
$O + C(CH_3)_4$	3.116×10^9	1.412	2886	6053
$OH + C_2H_6$	3.536×10^6	2.124	439.3	2012
$OH + C(CH_3)_4$	1.746×10^8	1.559	242.6	1284
$CH_3 + C_2H_6$	6.139×10^6	1.737	5257	11718
$CH_3 + C(CH_3)_4$	2.057×10^8	1.227	5606	12106

Note: $k(T) = AT^B \exp(-C/T)$; $\Delta H_0^{\ddagger}$ is shown for interest.

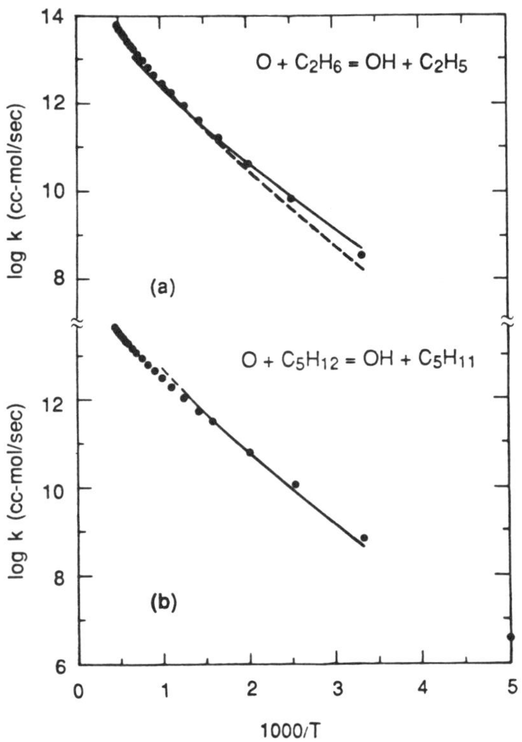

Figure 2. a) ●, Transition State Theory Calculations; —, Reference 40; ---, Reference 36. Transition State Theory calculations forced to agree with Reference 40 at 500 K. The temperature ranges displayed correspond to the range of data used in the sources cited.
b) ●, Transition State Theory Calculations; —, Reference 36, ---; Reference 41. Transition state theory calculations forced to agree with Reference 36 at 500 K. The temperature ranges displayed correspond to the range of data used in the sources cited.

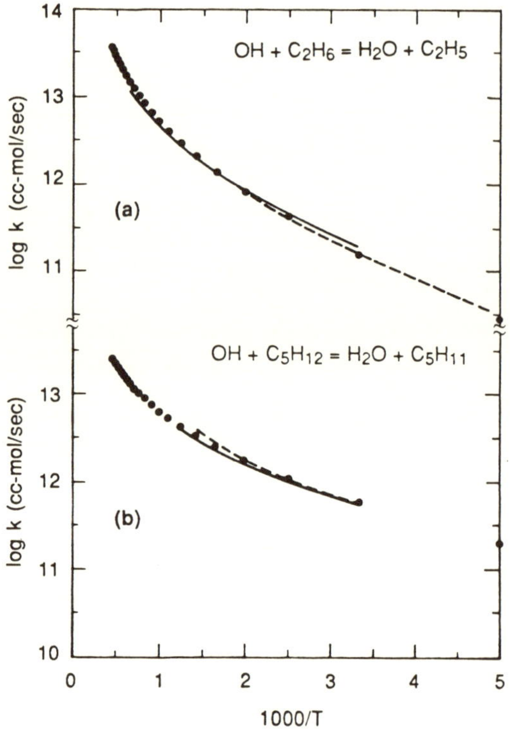

Figure 3. a) ●, Transition State Theory Calculations; —, Reference 40; ---, References 42. Transition state theory calculations forced to agree with References 42 at 300 K. The temperature ranges displayed correspond to the range of data used in the sources cited.
b) ●, Transition State Theory Calculations; —, Reference 36; ---, Reference 43. Transition state theory calculations forced to agree with Reference 43 at 300 K. The temperature ranges displayed correspond to the range of data used in the sources cited.

B. Single Channel Unimolecular Reactions

Falloff and the Lindemann Mechanism

If the simple Lindemann mechanism is used to describe a unimolecular decomposition:

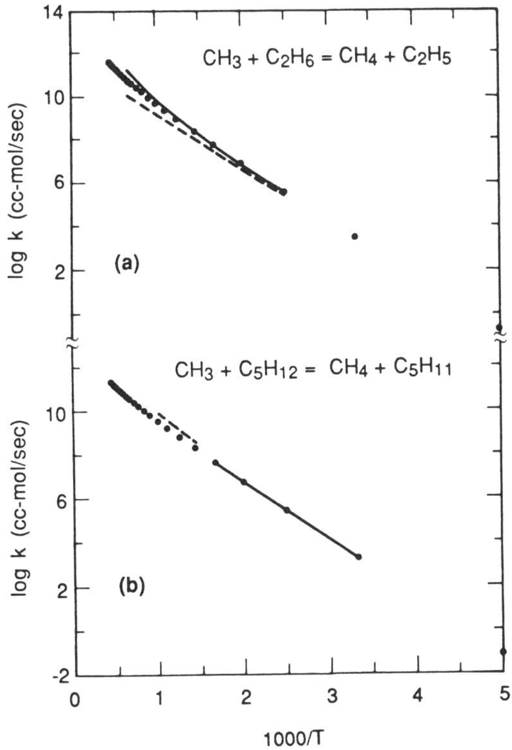

Figure 4. a) ●, Transition State Calculations; —, Reference 40; ---, Reference 44. Transition state theory calculations forced to agree with Reference 40 at 500 K. The temperature ranges displayed correspond to the range of data used in the sources cited.
b) ●, Transition State Calculations; —, Reference 44; ---, References 45, 46. Transition state theory calculations forced to agree with Reference 44 at 500 K. The temperature ranges displayed correspond to the range of data used in the source cited.

$$XY + M \underset{k_{-1}}{\overset{k_1}{\rightleftarrows}} XY^* + M \tag{48}$$

$$XY^* \xrightarrow{k_{-a}} X + Y \tag{49}$$

and we write $k_{-1}[M] = \beta\omega$ where ω is collision frequency and β $(0 < \beta \leq 1)$ is the collision efficiency:

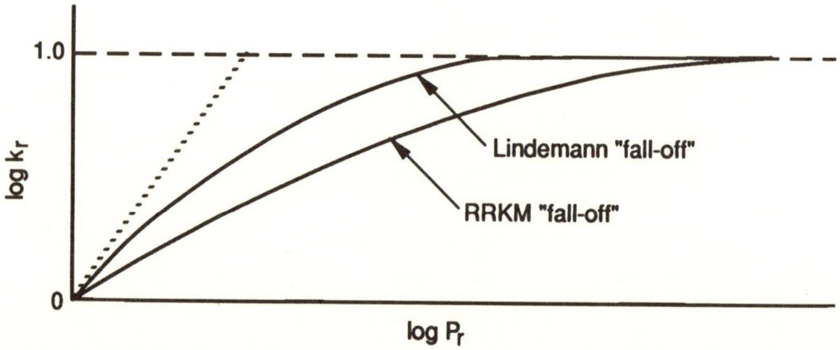

Figure 5. Reduced rate constant versus reduced pressure for a unimolecular process.

$$k_{uni} = \frac{K_1 \beta\omega\, k_{-a}}{\beta\omega + k_{-a}}\text{ ; where } K_1 = k_1/k_{-1} \tag{50}$$

In the limits $\omega \rightarrow \infty$ and $\omega \rightarrow 0$,

$$k_{uni}^{\infty} = K_1 k_{-a}; \qquad k_{uni}^{o} = K_1 \beta\omega \tag{51}$$

Following Troe[48–51] we define the reduced pressure:

$$P_r \equiv k_{uni}^{o}/k_{uni}^{\infty} \tag{52}$$

and the reduced rate constant,

$$k_r \equiv \frac{k_{uni}}{k_{uni}^{\infty}} = \frac{P_r}{1 + P_r} \tag{53}$$

If we were to plot log k_r vs log P_r, we would find the universal function plotted in Figure 5 and labeled "Lindemann fall-off."

Energy-Dependent Mechanisms

The Lindemann mechanism is too simple, however, in that it assumes that all excited molecules, XY^*, are the same, and that the product forming step, k_{-a}, proceeds with a single rate. In fact, k_1, k_{-1}, and k_{-a} are all functions of the energy of XY^* (at a still higher level of approximation, neglected here, these quantities also depend on J, the rotational quantum number). The distribution of $XY(E)$ is given by $K_1(E)$, which is simply the Boltzmann distribution function, $B(E)$.

Taking this into account, equation (50) becomes:

$$k_{\text{uni}} = \beta\omega \int_{E_{-a}}^{\infty} \frac{k_{-a}(E)}{\beta\omega + k_{-a}(E)} B(E)dE \tag{54}$$

From statistical mechanics, $B(E) = \dfrac{\rho(E)}{Q} \exp\ (-E/RT)$, where $\rho(E)$ is the density of states of XY at E and Q is the partition function of XY. Thus,

$$k^{o}_{\text{uni}} = \beta\omega \int_{E-a}^{\infty} B(E)dE \tag{55}$$

and

$$k^{\infty}_{\text{uni}} = \int_{E_{-a}}^{\infty} k_{-a}(E)B(E)dE \tag{56}$$

We may still define k_r and P_r as $k_{\text{uni}}/k^{\infty}_{\text{uni}}$ and $k^{0}_{\text{uni}}/k^{\infty}_{\text{uni}}$ respectively, and we notice that k_r may be expressed as in the Lindemann representation, but with an additional term, leading to the function in Figure 5 labeled "RRKM fall-off."

$$k_r = \frac{P_r}{1 + P_r} \cdot F_K \tag{57}$$

Still following Troe[48–51] we define F_K (the "broadening parameter") in terms of its central value F_c (i.e., its value at $P_r = 1$, where $k_r = 1/2\ k_\infty$ for the Lindemann model):

$$F_K \equiv F_c^X \tag{58}$$

Using simplest assumption of symmetric broadening about $P_r = 1.0$,

$$X = [1 + (\log P_r)^2]^{-1} \tag{58a}$$

It is easy to see that if one uses available unimolecular rate theories, such as RRKM, one can calculate F_c; since at $P_r = 1$, $F_c = 2k_{\text{uni}}/k^{\infty}_{\text{uni}}$. Thus one calculates k_{uni} at pressures corresponding to $P_r = 1$ using RRKM.

Numerical Representation of Pressure and Temperature Dependence

The pressure dependence of the rate has thus been characterized in the form of three parameters, k^{0}_{uni}, k^{∞}_{uni}, and F_c. If the analysis is carried out over a temperature range, expressions for $k^0_{uni}(T)$, $k^\infty_{uni}(T)$, and $F_c(T)$ can be obtained. We[52] have suggested the forms:

$$k^\infty_{uni} = aT^b \exp(-c/T), \tag{59}$$

$$k^o_{uni}[M] = a' \, T^{b'} \exp(-c'/T), \text{ and} \tag{60}$$

$$F_c = a'' \exp(-b''/T) + \exp(-T/c''). \tag{61}$$

The entire complex pressure and temperature dependence is thus accurately described by nine parameters. In terms of computer modeling, this is a trivial number of coefficients, yet allows the confident extrapolation of the model over a wide range of conditions (at least as it pertains to such reactions). Notice that these parameters are pertinent to strong collision conditions ($\beta = 1$). In fact, since gases are generally weak colliders, and the value of β varies with the gas, we will have to adjust these values slightly to fit the conditions. This will be discussed shortly.

Of course, k_{bi}, the rate constant for the reverse (association) reaction, may be computed from k_{uni} and the equilibrium constant.

In order to obtain descriptions of reactions in the form of equations (59–61), models must be developed which can be fit to the available experimental data of k_0, k_∞, and the falloff regime. Since these are empirical fits, not a priori calculations, overly sophisticated models are not necessary. They should be rigorous enough to adequately describe the behavior, and thus allow extrapolation, but the details are often not critical.

Restricted Gorin Models

A central problem in calculating the pressure dependence of a unimolecular reaction is building up an adequate picture of the transition state. We need to know its thermodynamic properties, hence its mass, vibrational frequencies, moments of inertia, and so on (section II).

We begin by assuming that, at 0 K, the barrier to dissociation of a closed shell species is equal only to the endothermicity of the reaction (i.e., that the recombination reactions of radicals proceed with no barrier at this temperature). Various simple functions, for example, Morse or Lennard-Jones potentials, can then be used to describe the dissociation. All of these reach a maximum at infinite nuclear separation, creating a quandary as to where to locate the transition state. At higher temperatures, however, where molecules have rotational energy, an effective potential term arising from the conservation of angular momentum must be added. This creates a maximum in the potential, the so-called centrifugal barrier, and it is here that we locate the transition state.

If a Lennard-Jones potential is used and it is assumed that the average transition state has an average rotational energy of 1/2RT, the position of the maximum is[3]:

$$\frac{r}{r_0} = \left(\frac{6D_0}{RT}\right)^{1/6} \tag{62}$$

where r is the internuclear distance, r_0 is the equilibrium distance, and D_0 is the enthalpy change at 0 K plus the zero point energy of the molecule. Notice that this leads to a temperature dependent transition state, with the maximum occurring at ever smaller internuclear distances as the temperature increases.

An Example:

$$CH_4 \rightarrow CH_3 + H$$

From the above discussion we locate the transition state at the centrifugal maximum. As the molecule dissociates its geometry and vibrational modes must shift smoothly from those of CH_4 to those of CH_3 + H. Gorin models assume that the transition state is so "loose" that each end "looks like" the incipient radicals. Thus the CH_3 end is flat and has six vibrational frequencies equal to those of methyl radical. Of the three remaining frequencies, one is the reaction coordinate and the other two correspond to the bending motions of the H relative to the CH_3. Because the $[H_3C{-}H]^{\dagger}$ bond is so stretched these are better described as a two dimensional rotation of the methyl rather than as vibrations. This rotation will be restricted, however, since if the methyl is rotated about the axes perpendicular to the hydrogen a steric interaction will ensue (reduced orbital overlap could also play a role[53]). The degree of hindrance is expected to increase as the transition state moves inward at higher temperatures. Hence we treat those transition modes that are the incipient, diatomic-like, free rotations in the products, and presumably in the transition state at absolute zero, like particles in an ever decreasing two dimensional box. One way to account for this analytically is to change the product of the moments of inertia in the transition state to some effective value according to $\Pi\, I^{\ddagger}_{eff} = \Pi\, I^{\ddagger}(1 - \eta/100)$, where η is the percent hindrance.[54] We find that the

Table 5. Data for RRKM Calculations

Frequencies of CH_4:	3019 (3), 2916, 1534 (2), 1306 (3) cm^{-1}
Moments of Inertia for CH_4:	5.33, 5.29, 5.29 $\times$ 10^{-40} g cm^2
Heat of Formation at 0 K CH_4:	-15.99 kcal mol^{-1}
Active External Rotor in CH_4:	1,I $-$ 5.313 $\times$ 10^{-40} g cm^2; σ = 3
Complex CH_3 $-$ H	
Frequencies of the complex:	3162 (2), 3044, 1396 (2), 580 cm^{-1}
Moments of inertia of the complex:	I^+/I = 18.8 at 300 K, 6.2 at 2500 K
Active external rotors in the complex:	1, I = 5.86 $\times$ 10^{-40} g cm^2; σ = 3
Active 2-dimensional rotors in the complex:	1, I = 2.93 $\times$ 10^{-40} g cm^2; σ = 1
Barrier at 0 K for the complex:	103.0 kcal mol^{-1}
Hindrance Parameter:	$\eta = 146.3 - 193.8T^{-1/6}$

temperature dependent form $\eta = a + bT^{-1/6}$ is reasonable[55] and fits most data. In methane dissociation we found[55]: $\eta = 146.3 - 193.8T^{-1/6}$. The suggested parameters for use in determining the strong collision values of this rate constant are in Table 5.

The model was chosen so as to reproduce the high pressure-limiting rate constant from 300–1100K given by[56] and, via an equilibrium constant, the data of references 40 and 41. Values for k_o, k_∞, and F_c were then calculated from 300–2500K and fit to three-parameter expressions as outlined above. These parameters are listed in Table 6. k_{uni} for methane dissociation has been measured between 995 K–1103 K and 25–700 torr by Chen et al.,[56] with methane as the bath gas. The pressure dependence of the rate constant was modeled using the parameters listed in Table 6. However, since methane is not a unit-

Table 6. Parameters for $CH_4 \rightarrow CH_3 + H$

$$k^{sc} = 10^a T^b \exp(-c/T)$$
$$F_c^{sc} = a \exp(-b/T) + \exp(-T/c)$$

	$(k_0^{sc}/[M])$ $cm^3\ mol^{-1}\ s^{-1}$			k_∞/s^{-1}			F_c^{sc}	
	a	*b*	*c*	*a*	*b*	*c*	*a*	*b*
$CH_4 \rightarrow CH_3 + H$	33.111	-3.732	53596	17.569	-0.558	52782	0.45	797

Equilibrium constant $K_c/mol\ cm^{-3} = 10^{\alpha}T^{\beta} \exp(-\gamma/T)$

	α	β	γ
$CH_4 \leftrightarrows CH_3 + H$	1.163	0.26	52623

efficient strong collider, corrections for weak collision effects were carried out as follows:[48–51]

$$k_o^{wc}(T) = k_o^{sc}(T)\beta_c(T) \tag{63}$$

$$F_c^{wc}(T) = F_c^{sc}(T)\beta_c(T)^{0.14} \tag{64}$$

β_c was calculated from the expression:[48–51]

$$\beta_c/(1 - \beta_c^{1/2}) = \langle\Delta E\rangle/(F_E RT) \tag{65}$$

F_E, the energy dependence of the density of states, is calculated according to Troe.[48–51] The fit to data shown in Figure 6, requires a value for $\langle\Delta E\rangle$ of 3000 cals/mol, giving $\beta_c \simeq 0.44$ at 1000 K. The pressure dependence of k_{66}, as determined by Brouard et al.[57] at several temperatures using helium as a bath gas, was modeled in a similar manner using $\langle\Delta E\rangle$ = 155 cals/mol. This is shown in Figure 7. Using a value for β_c(Ar) of ~ 0.03, we are able to produce values for

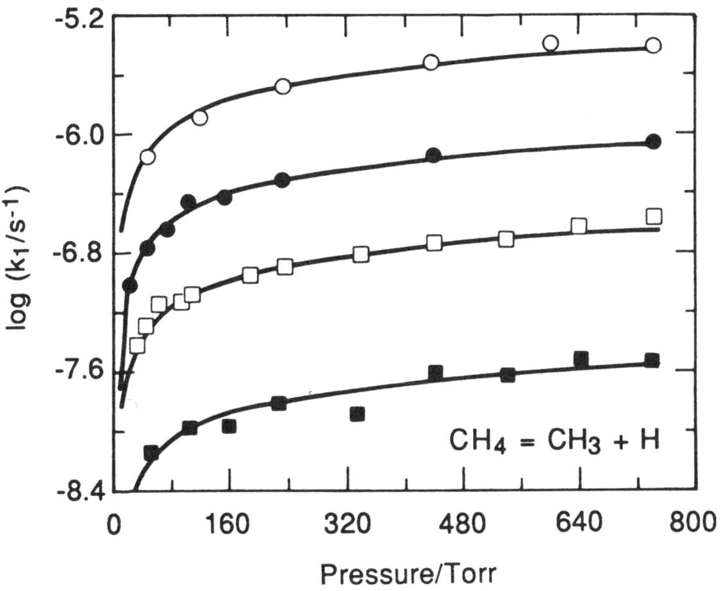

Figure 6. Rate constant versus pressure for the reaction $CH_4 \rightarrow CH_3 + H$ over the temperature range 995–1103 K. Lines are calculated from data in Table 6. Data points are from Reference 56: ■, 995 K; □, 1038 K; ●, 1068 K; ○, 1103.

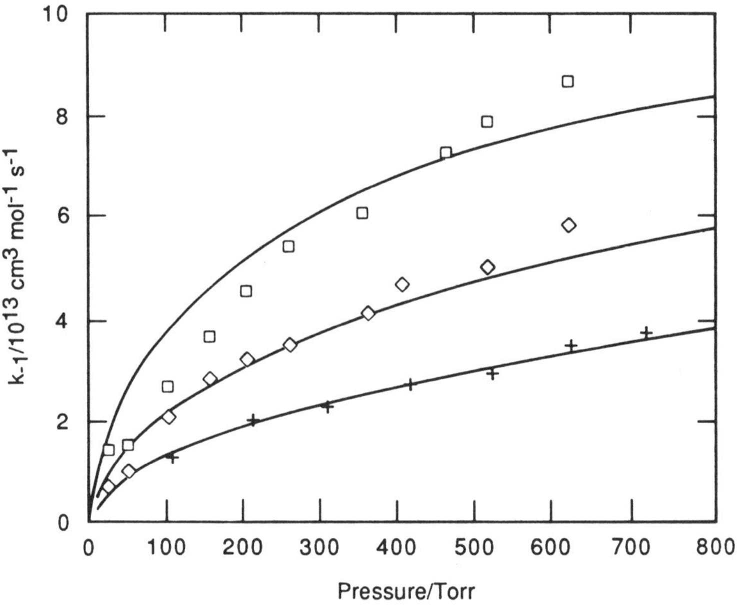

Figure 7. Rate constant versus pressure at 504 K for the reaction $CH_3 + H \rightarrow CH_4$. Solid lines calculated using the parameters in Table 6. Data points are from Reference 57: □, 401 K; ◇, 504 K; +, 601 K.

k_o^{wc}, which were in excellent agreement with those of Tabayashi and Bauer[58] between 1900–2770 K.

$$CH_3 + H + M = CH_4 + M \qquad (66)$$

Clearly, with a suitable choice of either $<\Delta E>$ or β, the parameters listed in Table 6 describe very well the pressure and temperature dependence of the rate constant for methane decomposition.

C. Multichannel Unimolecular Reactions

Thermally Stable Intermediates

These types of processes are characterized in terms of potential surfaces, such as represented in Figure 8c. Here, the association reaction produces the excited species XY^*. This species will decompose if proper localization of the excess energy occurs prior to colli-

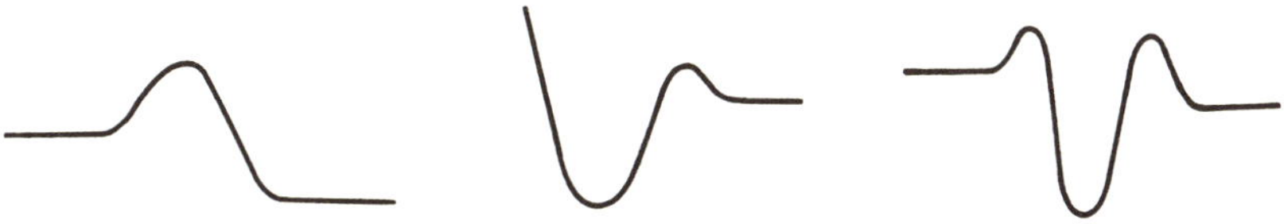

Figure 8. Representative reaction coordinate diagrams.

sional stabilization. As there may be several reactive surfaces, the decomposition products may be other than the starting species. Since these other reactions are energetically accessible due to the release of energy in the initial association, they are known as chemical activation processes. For a system with two reactive surfaces, and ignoring thermal dissociation of XY (that is $k_{-68} = 0$), the Lindemann mechanism is:

$$X + Y \underset{k_{-a}}{\overset{k_a}{\rightleftarrows}} XY^* \tag{67}$$

$$XY^* + M \overset{\beta\omega}{\rightarrow} XY + M \tag{68}$$

$$XY^* \overset{k_b}{\rightarrow} C + D \tag{69}$$

Notice that the value of β in (68) does not have to be the same as for thermal processes. It may be thought of as the efficiency for lowering the initial energy to a point such that k_{-a} and k_b are much less than the collision rate ω.

Experimentally, the rate constant measured in a multichannel system depends on whether loss of X and/or Y is monitored or whether appearance of C and/or D or XY is observed. The rate constant for loss of X (and/or Y) is:

$$k_{\text{loss}} \equiv \frac{1}{[X]\,[Y]} \left[\frac{d[C]}{dt} + \frac{d[XY]}{dt} \right] \tag{70}$$

The Association Reaction

Pursuing the Lindemann formulation further, $k_{\text{bi}}^{\text{assn}}$ may be written by analogy to the previous discussion with the addition of the second pathway.

$$k_{\text{bi}}^{\text{assn}} = \frac{k_a \, \beta\omega}{k_{-a} + k_d + \beta\omega} \tag{71}$$

$$P_r^{\text{assn}} = \frac{\beta\omega}{k_{-a} + k_b} \tag{72}$$

$$k_r^{\text{assn}} \equiv \frac{k_{\text{bi}}^{\text{assn}}}{k_{\text{bi}}^{\text{assn}\infty}} = \frac{\beta\omega}{k_{-a} + k_b + \beta\omega} = \frac{P_r^{\text{assn}}}{1 + P_r^{\text{assn}}} \tag{73}$$

Actually, of course:

$$k_{\text{bi}}^{\text{assn}} = K_{XY\,\beta\omega} \left\langle \frac{k_{-a}(E)}{k_{-a}(E) + k_b(E) + \beta\omega} \right\rangle_B \tag{74}$$

where $< >_B$ signifies averaging with respect to the Boltzmann distribution.

$$K_{\text{XY}} = \frac{[XY]_{\text{eq}}}{[X]_{\text{eq}}[Y]_{\text{eq}}} \tag{75}$$

$$\text{Thus, } k_r^{\text{assn}} = \frac{P_r^{\text{assn}}}{1 + P_r^{\text{assn}}} \, (F_c^{\text{assn}})^x \tag{76}$$

Chemical Activation

For the chemical activation process, the Lindemann mechanism yields:

$$k_{\text{bi}}^{\text{ca}} = \frac{k_a \, k_b}{k_{-a} + k_b + \beta\omega} \tag{77}$$

$$P_r^{ca} = \frac{\beta\omega}{k_{-a} + k_b} \tag{78}$$

and

$$k_r^{\text{ca}} = \frac{\beta\omega}{k_{-a} + k_b + \beta\omega} = \frac{P_r^{ca}}{1 + P_r^{ca}} \tag{79}$$

Again, the actual expression is:

$$k_{\text{bi}}^{\text{ca}} = K_{XY} \left\langle \frac{k_{-a}(E) \, k_b(E)}{k_{-a}(E) + k_b(E) + \beta\omega} \right\rangle_B \tag{80}$$

$$k_r^{ca} = \frac{P_r^{ca}}{1 + P_r^{ca}} \left(F_c^{ca}\right)^x \tag{81}$$

Notice that for this case the low-pressure limit of the bimolecular rate constant is independent of pressure, while the high-pressure limiting bimolecular rate constant is best expressed as a first-order constant by multiplying by the pressure.

The rate constant for production of C (and/or) D is just the chemical activation rate constant, k^{ca}, if XY does not thermally decompose.

$$k_c = \frac{1}{[X][Y]} \frac{d[C]}{dt} = k_{bi}^{ca} \tag{82}$$

Some general considerations to be remarked on are that the rate of the chemical activation step(s) relative to collisional stabilization can be heavily dependent upon where in the energy manifold of $XY^*(E)$ a particular molecule is located. The more energetic molecules, of course, have more pathways available to them and must undergo more collisions before they are deactivated. A consequence of this is that going up in temperature or down in pressure can strongly affect the branching ratios. Similarly, these changes may result in new pathways becoming energetically accessible or important.

Thermally Unstable Intermediates

If the adduct, XY, is weakly bonded, it may also thermally decompose. This has been discussed[59] in the case of the HO_2 self-reaction. The Lindemann mechanism must contain another step (i.e., $XY + M \xrightarrow{k_{-68}} XY^* + M$), that is, the reverse of the collisional stabilization step. Now,

$$k_{\text{loss}} = k_{\text{bi}}^{\text{ca}} + k_{\text{bi}}^{\text{assn}'} \tag{83}$$

where

$$k_{\text{bi}}^{\text{assn}'} = k_{\text{bi}}^{\text{assn}} - \frac{k_{-a}k_{-s}[M]}{(k_{-a} + k_r + \beta\omega)} \frac{[XY]}{[X]\,[Y]} \tag{84}$$

$[XY]$, $[X]$, $[Y]$ are time-dependent concentrations.

The rate of production of C (and/or D) when thermal decomposition of the adduct plays a role has contributions from both the chemical activation and the thermal decomposition of the adduct. In the simple Lindemann form:

$$k_c = \frac{k_a k_{-a} k_b}{(k_{-a} + k_b + \beta\omega} + \frac{[XY]_t}{[X]_t[Y]_t}$$

$$\frac{k_{-s}[M]}{\beta\omega} \quad \frac{\beta w\ k_b}{(k_{-a} + k_b + \beta\omega)} \tag{85}$$

$$k_c = k^{ca} + k_{bt} \frac{[XY]_t}{[X]_t\ [Y]_t} \tag{86}$$

In general, a model for this system which contains k_1, k_{-1}, k_2, k_{-2}, and k^{-ca} has to be used to solve for the appropriate time-dependent concentrations. Such a system, then, should be represented for modeling purposes (over all P and T) by tabulating entries for each rate constant, as in Table 6.

While this may seem to be a large number of parameters to define this system, it must be remembered that they may be derived from more than one type of experiment and that not all problems are equally sensitive to all these parameters. In any case, no current representation of data for systems such as these adequately represents the values of the rate constant over wide temperature and pressure ranges.

D. Applications

$HO + CO \rightleftarrows [HOCO] \rightarrow H + CO_2$

The oxidation of carbon monoxide by the hydroxyl radical to form carbon dioxide and a hydrogen atom is an important reaction in both combustion and atmospheric chemistry. Accurate parameterization of its pressure and temperature dependence over a wide range of conditions is thus needed. The observation that the rate coefficient of this exothermic reaction (-25 kcal/mol) increases with pressure suggests that the reaction proceeds through a bound HOCO intermediate in a chemical activation step.

We have performed two-channel RRKM calculations for the OH(D) + CO reaction with the aim of developing a self consistent model which can be used to extrapolate the available experimental data.[60] The model parameters for the OH case are given in Table 7. Barriers were chosen to fit the OH data[61] at low pressure over the temperature range $200 \leq T/K \leq 2200$. RRKM calculations were then fit to the Troe equation and values of $F_c(T)$ extracted. As suggested above, three-parameter equations were used to fit $k_0(T)$, $k_\infty(T)$, and $F_c(T)$. These coefficients, listed in Table 8, lead to the calculated values shown in Figures 9–11.

Table 7. Parameters used in RRKM Calculations to Describe HOCO (DOCO) Decomposition

HOCO complex	
Frequencies:	3460, 1830, 1260, 1070, 620, 620 cm^{-1}
	(2560, 1820, 1120, 890, 560, 460)
Moments of Inertia: adiabatic	49.6, 47.2 amu Å^2
	(52.9, 50.1)
nonadiabatic	2.40
	(2.77)
Heat of formation at 0 K:	−52.5 kcal mol^{-1}
	(−53.7)
TS 1. HO—CO (trans-Bent)	
Frequencies:	3650, 1900, 1000, 510 cm^{-1}
	(2600, 1890, 850, 490)
Moments of Inertia:	
internal:	0.5 amu Å^2
	(1.0)
external: adiabatic	57.3, 55.1 amu Å^2
	(62.2, 60.0)
nonadiabatic	2.1
	(2.1)
Heat of formation at 0 K:	−17.39 kcal mol^{-1}
	(−18.23)
TS 2. H-OCO (cis-Bent)	
Frequencies:	2700, 2100, 1200, 900, 650 cm^{-1}
	(2200, 2070, 1250, 700, 600)
Moments of Inertia:	
external: adiabatic	46.5, 44.8 amu Å^2
	(49.4, 46.7)
nonadiabatic	1.7
	(2.7)
Heat of formation at 0 K:	−17.65 kcal mol^{-1}
	(−17.75)
Lennard-Jones Diameter:	4 Å
Lennard-Jones Well Depth:	200 K

Figure 9 compares the temperature dependence of k_0 predicted by the model with that recommended by Baulch et al.[61] Figure 10 shows the agreement between the modeled pressure dependence of the rate of OH + CO in various bath gases and that found by Paraskevopoulos and Irwin[62] at 298 K.

After making appropriate isotopic corrections to the frequencies and ZPE, this same model was used to fit OD + CO data. The experimental[60] and modeled pressure dependence of the total rate constant at various temperatures in several bath gases is shown in Figure 11. Again the agreement is shown to be very good.

Table 8. Parameters Obtained from Two-Channel RRKM Calculations for OH (OD) + CO

$$k = 10^{a}T^{b}e^{-c/T}$$
$$F_c = ae^{-b/T} + e^{-T/c}$$
$$K = 10^{\alpha}T^{\beta}e^{-\gamma/T}$$

*(a) Unimolecular Reactions**

	$(k_o/[M])/cm^3mol^{-1}s^{-1}$			k_∞/s^{-1}			F_c		
	a	*b*	*c*	*a*	*b*	*c*	*a*	*b*	*c*
HOCO → OH + CO	24.70	−2.267	18700	14.11	0.034	18161	0.854	1600	1420
DOCO → OD + CO	26.55	−2.785	18990	14.76	−0.176	18410	0.794	818	731
HOCO → H + CO_2	25.89	−2.875	18500	12.59	0.178	17811	0.665	623	422
DOCO → D + CO_2	25.04	−2.671	19190	12.99	0.034	18350	2.615	6180	2890

(b) Chemical Activation Process

	$k_o^{ca}/cm^3mol^{-1}s^{-1}$			$[M]k_\infty^{ca}/s^{-1}$			F_c		
	a	*b*	*c*	*a*	*b*	*c*	*a*	*b*	*c*
H + CO_2 → OH + CO	13.354	0.147	12471	3.311	2.504	12247	1.321	1892	1677
DO + CO → D + CO_2	−5.845	1.533	−868.9	−21.68	5.552	−3303	0.887	103	252

(c) Equilibrium Constant

	$K/mol\ cm^{-3}$		
	α	β	γ
HOCO = OH + CO	5.809	−1.239	17370
DOCO = OD + CO	6.422	−1.375	17700
HOCO = H + CO_2	−0.036	−0.214	4506
DOCO = D + CO_2	1.190	−0.500	5649
OH + CO = H + CO_2	−5.845	1.025	−12865
OD + CO = D + CO_2	−5.232	0.875	−12056

Note: *Strong collision values (β = 1.0).

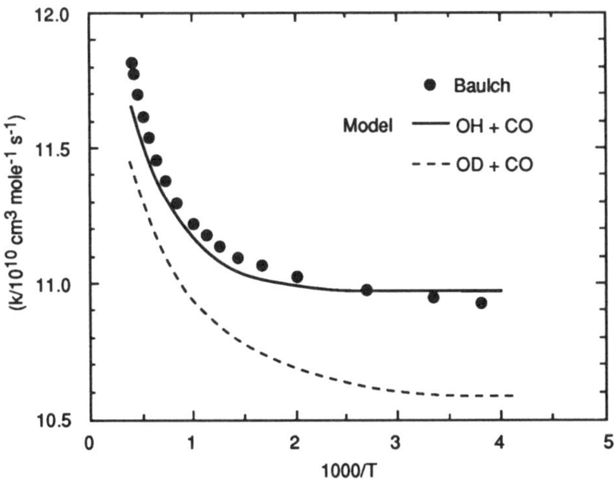

Figure 9. An Arrhenius plot for the rate constant of OH(D) + CO at zero pressure. Comparison of the recommended temperature dependence of Reference 61 for OH + CO, (●,), to our model predictions for OH + CO (—) and OD + CO (--).

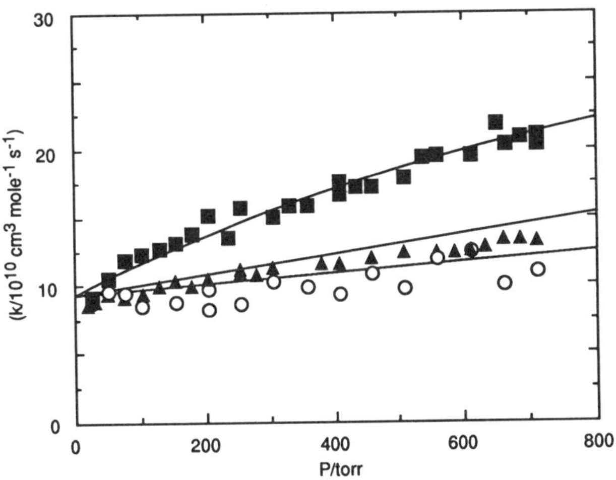

Figure 10. Comparison of the experimental[62] pressure dependence at 298 K of the rate constant for OH + CO in several bath gases to the model. The experimental data points in He (○), N_2(Δ), and SF_6 (■) are compared to the lines representing the model prediction for the different bath gases.

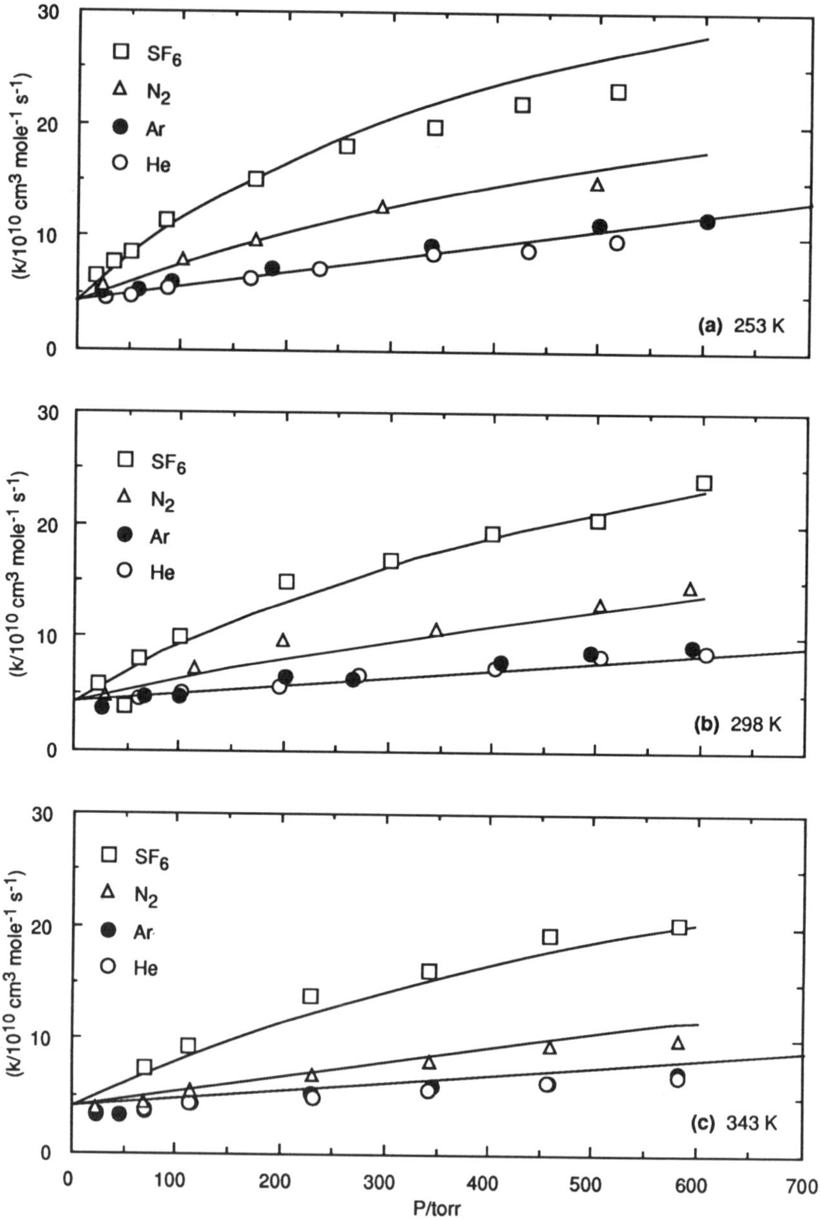

Figure 11. Comparison of the pressure dependence of the rate constant for OD + CO to the model. Experimental data[60] in He (○), Ar (●), N_2 (Δ), and SF_6 (□) bath-gases are compared to the model predictions at several temperatures: a) 253 K; b) 278 K; c) 298 K; d) 343 K.

$$2\ CH_3 \rightarrow C_2H_5 + H$$

Methyl radicals play a major role in the combustion of methane, ethane, and higher hydrocarbons. In contrast to low temperature behavior, where methyl recombination leads to the relatively stable ethane molecule, under combustion conditions chemically activated fission of the bound intermediate can result in hydrogen atom and ethyl radical (which in turn rapidly decomposes to C_2H_4 + H). The occurrence of this step is very important since it converts a chain termination into a chain branching step.

In order to describe this system, proceeding via a bound ethane intermediate, a treatment similar to that outlined for the HOCO case was employed.[63] The model is given in Stewart et al.,[63] and the parameters obtained are listed in Table 9. The experimentally determined pressure dependence of the methyl + methyl association reaction[64] was modeled using the parameters in Table 9 and the results are shown in Figure 12. The bath gas was argon, and $<\Delta E>$ was taken as 320 ± 100 cals/mol. The pressure and temperature dependence for the loss of methyl radicals is shown in Figure 13. The bath gas was again taken as argon with $<\Delta E>$ equal to 320 cals/mol. The peculiar shape of Figure 13 is attributable to the combined effects of the slight negative temperature dependence of the association reaction, the increasing pressure dependence of this reaction with temperature, and the increase with temperature of the chemical activation step.

The dangers of simple extrapolation to temperatures and pressures outside those of experiment should be graphically obvious by simple inspection of Figure 13. The general form of this graph will be common to all reactions proceeding through a bound intermediate. Care should be taken to recognize these reactions and to act accordingly. Of course, should parameters such as those listed in Tables 8 and 9 be available to combustion modelers for *any* desired reaction then it is a relatively trivial matter to estimate with confidence the rate constant for any given temperature and/or pressure. The ability of such a parameterization to predict the temperature and pressure dependence of reactions has been amply justified by the examples above. Of course, our parameters are only as good as the data on which they are based. More accurate determinations of these rate constants may result in revision of the parameters of Tables 8 and 9. It is the authors' hope that eventually all appropriate reaction rate constants will be parameterized in this manner.

Table 9. Parameters for $CH_3 + CH_3$

	$k_0/[M]$ ($cm^3\ mol^{-1}\ s^{-1}$)			k_∞ (s^{-1})			F_c		
	a	*b*	*c*	*a*	*b*	*c*	*a*	*b*	*c*
	Unimolecular Reactions: $k = 10^a T^b exp(-c/T)$; $F_c = a\ exp(-b/T) + exp(-T/c)$								
$C_2H_6 \rightarrow 2CH_3$	58.257	−10.609	49601	26.027	−2.792	46858	0.31	518	445
$C_2H_6 \rightarrow C_2H_5 + H$	42.839	−6.431	53938	20.947	−1.228	51439	47.61	16182	3371
	Chemical Activation Process: $k = 10^a T^b exp(-c/T)$; $F_c = a\ exp(-b/T) + exp(-T/c)$								
$2CH_3 \rightarrow C_2H_5 + H$	12.698	0.099	5335	−6.420	4.838	3880	1.641	4334	2725

	Equilibrium Constant: $K_c/mol\ cm^{-3} = 10^\alpha T^\beta exp(-\gamma/T)$		
	α	β	γ
$C_2H_6 \leftrightarrows 2CH_3$	+9.70	−1.825	+46546
$C_2H_6 \leftrightarrows C_2H_5 + H$	+3.23	−0.24	+50648

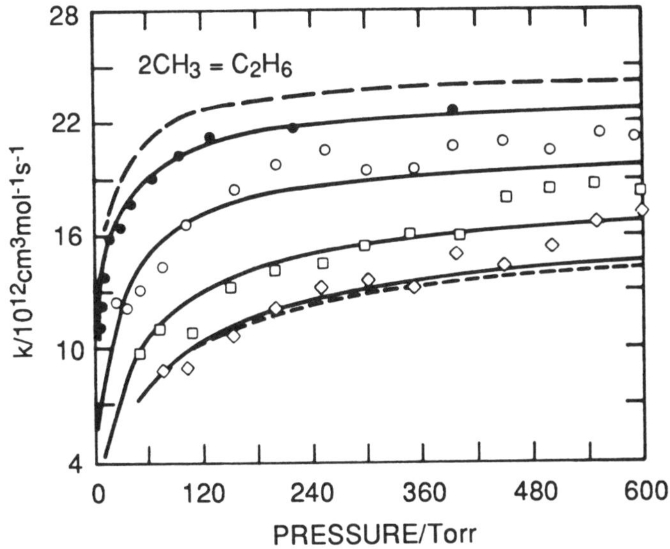

Figure 12. Rate constant for methyl radical association versus pressure. Solid lines are calculated from the data in Table 9. Data points are taken from Reference 64. ●, 577 K; (○), 700 K; (□), 810 K; (◇), 906 K; dashed lines are falloff curves derived from the parameters of Wagner and Waldlaw, Reference 65; — —, 577 K; _ _ _, 906 K.

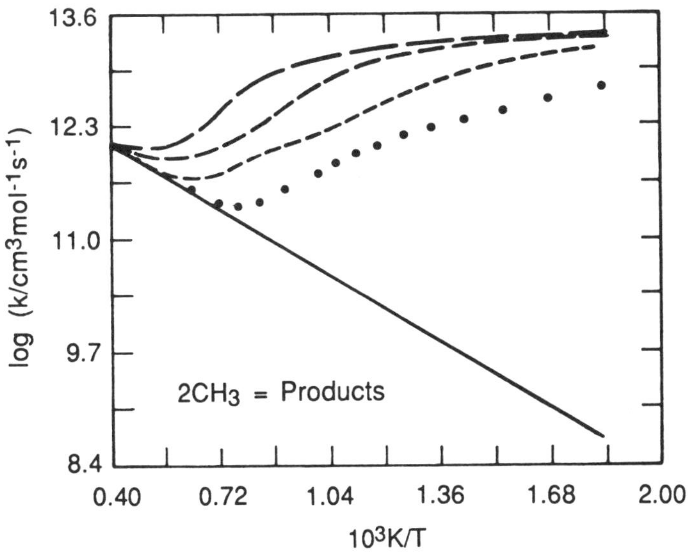

Figure 13. Arrhenius Plot for 2 $CH_3 \rightarrow$ Products. Lines correspond to pressures of 0 (solid line), 1, 10, 100, and 760 Torr of Argon.

IV. MEASURING RATE CONSTANTS

A. Gas Phase Applications and Experimental Techniques

An experimental technique developed in our laboratories, and now used widely, involves the use of a Knudsen cell reactor, that is, a reactor in which pressure in the millitorr range ensures that gases are in the molecular flow regime and that their lifetimes in the reactor are governed by the size of the escape aperture. This puts all measurements on an absolute basis. Our earliest experiments were studies of unimolecular decompositions, but fast metathesis reactions at temperatures where free radicals could be thermally generated were examined as well. Products were detected using residual gas analysis via mass spectrometry (now usually replaced by modulated molecular beam sampling, which yields a better signal/noise ratio). The experimental technique was called VLPP (Very Low Pressure Pyrolysis). It has been our habit, as variations on the technique have been developed, to change the initials used to describe the experimental technique. Thus the use of photolytic in situ radical production became VLPØ (Very Low Pressure Photolysis) and the use of this type of experiment to study gas-surface interactions has been called VLP/S (Very Low Pressure Surface). Others have coined the acronym VLPR (Very Low Pressure Reactor).

VLPP—Very Low Pressure Pyrolysis

Knudsen Cell Kinetics. In a simple VLPP experiment, molecules of reactant gas are allowed to flow under steady state conditions into a Knudsen cell reactor (Figure 14) made of fused silica (or other appropriate material) which is electrically heated in a controlled and uniform way. The flow rate is confined to a range such that the mean free path of molecules within the cell exceeds the exit aperture diameter by a factor of ca. ten. This guarantees molecular effusion in the exit beam. Usually the mean free path is kept longer than the vessel diameter so that most collisions of reactant molecules are with the walls. Under these conditions, "stirring" is very rapid and the reactor fulfills the kinetics requirements for behaving like a well-stirred reactor. That is, mixing of the entering gas with residual gas takes a time which is very small compared to t_r, the total residence time in the system. Each

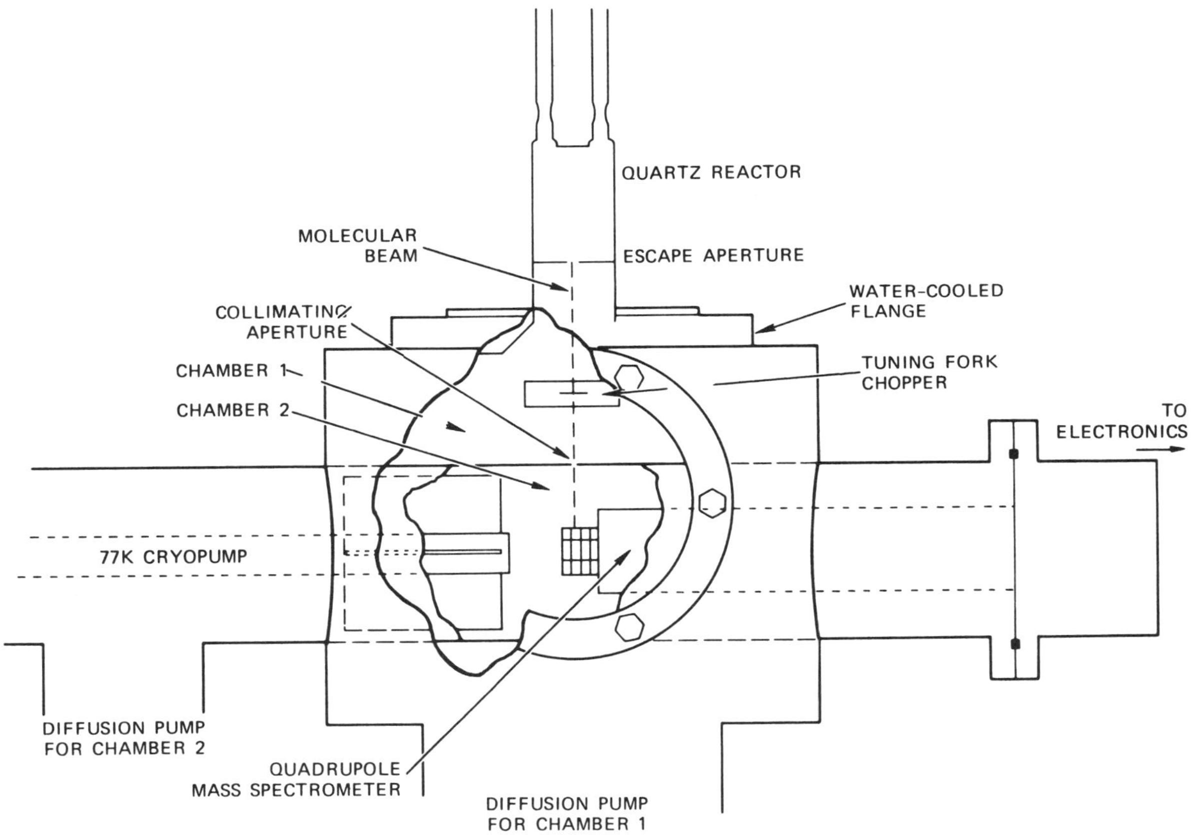

Figure 14. Schematic of GC-MS monitored Knudsen cell.

Knudsen cell reactor of the type depicted in Figure 14 is characterized by a collision number,

$$Z_w + A_w/A_h \tag{87}$$

where A_w is the area of the walls of the reactor, A_h is the area of the escape aperture, and Z_w is the average number of collisions with the walls made by a molecule during its residence in the cell. The residence time, t_r, of the average molecule of mass M in the reactor is given by the Knudsen equation as:

$$t_r = \frac{4V}{\bar{c}A_h} \; s^{-1} \tag{88}$$

where V is the volume of the reactor in cm^3, and $\bar{c}$ is the mean unimolecular speed given by kinetic theory as:

$$\bar{c} = \left(\frac{8RT}{\pi M}\right)^{1/2} = 1.46 \times 10^4 \left(\frac{T}{M}\right)^{1/2} \text{cm/s} \tag{89}$$

It is useful to define a first-order rate constant, k_{em}, which characterizes the rate of escape of a species of molecular weight M from the reactor:

$$k_{em} \equiv \frac{1}{t_r} = 3.65 \times 10_3 \frac{A_h}{V} \left(\frac{T}{M}\right)^{1/2} \tag{90}$$

The frequency of collision of the average molecule with the reactor walls, ω, is given by:

$$\omega \equiv \frac{Z_w}{t_r} = Z_w k_{em} = \frac{\bar{c}A_w}{4V} \tag{91}$$

which depends only on the reactor geometry and not on its aperture size.

If F_M is the flux of molecules M into the reactor cell, then under steady state conditions, with no reaction going on, the steady state concentration of M in the reactor is given by:

$$[M]_{ss} = \frac{F_M}{Vk_{em}} = \frac{R_M}{k_{em}} \tag{92}$$

V is the volume of the reactor and $R_M = F_M/V$ is the rate of introduction of M into the reactor, while F_M is the flux. Typical fluxes have ranged from 10^{14} to 10^{17} molecules/s; typical reactor volumes are ca 30–500 cm^3.

Unimolecular Decompositions. Let A be a molecule which can undergo an irreversible unimolecular decomposition whose stoichiometry is:

$$A \xrightarrow{1} pP + qQ \tag{93}$$

If k_1 is the first-order rate constant for this process, and k_{ea} the first-order rate constant for A to escape from the vessel, then the steady state concentration of A is:

$$[A]_{ss} = \frac{R_A}{k_1 + k_{ea}} \tag{94}$$

and for a typical stable product, P, the steady state concentration is:

$$[P]_{ss} = \frac{pk_1[A]_{ss}}{k_{ep}} \tag{95}$$

Defining: $[A]_o \equiv \dfrac{R_A}{k_{ea}} = [A]_{ss}$ when $k_1 = 0$

$$f \equiv \frac{[A]_o - [A]_{ss}}{[A]_o} = \text{fraction of } A \text{ decomposed} \tag{96}$$

$$F \equiv \frac{f}{1-f} = \frac{[A]_o - [A]_{ss}}{[A]_{ss}} = \frac{k_1}{k_{ea}} \tag{97}$$

The essence of the technique can be ascertained from the understanding that equation (94) describes $[A]_{ss}$ as resulting from the competition between chemical reaction (k_1) and escape from the reactor (k_{ea}). VLPP is thus a relative rate measurement technique, wherein one of the processes is not a chemical reaction, but rather a physical process whose rate constant can be computed from easily known parameters of the experiment, or measured as pressure in the reactor, thus yielding absolute values of chemical rate constants.

The signal measured at the mass spectrometer detector can be calibrated as a function of the flow rate R_M for each reactant and product species M, and at constant temperature should just be proportional to the flux of M reaching the detector. If this signal is I_M, then:

$$I_M = \alpha_M k_{em}[M]_{ss} \tag{98}$$

and

$$I_A^{\circ} = \alpha_A k_{ea}[A]_o = \alpha_A R_A \tag{99}$$

and from equation (97):

$$F = \frac{[A]_o - [A]_{ss}}{[A]_{ss}} = \frac{I_A^o - I_A}{I_A} = \frac{k_1}{k_{ea}} \qquad (100)$$

F is an experimentally measured quantity, as is k_{ea}. The constant k_1 is then calculated from equation (100). The precision of measurement is greatest for $F = 1$ and least (unless measuring product) near $F = 0$ or when F is very large. We usually confine our measurements to $1/9 \leq F \leq 9$ (i.e., $0.1 \leq f \leq 0.9$).

Relationship of k_1 to $k_1\infty$. The rate constants k_1 (and, of course, those for the reverse reactions) will be a function of ω, the frequency of collision of the average molecule with the walls of the reactor. This is a result of the fact that the frequency of wall collisions is not sufficient to maintain a Boltzmann distribution of energy in the reacting molecules and thus the rate constant measured is lower than $k_{1\infty}$, the rate constant for $\omega \rightarrow \infty$.

The relationship between k_1 and $k_{1\infty}$ is given by unimolecular rate theory, and it is possible to predict k_1 if the Arrhenius parameters of $k_{1\infty}$ are known. Accordingly, if the value of the Arrhenius A-factor is known, the value of the activation energy may be determined from a fit of k_1 versus T to a computed curve. Details of such determinations of activation energies are given in several publications.

In general, A-factors for unimolecular processes can be estimated with considerable accuracy if the process is unambiguous. When, on the other hand, doubt exists as to whether a given process is occurring via, say, bond homolysis or molecular elimination, estimated A-factors may differ by ca $10^{3\pm1}\ s^{-1}$. Because we are in the fall-off regime, either model will often give a good fit to the experimental k versus T curve, albeit, of course, for different values of E_∞. Often VLPP can answer mechanistic questions of this type via mass spectral detection of intermediates, since few, if any, secondary reactions will have occurred.

There are many examples, from SRI and other laboratories, of the use of VLPP for simple unimolecular studies. We have not pursued such studies for several years, our most recent work having been the determination of the alkyl C-H bond dissociation in diphenylmethane (82.3 kcal/mole) and thus the extra stabilization energy in diphenyl methyl radical when compared with α-phenylethyl radical (4 kcal/mole).[66]

d. Radical-Molecule Reactions. Since the VLPP experiment is simply a way of measuring the competition between first order loss by chemistry and by escape, bimolecular reactions can be measured as well, if a pseudo first-order loss rate competes with escape. This condition can be achieved for reactions with rate constants between collision frequency and about 10^{-5} of collision frequency. This restricts the technique to fairly rapid reactions, including the reactions between free radicals and hydrogen (or deuterium) halides. These reactions are very important, since they have been used to determine prototypical bond strengths and radical heats of formation, the values of which have great bearing on many current chemical problems, from combustion to atmospheric chemistry.

When we first studied these reactions, radicals were produced using thermal precursors. This limited our experiments to temperature ranges high enough to pyrolyze species such as azo compounds, but low enough to prevent radical decomposition.

The mechanism and associated algebra are as follows:

R	$\rightarrow$	Flow In	F_R
R	$\rightarrow$	Escape	k_{er}
HX	$\rightarrow$	Flow In	F_{HX}
HX	$\rightarrow$	Escape	k_{ehx}
$R + HX$	$\rightarrow$	RH + X	k_a
RH	$\rightarrow$	Escape	k_{erh}

Solving for the steady state concentrations of the various species:

$$[RH] = k_a[R][HX]/k_{erh} \tag{101}$$

$$[R] = F_R/(k_{er} + k_a[HX]) \tag{102}$$

$$[HX] = F_{HX}/(k_{ehx} + k_a[R]) \tag{103}$$

Substituting for $[R]$ and $[HX]$ leads to:

$$[RH] = \frac{k_a F_R F_{HX}}{k_{erh}\,\{k_{ehx}(k_{er} + k_a\,[HX]) + k_a F_R\}} \tag{104}$$

The initial radical concentration (before the introduction of HX) is:

$$[R]_0 = F_R/k_{er} \tag{105}$$

$$\frac{[R]_0}{[RH]} = \frac{k_{erh}}{k_{er}} \left[\frac{k_{er}\, k_{ehx}}{k_a F_{HX}} + 1 + \frac{F_R}{F_{HX}} \right] \tag{106}$$

Since the flux out of the reactor is given by $F_Y = k_{ey}[Y]$ for any Y, and since F_R/F_{HX} is about zero:

$$\frac{F_R}{F_{RH}} \equiv \frac{1}{f} = \frac{k_{er}k_{ehx}}{k_a F_{HX}} + 1 \tag{107}$$

The quantity F_R may be measured from the flux out of the reactor of a cogenerated species that undergoes no subsequent chemistry or by adding enough HX to completely titrate the radicals, while measuring the flux of RH out of the reactor, viz:

$$[RH]_\infty = \frac{F_R}{k_{erh}} \quad or \quad F_{RH_\infty} = F_R \tag{108}$$

This analysis suggests that a plot of $1/f$ versus $1/F_{HX}$ will yield a straight line with the slope of which will yield k_a, since the escape constants are known. An early study of this type[67] using 2,2′-azoisobutane as a thermal source of t-butyl radicals was restricted to the temperature range 644–722 K. The lower end corresponds to the lowest temperature at which we could produce sufficient radical concentration and the upper end represents the temperature at which the radicals begin to thermally decompose. In more recent experiments, we have produced the radicals photolytically and have thus decoupled the radical production from the ambient temperature of the reactor. (See below.)

Radical–Radical Reactions. Reactions between radicals are usually rapid and occur on the correct time scale for VLPP measurement. We have approached these reactions in the following two ways:

1. In cases where a unimolecular process is a reversible bond scission reaction (e.g., 1,5-hexadiene [biallyl{BA}] ⇄ 2 Allyl Radicals) the equation for a unimolecular rate constant is modified to:

$$\frac{k_{\text{decomp}}}{k_{eba}} = \frac{F}{1 - 2K[BA]F} \tag{109}$$

where $K = k_{\text{decomp}}/k_{\text{recomb}}$ and rearrangement gives:

$$\frac{1}{F} = \frac{k_{eba}}{k_{\text{decomp}}} + 2K[BA]F \tag{110}$$

A plot of $1/F$ versus $[BA]F$ should be a straight line from whose slope and intercept K and k_{decomp} can be determined. Of course this yields k_{recomb} as well. The value of k_{recomb} determined this way has to be corrected via a transition state model and RRKM theory in order to deduce the high pressure limiting value. We found[68] that allyl radical combination was rapid and seemingly unaffected by the delocalization of the unpaired electron in the radical.

2. In general if we can generate radicals from the precursor RXR, we can write a mechanism as follows:

RXR	$\rightarrow$	Flow In	F_{RXR}
RXR	$\rightarrow$	Escape	k_{erxr}
RXR	$\rightarrow$	$2R + X$	k_1
$2R$	$\rightarrow$	R_2	k_{rec}
$2R$	$\rightarrow$	RH + Olefin	k_{disp} (representative of disproportionation reactions)
R_2	$\rightarrow$	Escape	k_{er2}
R	$\rightarrow$	Escape	k_{er}

If f is the fraction of RXR decomposed, the $2fF_{RXR}$ is the initial flow of radicals into the reactor. Since radicals recombine, disproportionate and escape:

$$2fF_{rxr} = 2(k_{rec} + k_{disp})[R]^2 + k_{er}[R] \quad (111)$$

Given that the recombination product comes only from radical recombination, the flow out of the reactor of the dimer $= F_{R2} = k_{er2}[R_2] = k_{rec}[R]^2$, we can now write:

$$fF_{rxr}/F_{R2} = (k_{rec} + k_{disp})/k_{rec} + k_{er}/2k_{rec}^{1/2}F_{R2}^{1/2} \quad (112)$$

A plot of the left hand side of the above equation versus $F_{R2}^{1/2}$ will yield a straight line whose intercept and slope will allow the determination of both the recombination and disproportionation rate constants. This method, also limited to temperatures where the radicals may be thermally produced and not decomposed, has been used in several studies. Once again the use of photolytic radical production decouples the radical production from ambient temperature of the reactor.

We applied this method to the determination of the rate constant for the combination of perfluoromethyl radicals, using hexafluoroazomethane as the radical precursor.[69] We were able to perform measurements between 932 and 1107 K, which had to be corrected via a transition state model and RRKM theory to the high pressure values. We also incorporated available room temperature values into this study and, although they span almost one order of magnitude, we were able to distinguish between transition state models of the conventional vibrational type and those in which the transition modes are considered to be hindered rotations, showing that the latter made more physical sense.

VLPØ—Very Low Pressure Photolysis

More than ten years ago we realized that the Knudsen cell experiments of VLPP were well suited for modification to make use of the fact that radicals of interest could be generated with both infrared and excimer lasers. The kinetic schemes discussed above with regard to thermal radical generation, are equally applicable to photolytic generation. In addition, we take advantage of the fact that radicals generated in a Knudsen cell reactor make thousands of collisions with the reactor surface prior to reacting. Thus, even though they may have been created with excess internal energy, they are accommodated to the ambient reactor wall temperature before reacting. Of course, if the walls were reactive in this type of experiment, homogeneous processes could be obscured. We have taken pains to alternatively ensure that these wall reactions are absent or to study specific heterogeneous processes.

Activation Energies of R + HX Reactions. In an experiment[70] that extended the earlier study mentioned above on the reaction of *t*-butyl radicals with DI, we photolyzed 2,2′-azoisobutane with radiation at 351nm using a XeF excimer laser delivering in excess of 100 mJ/pulse. The pulse repetition frequency of the laser was sufficiently rapid (10–50/sec) so that the experiment effectively reached a steady state, since the escape frequency is of the order of seconds. The reactor could be heated as well and, since radical production is unrelated to wall temperature and wall collisions ensure accommodation to the wall temperature, we were able to make rate constant measurements between room temperature and about 410 K. These were then combined with our earlier measurements near 700 K to extract Arrhenius parameters for the D-atom metathesis of:

$$\log (k^D/M^{-1}s^{-1}) = 9.60 - 0.45/2.30T \tag{113}$$

A small correction (as well as some measurements with HI) leads to slightly different parameters for the isotopic reaction:

$$\log (k^{\mathrm{H}}/M^{-1}s^{-1}) = 9.73 - 0.40/2.30T \qquad (114)$$

The important fact is that we find that reactions of radicals with H(D)I have small positive activation energies. There are recent studies[71] that claim the activation energies for reactions such as these with both HI and HBr are negative. We have therefore recently repeated our work using DBr[31] and still conclude the opposite.

A good deal of the differences in this area have to do with the nature of heterogeneous interferences and with the sampling methods. We feel that our results reveal some heterogeneous first-order radical loss that we can actually measure in the overdetermined situation that comes about from having multiple aperture sizes. We also show that one has to be very careful to avoid a second-order heterogeneous pathway for the reaction of interest, even in a fluorocarbon coated reactor. (See Figures 3 and 4 of Reference 31 and discussion therein.) Another concern is that other studies may be complicated by the internal energy imparted to the radicals when they are photolytically produced. This is particularly true when 248 and 193 nm radiation is used to produce the radicals. The VLPØ measurement occurs after thermal accommodation and so is unaffected by this problem. With the hope of further clarifying the situation, we are planning to study some of these processes with in situ detection of the radical, probably using REMPI (Resonance Enhanced Multi-Photon Ionization).

Radical Radical Reactions. The rate constant for perfluoromethyl radical combination was determined near room temperature by using either infrared photolysis by a CO_2 laser at 1076 cm^{-1} or an excimer laser at 248 nm in conjunction with equation 4.26 above.[72] Values were independent of the nature of the photolysis and, after a small correction for pressure dependence, still fall at the lowest end of the spectrum of measured values.

Photophysics. During the flurry of excitement that accompanied the realization in the late 1970s that molecules could be collisionlessly dissociated by infrared multiphoton dissociation, we turned our attention to some of the photophysical events associated with these phenomena.[73] VLPØ was a technique wherein the molecular consequences of infrared photolysis could be detected mass spectrometrically. We investigated questions having to do with distribution of reacting species and energy transfer, but have not pursued these further since the editor of this volume moved from SRI to the University of Michigan.

VLPØ With In Situ *Real-Time (REMPI) Detection.* The various VLPX methods discussed above rely on steady state experiments with detection of effusing species. We have recently added the capability of measuring certain transient species in the reactor using Resonance Enhanced MultiPhoton Ionization (REMPI).[74] This has been applied to CF_3 radicals in a study where the radicals were produced from the IRMPD of hexafluoroacetone and the radical concentration was followed in real-time between laser pulses. Data from these experiments yields a value for the combination reaction of CF_3 radicals in excellent agreement with that our 1986 results.[72] This indicates that the rate of wall loss for these radicals has a relatively small value (ca. 1 s^{-1}) for this reactor with its gold-coated walls and quartz and salt windows.

Laser Pyrolysis

There are two very important limitations of the VLPX technique for the determination of homogeneous thermal rate constants. These are the intercession of heterogeneous processes and the necessity of correcting unimolecular data through RRKM theory. In order to circumvent these difficulties, we have spent some time in developing the technique that is usually called Laser Pyrolysis, although we have also coined the term Laser Powered Homogeneous Pyrolysis (LPHP).

The essence of the technique involves the heating of an inert infrared absorber, usually a few torr of SF_6 or SiF_4, in the presence of about 100 torr of a "bath gas" such as Ar, CO_2, CH_4, and so on. The absorber exchanges energy with the bath gas on a microsecond time scale, and a heated zone in the center of the reactor is created for about ten microseconds before diffusional cooling on the millisecond time scale dissipates it. If a substrate of interest is also present in the mixture, it will react in the heated zone removed from the reactor surfaces.

GC/MS. In the earliest versions of the laser pyrolysis technique in our laboratory the analysis was performed by batch sampling with either gas chromatographic or mass spectrometric detection of products. In more recent studies gc/ms detection has been utilized. In these experiments the relative rate technique, used often by Tsang[75] in single pulse shock tube experiments, is employed. We add to the sample mixture a small amount of a substrate whose decomposition Arrhenius parameters are well known and such that it will decompose in the same temperature range as the compound under investigation. The analysis of decomposition products of the standard and the substrate of interest yield the rate parameters being sought. If we expect chemical problems from radicals that may be formed during the decomposition, we add radical scavengers, such as toluene, to obviate fast chain reactions. A schematic of a typical apparatus is shown in Figure 15.

Bond Energies of Metal Carbonyls. We have used the laser pyrolysis technique with some success in determining the bond dissociation energies of several metal carbonyls, compounds which are notorious for their surface catalyzed instability.[76] An apparatus similar to that shown in Figure 15 was used, although no GC was attached in this study. A slow flow through the cell of a few torr of SF^6 in 100 torr of N_2 bath gas, together with the carbonyl and the standard, was set up and maintained. The gas was then irradiated with a pulsed CO_2 laser, creating a rapid cycle of heating and diffusional cooling, with the reaction occurring in the temperature regime of 750–850 K. The pulses lead to a steady state of reactants and products downstream, which can be monitored mass spectrometrically. The reaction yield, kt_r, is given by:[76]

$$kt_r = -\ln [1 - (A_0/A - 1)(V_T/V_R)(t_L/t_F)] \tag{115}$$

where A_0 and A are the initial and irradiated steady-state concentrations of substrate determined mass spectrometrically, V_T/V_R is the ratio of cell to irradiated volume, and t_L/t_F is the ratio of the time between laser shots and the measured flow lifetime. Given the Arrhenius form for the standard and substrate rate constants, $k = A\ e^{-E/RT}$, and plotting log k_1t_r versus log k_2t_r, we get a line described by:

$$\log k_1t_r = \log A_1 + (1-E_1/E_2) \log t_r - (E_1/E_2) \log A_2 + (E_1/E_2) \log k_2t_r \tag{116}$$

The slope of this comparative rate plot is E_1/E_2, so that knowledge of the activation energy for the standard E_2, furnishes that of the unknown. From the intercept and known values of t_r and A_2, the value of A_1 can also be determined.

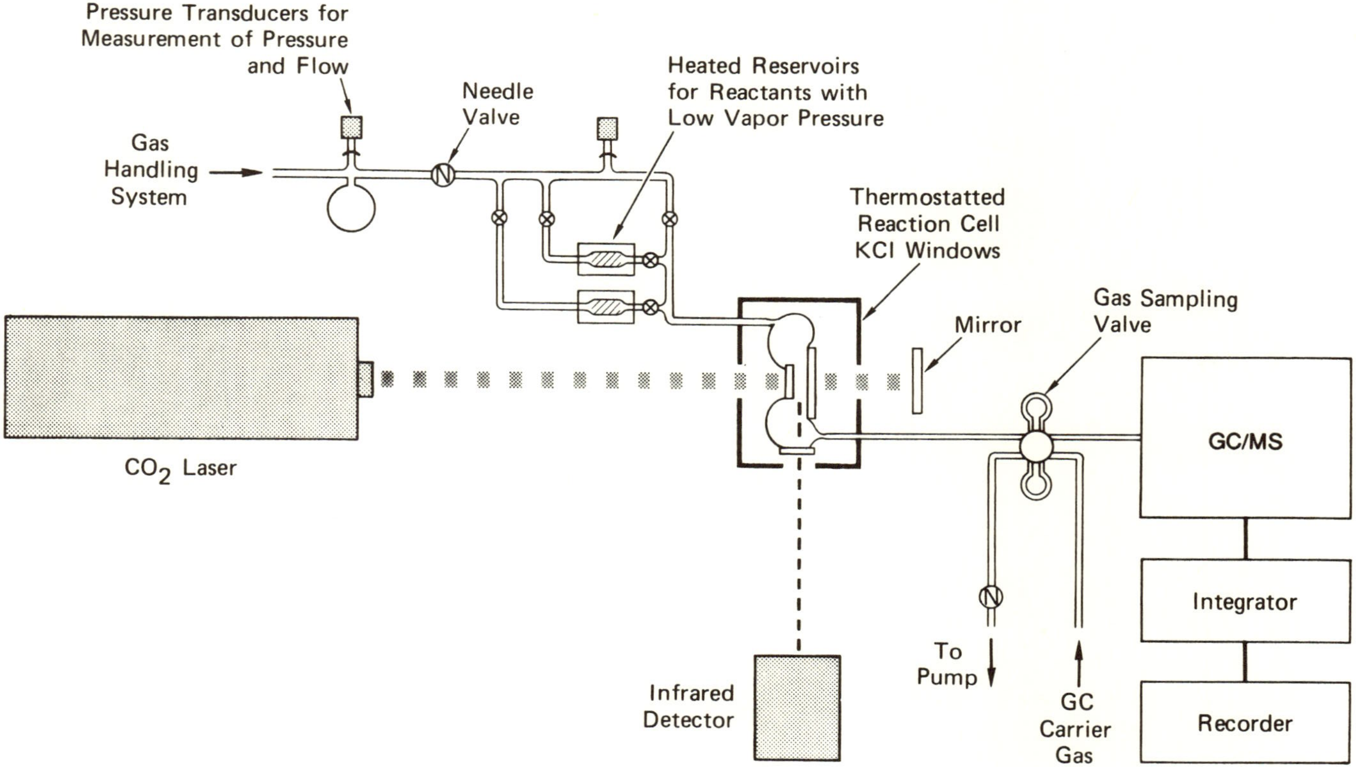

Figure 15. Schematic of GC-MS monitored LPHP apparatus.

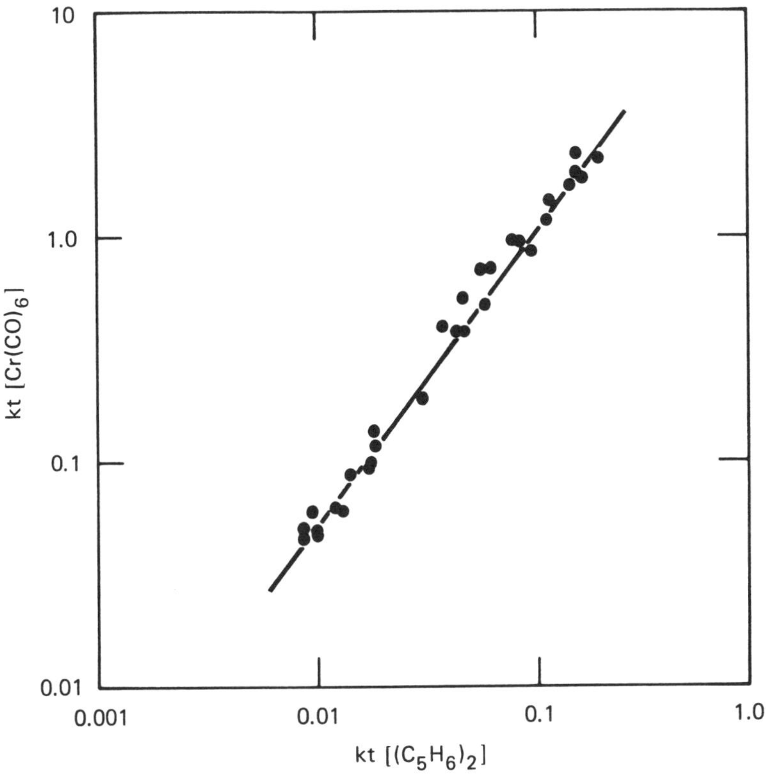

Figure 16. Logarithmic comparative rate plot for chromium hexacarbonyl vs. dicyclopentadiene composition. (Line is least-squares fit.)

In our experiments, $Cr(CO)_6$ decomposition was measured relative to the reverse Diels–Alder decomposition of dicyclopentadiene, while the other carbonyls were measured relative to the chromium compound. Typical data is shown in Figures 16 and 17 while Tables 10 and 11 summarize the results.

Energetic Materials: Decomposition of Dimethylnitramine [$(CH_3)_2$ N-NO_2]. To help gain an overall understanding of what controls the decomposition of cyclic nitramines and other high energy materials, we have more recently applied LPHP to the decomposition of dimethylnitramine (DMNA).[1] The apparatus is shown in Figure 15, and the experimental procedure similar to that described above.

The initial step in this reaction might reasonably be expected to be

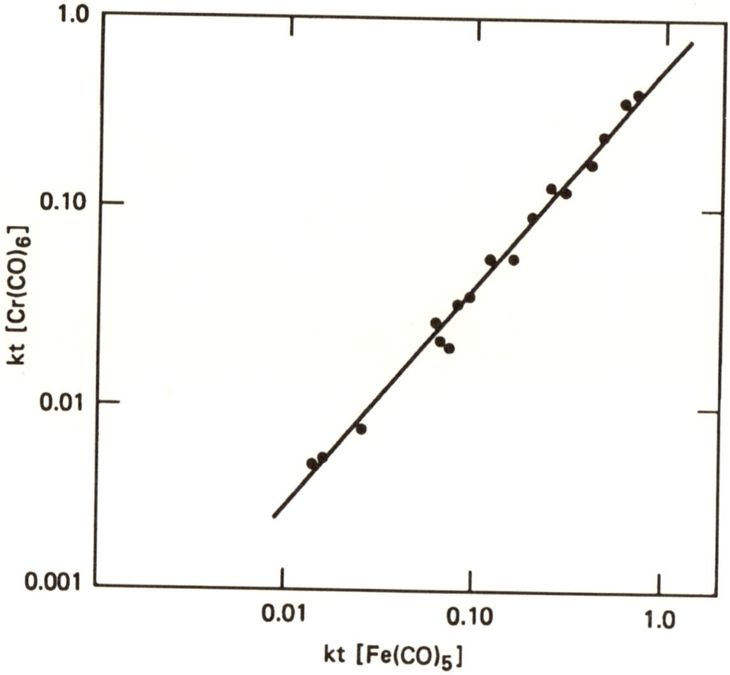

Figure 17. Logarithmic comparative rate plot for iron hexacarbonyl vs. chromium hexacarbonyl decomposition. (Least-squares line slope is 1.15 ± 0.03.)

N-N bond fission: $(CH_3)_2N\text{-}NO_2 \rightarrow (CH_3)_2N + NO_2$. Indeed, experiments conducted with added NO showed dimethylnitrosamine, $(CH_3)_2$ N-NO (DMNO), as a major product, consistent with trapping of the dimethylamino radical with NO. In order to determine and limit the importance of secondary processes, reactions were also carried out

Table 10. Summary of Laser Pyrolysis Results

Compound	*T, K*	E_a, *kcal/mol*	*log A*	k/k_∞	*log* A_∞	E_a^∞ *kcal/mol*	$\Delta H°$ *kcal/*
$Cr(CO)_6$	740–820	45.5 ± 1.8	17.2				
		(35.3)	(15.4)	(0.75)	(15.5)	(35.3)	(36.
$Mo(CO)_6$	670–760	38.9 ± 2.4	15.4 ± 0.8	0.80–0.79	15.6	39.0	(40
$W(CO)_6$	745–810	44.3 ± 2.8	15.5 ± 0.7	0.90–0.86	15.6	44.5	(46
$Fe(CO)_5$	670–780	39.6 ± 2.3	15.7 ± 0.8	0.88–0.80	15.8	40.0	(41

Note: Parameters in parentheses correspond to the first $Cr(CO)_6$ bond fission based on an estimated log A. See Referen for details.

Table 11. Metal Carbonyl Bond Dissociation Energies, kcal/mol

Compound	ΔH°_{298}	$\bar{E}$ *(Ref. 77)**	*Source*
$Cr(CO)_6$	36.8	26	LPHP
$Mo(CO)_6$	40.5	36	LPHP
$W(CO)_6$	46.0	43	LPHP
$Fe(CO)_5$	41.5	28	LPHP
	55 ± 12		Ref. 78
$Ni(CO)_4$	25 ± 2	35	Ref. 79
	22		Ref. 80

*$\bar{E}$ refers to the average CO—M bond strength in the molecule.

with varying concentrations of excess radical scavenger (toluene or cyclopentane), both with and without added NO. The decomposition rate was measured relative to the reaction 4-hydroxy-4-methyl-2-pentanone → 2 $(CH_3)_2CO$, and the results are shown in Figure 18. (The mass balance without added NO was not complete, and these parameters, although similar, are not considered as reliable.) Surprisingly, the

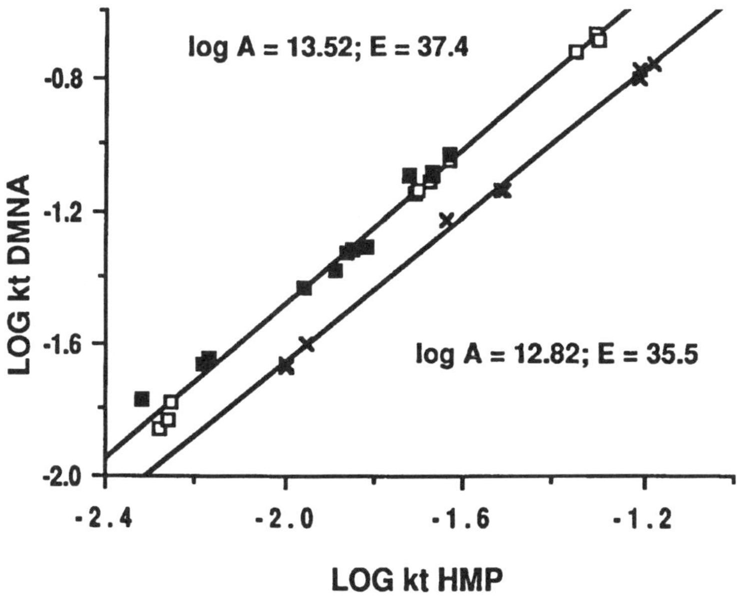

Figure 18. Comparative rate plot of dimethylnitramine (DMNA) versus 4-hydroxy-4-pentanone (HMP): (■) 50-fold excess of NO; (□) 100-fold excess of NO; (×) no additional NO, toluene as scavenger.

low Arrhenius parameters of DMNA decomposition are not consistent with simple bond fission. Further evidence for a more complex mechanism was that DMNO was found to be the major product even in the absence of NO and radical scavenger.

Since DMNO was a major product, the reaction of this compound was also studied, Figure 19. In contrast with DMNA, the rate parameters for this substrate are indicative of simple bond fission, $(CH_3)_2N\text{-}NO \rightarrow (CH_3)_2N + NO$. In order to explain the obviously more complex reaction of DMNA, we developed a numerical model with the key steps shown in Figure 20. The critical elements that allow the model to explain the observed product distributions and low Arrhenius parameters are the following:

1. N-NO_2 bond scission and NO_2-ONO rearrangement that are roughly competitive at 900 K (reactions 1 and 3).
2. Rapid, but reversible, thermolysis of the very weak O-NO bond in the N-nitrite (reactions 5 and 6).
3. Reverse of the nitro-nitrite rearrangement (reaction 4).

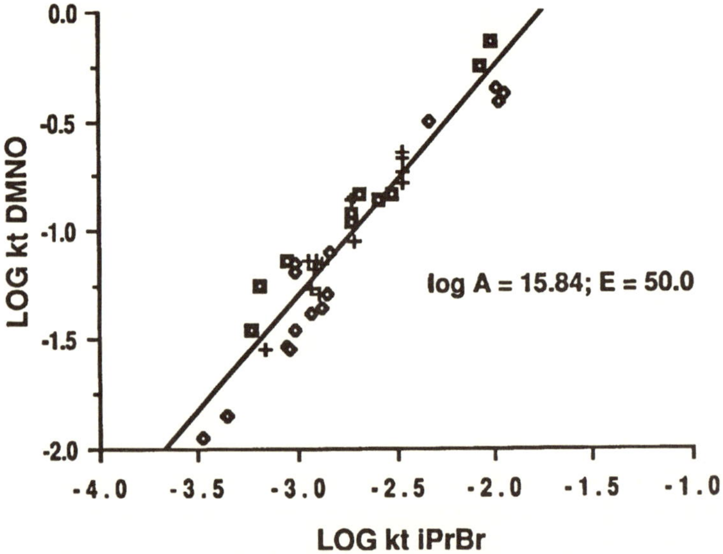

Figure 19. Comparative rate plot of dimethylnitrosamine (DMNO) vs. isopropyl bromide (*i*-PrBr). The different symbols represent experiments on different days.

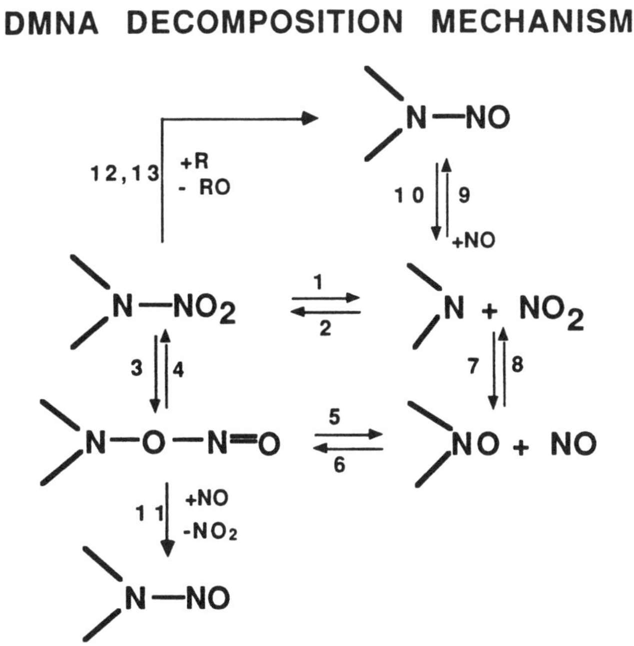

Figure 20. Schematic representation of the mechanism at DMNA decomposition in the presence of excess NO.

4. Attack of NO on the N-nitrite to displace NO_2 and directly yield the nitrosamine (reaction 11).

Our results thus suggest competing initial reactions and facile secondary reactions in the decomposition of the simple nitramine DMNA. Such information is clearly necessary to understand under what conditions the decomposition of this and more complex nitramines can become thermally self-sustaining.

MBMS (Molecular Beam Mass Spectrometry). The simple procedure described above does not allow the direct detection of transient intermediates. We have therefore built an apparatus designed to sample directly from a laser pyrolysis zone through a nozzle and skimmer into a mass spectrometer. Quantitative rate data is forgone in this arrangement, but transients are detectable and mechanistic information may be thus obtained. We have applied this system to dimethylnitramine, and were able to show[81] that NO, NO_2, dimethylamino radical, and dimethylnitroxyl radical were nascent products in the decomposition of DMNA. This further supports our mechanistic conclusions from the LPHP studies.

Laser Induced Fluorescence (LIF). In the modes discussed above we have followed unimolecular reactions with the help of a standard reaction, which obviates the need to measure the temperature and residence time. With the use of real-time techniques and a suitable method of temperature measurement, however, it is possible to study bimolecular reactions via LPHP.

High Temperature Abstractions by OH Radicals. Others at SRI have used laser pyrolysis to make OH radicals, which they have observed reacting with hydrocarbons by following the decay of the Laser Induced Fluorescence (LIF) signal from the radical.[82] In these studies the setup is as in Figure 21. The pulsed laser heats the slowly flowing mixture to 800–1400 K, depending on the laser energy and SF_6 concentration, and some OH is generated via thermal decomposition of H_2O_2. A dye laser perpendicular to the IR laser and tuned to the OH band at 310.7 nm is fired at an adjustable delay time Δt after the CO_2 laser. The time-dependent decay of the OH fluorescence is then monitored both with and without added RH. The variation in the first-order loss rate with reactant concentration yields the bimolecular rate constant, Figure 22.

The rate constants so determined must be associated with a temperature. This was accomplished by scanning the dye laser through various OH lines, yielding the rotational level populations from which a Boltzmann rotational temperature can be deduced. The vibrational temperature can also be determined by examining the relative populations of the $v = 0$ and $v = 1$ states. Both methods give the same result within experimental error, indicating that the OH is fully thermalized.

The above procedure was used to obtain direct high temperature measurements of the reaction of OH with methane, propane, and propene.[82] The results with propene, Figure 23, are particularly interesting. At low temperatures the dominant mechanism is addition, while, due to falloff, abstraction is the important pathway at higher temperatures. Isolation of this latter channel, difficult at low temperatures, is easily accomplished via laser pyrolysis and is illustrative of

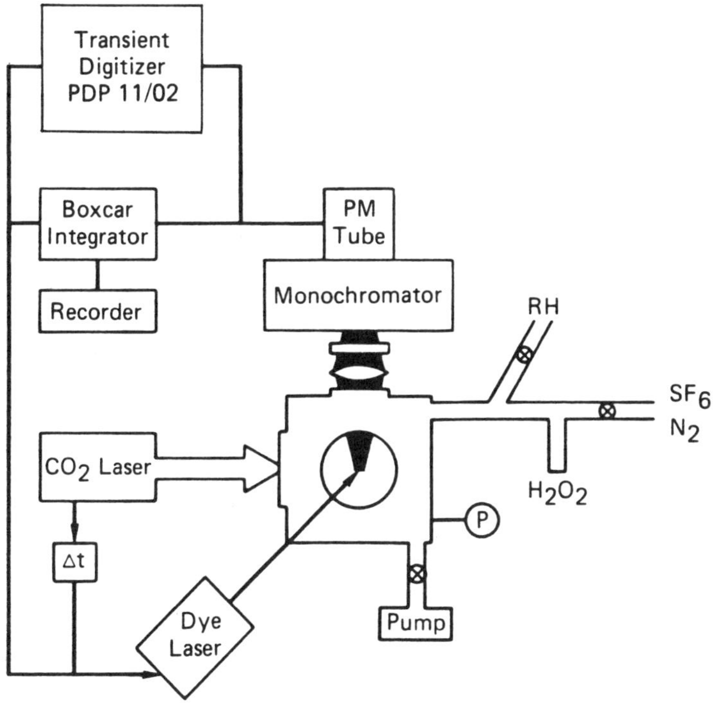

Figure 21. Schematic of the laser pyrolysis/laser fluorescence experiment. *P* represents a pressure gauge, circles enclosing crosses are flow control values, and RH is the chosen hydrocarbon reactant.

the potential of this method for the study of reactions at high temperatures.

In later studies, the reaction of OH with several butenes was examined,[86] with the somewhat surprising result that for these substrates the rate per allylic hydrogen is significantly faster than for propene.

REMPI. We are currently using Laser Pyrolysis to generate methyl radicals from precursors such as di-*t*-butyl peroxide and azomethane, while detecting the radical via Resonance Enhanced Multi-Photon Ionization (REMPI). As this is being written we hope to be studying the again controversial reaction, or lack thereof, between methyl radicals and oxygen molecules at about 1000 K.

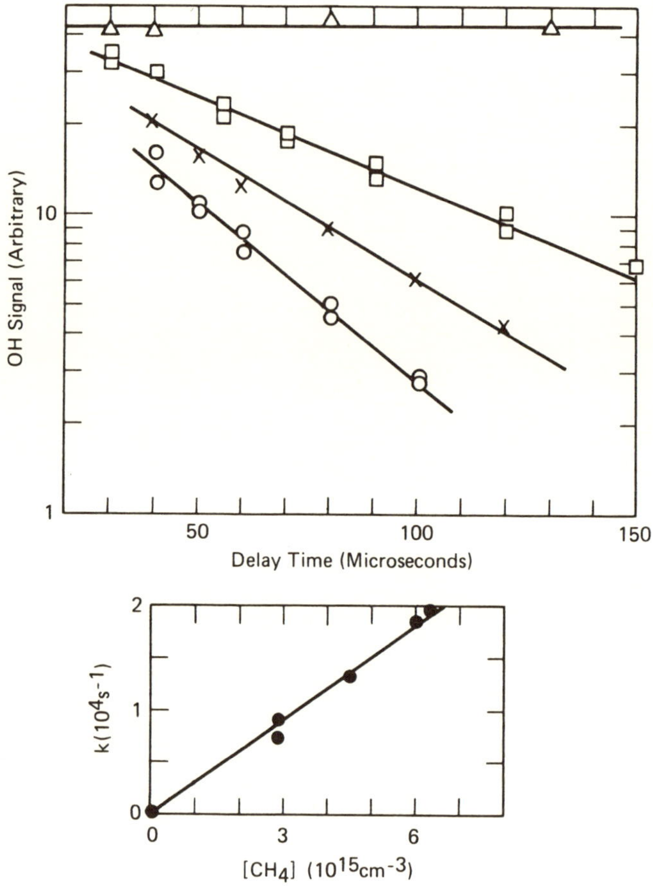

Figure 22. (top) Logarithmic plot of the decline in OH LIF signal vs. delay time for various partial pressures of added CH_4: (Δ), no CH_4; (□), two separate runs at 0.37 torr; +, 0.57 torr; (○), runs at 0.74 and 0.80 torr. The experiment was performed at 1240 ± 50 K in 47 torr of N_2 and 10 torr of SF_6. (bottom) Decay rate of OH vs. CH_4 concentration from the above data, with least-squares-fitted line.

B. Gas Surface Studies

Uptake and Sticking Coefficients

VLP/S. We have discussed the fact that VLPX techniques allow the measurement of the first order loss of some species in competition

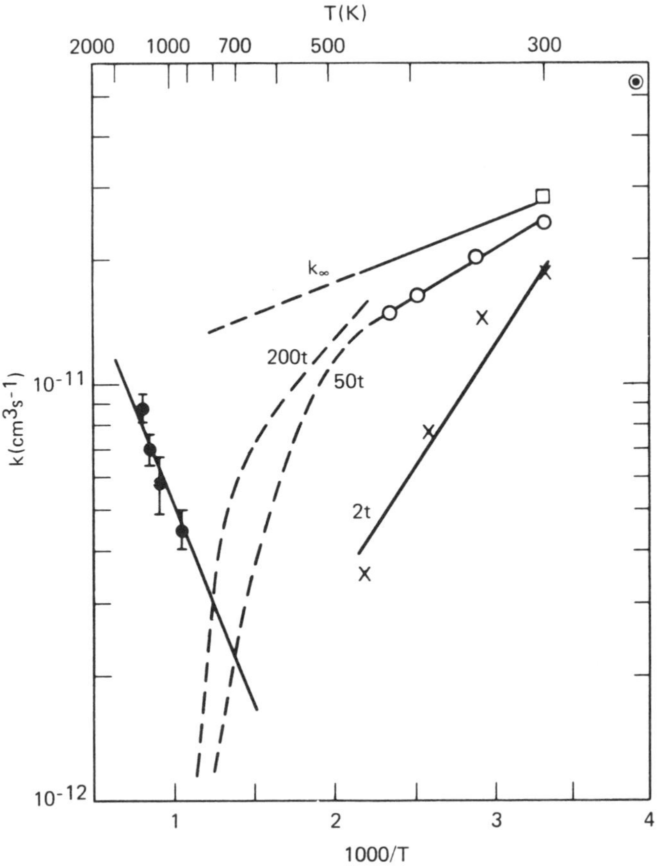

Figure 23. Arrhenius plot of the OH + C_3H_6 reaction: solid circles with errors bars giving 1 σ statistical uncertainty, this work; square, Reference 83; open circles, in 50 torr of He, Reference 84; crosses in 2 torr of He, Reference 85. The lines drawn are the results of the calculations described in the text.

with its escape through the aperture. In the same way, we can measure loss to a surface in the Knudsen cell. If the surface area is known, we can determine the loss per unit area of surface, a quantity that we call the "uptake" coefficient. We have sometimes called this the VLP/S technique.

A simple mechanism that is representative of the VLP/S experiment is shown in Table 12.

Table 12. Simplified Kinetic Scheme for the Reaction of a Gas on a Surface in a Knudsen Cell

	Process	*Rate*	*Description*
1.	$\rightarrow G$	F^i	Gas flow in
2.	$G + S \rightarrow G \cdot S$	$k_{trap}[G][S]$	Trapping
3.	$G \cdot S \rightarrow G + S$	$k_{evap}[G \cdot S]$	Evaporation
4.	$G \cdot S \rightarrow$ Prod.	$k_{rxn}[G \cdot S]$	Reaction
5.	$G \rightarrow$	$k_e[G]$	Escape to mass spectrometer

In this set of reactions, G represents a gas phase molecule, S an available surface site, and $G \cdot S$ a molecule trapped in a precursor state on the surface. Molecules trapped in a precursor state can either react on the surface to form products, or evaporate back to the gas phase. The pseudo–first-order rate constant for trapping, $k_{trap}[S]$, is given by $\alpha\omega$ where α is the "trapping" efficiency (the average likelihood that a molecule striking the surface will be bound in some manner), and ω is the total gas-surface collision frequency, ω(col/sec), obtained from kinetic theory as:

$$\omega = 1/4 <v> A_s(N/V) \tag{117}$$

where v is the average molecular speed (cm/s), A_s is the area of the surface (cm^2), and N/V is the reactant gas number density (molec/cm^3). The number density in a Knudsen cell is related to the reactant flux, F, and the size of the escape orifice, A_h, by equation (118).

$$N/V = 4\ F/A_h <v> \tag{118}$$

The experimental quantity that is measured is the net uptake efficiency, γ, defined by equation (119):

$$\gamma = \frac{\text{No. of reactant molecules lost (molec/s)}}{\text{No. of gas-surface collisions (col/s)}} \tag{119}$$

The number of reactant molecules lost is measured in our Knudsen cell as the quantity ($F^\circ - F$) where F° and F are the reactant fluxes (molec/s) out of the cell in the absence and presence of the surface, respectively.

Combining equations (117–119) the uptake coefficient may thus be defined:

$$\gamma = A_h/A_s\{(F^\circ - F)/F\} \tag{120}$$

In the literature, one often sees both α, the trapping efficiency, and γ referred to as the "sticking coefficient." We have avoided this ambiguous nomenclature. Applying a steady-state analysis to the equations in Table 12, equation (121) is obtained, where $[G^0]$ and $[G]$ are the steady state concentrations

$$\{ ([G^0] - [G])/[G] \} A_h/A_s = \gamma = \alpha \{k_{rxn}/(k_{\text{evap}} + k_{rxn})\} \quad (121)$$

of the gas in the absence and presence of the surface, respectively. It can be seen that $\gamma = \alpha$ only for the case where no reevaporation occurs, in keeping with the definition of γ as the *net* rate of sticking, while α refers to the probability of sticking on initial contact (generally an immeasurable quantity). When there is no barrier to adsorption via a precursor state, α is often taken to be unity. For systems where reevaporation does occur, the measured uptake efficiency will always be less than the trapping efficiency.

Surface Interactions. Interactions of molecules with surfaces may be strong or weak and may loosely be described as reactive (involving rearrangement of the atoms) or unreactive (involving physisorption or diffusion into the bulk):

Non-reactive—We have seen examples of very low values of the uptake coefficient, which presumably may be explained by rapid reevaporation compared to reaction or diffusion into the bulk. In such cases, of course, we can only say that the degree of reactivity is lower than our analytical sensitivity to loss of gaseous species. We have also examined situations where the gaseous species may diffuse into the bulk, controlled by their solubility, and have measured Henry's Law Coefficients.[87]

Reactive—Surface induced chemistry may involve the surface as catalyst for the decomposition of closed shell molecules or radicals, or the catalysis of effective bimolecular processes between closed shell and/or radical species, or reactions in which the surface is one of the reactive components. We have examined several of these possibilities that may be important in heterogeneous catalysis, etching, and heterogeneous atmospheric chemistry.

Steady-State Experiments. In order to easily measure the difference between mass spectral signals due to molecules that have encountered an active surface and those that have not, we use Knudsen cells that contain two chambers that may be isolated from each other. Figure

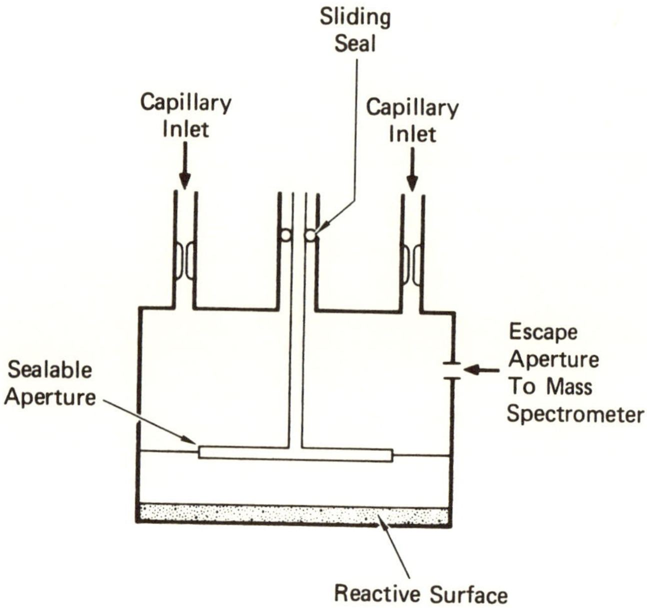

Figure 24. Schematic of a typical two chamber Knudsen cell used to study surface reactions.

24 depicts a schematic representation. The upper chamber is coupled to a differentially pumped mass spectrometer through an escape orifice with variable area. Reactant gases are routed through the upper chamber, initially bypassing the surface. The valve is then opened and, since the mass spectrometer intensity for a given molecule is proportional to its flux, F (molec/s), out of the Knudsen cell, the change in signal that occurs measures the flux of reactant gas that is lost to the surface. These experiments allow the direct measurement of uptake coefficients through the use of Equation 120. A reference experiment in which the surface of interest is absent is also done, of course, to make sure that interference from other reactor surfaces is not occurring. The mass spectrometer can also be used to detect and identify reaction products released into the gas phase.

Heterogeneous Reactions in the Atmosphere

In recent years many people have sought to understand the chemistry of the atmosphere, with an eye to predicting the effects of chemical additives on global warming, photochemical smog, ozone depletion, and so on. By the mid 1980s various failures in these models led people to suggest that it might be necessary to include heterogeneous processes.

Odd Nitrogen. One of the central questions concerns the fate of odd nitrogen. For instance, recombination of NO_2 and NO_3 radicals to give dinitrogen pentoxide, N_2O_5, is important throughout the atmosphere. Hydrolysis of N_2O_5 to nitric acid (equation 122), if rapid,

$$N_2O_5 + H_2O \rightarrow 2\ HNO_3 \tag{122}$$

could make this an important sink for NO_x. Although it is no surprise that this process proceeds as written, a direct reaction between closed-shell species in the gas phase is extremely slow. We were interested in the possibility that the hydrolysis could be surface catalyzed by small aerosol particles in the atmosphere.

To this end we have studied this process on several surfaces at room temperature and below. In early work,[88] we introduced carbonaceous material, in the form of commercial charcoal, into the second chamber of a two chamber Knudsen cell (see Figure 24).

If dry N_2O_5 is flowed through the upper chamber and the gate valve then opened, the N_2O_5 mass spectral intensity initially drops to about 25 percent of its starting value, but then returns to its original level after several minutes. This is indicative of physisorbtion followed by gradual saturation of the surface sites. At the same time, a slow increase in the flux of HNO_3 was observed in these experiments, most probably due to surface reaction of N_2O_5 with traces of water bound to the carbon. In contrast to the low production of HNO_3 in these experiments, when H_2O was flowed together with the N_2O_5, a rapid rise in HNO_3 production occurs upon exposure to the carbon surface. Since the homogeneous gas phase reaction is slow, this must be due to surface catalyzed hydrolysis.

In further experiments, the behavior of the product, HNO_3, with respect to the surface was examined and mass balances were obtained. We were able to show that several surface processes were important, including the stoichiometric conversion of H_2O and N_2O_5 to two nitric acid molecules (equation 123), physisorbtion of N_2O_5 and HNO_3 (equations 124 and 125), and surface-catalyzed redox reactions of HNO_3 to give NO and NO_2 (model reaction 126).

$$N_2O_5\ (g) + H_2O\ (g) \xrightarrow{\text{surface}} 2\ HNO_3\ (g) \quad (123)$$

$$N_2O_5 \leftrightarrows S{\cdot}N_2O_5 \quad (124)$$

$$HNO_3 \leftrightarrows S{\cdot}HNO_3 \quad (125)$$

$$HNO_3(g) \xrightarrow{\text{surface}} NO\ (g) + NO_2(g) \quad (126)$$

Thus, to the extent that atmospheric carbonaceous material is similar to the charcoal material used in these experiments, there are strong grounds for considering heterogeneous processes on such materials in atmospheric models.

The Antarctic Ozone "Hole." Another recent problem has been to explain the dramatic reduction in ozone levels observed in the spring over Antarctica. Chemical mechanisms for ozone destruction generally rely on the presence of "active" chlorine [e.g., atomic chlorine (Cl) and chlorine monoxide (ClO)] in the stratosphere. Homogeneous photochemistry alone, however, cannot produce the levels of active chlorine necessary to explain the observed ozone decrease. It seemed possible that surface-catalyzed reactions were the source of the required active chlorine. Reasonable candidates include reactions of chlorine nitrate with atmospherically available species such as H_2O or HCl (reactions 127 and 128).

$$ClONO_2 + H_2O \rightarrow HOCl + HNO_3 \quad (127)$$

$$ClONO_2 + HCl \rightarrow Cl_2 + HNO_3 \quad (128)$$

We began by studying these reactions on ice,[89] a surface present in the atmosphere of Antarctica in the form of polar stratospheric clouds (PSCs) (there are now thought to be at least two types of PSCs, Type 1 composed of mixtures of ice and nitric acid, and Type 2 which are largely ice). For this work, where we were interested in creating surfaces at polar stratospheric temperatures, we modified the reactor design to that shown in Figure 25. In this case an ice surface is formed in the lower chamber which is immersed in a cold methanol bath regulated to the desired temperature by a combination of liquid nitrogen cooling and resistive heating.

Figure 26 shows the mass spectra of the effluent of a flow of $ClONO_2$ both before and after exposure to ice at 185 K. In the before exposure spectrum, peaks at m/e = 30, 46, 51, and 53 correspond to $ClONO_2$, while others are due to water and the impurities Cl_2 (m/e = 70, 72, and 74) and Cl_2O (m/e = 51, 53, 86, 88, and 90). Upon

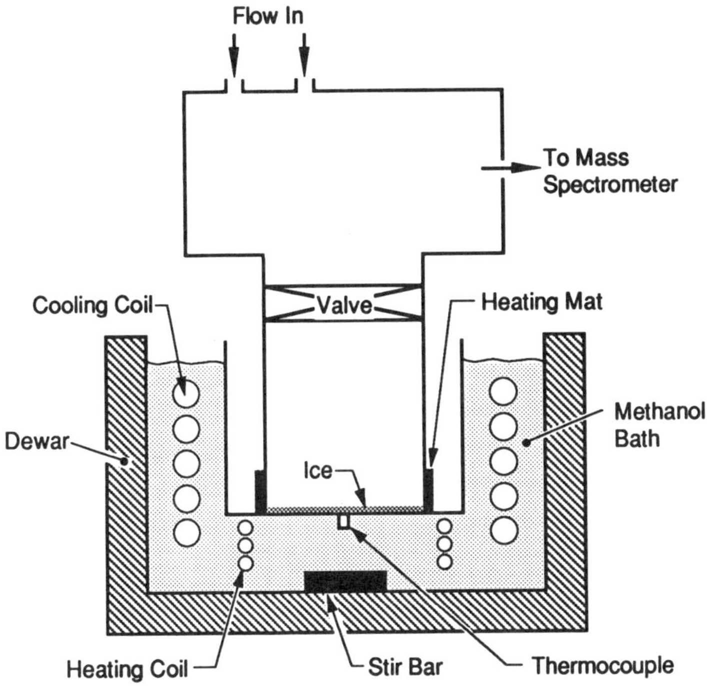

Figure 25. Schematic of apparatus used to study surface reactions at polar stratospheric temperatures.

exposure to ice, the near-disappearance of $m/e = 30$ and 46 shows that $ClONO_2$ is rapidly lost on ice. Characteristic peaks of the expected product of (117), HOCl ($m/e = 52, 54$), are present, but even more prominent is the growth in the Cl_2O signal. This suggests its formation via a secondary reaction. We have suggested:

$$HOCl + HOCl \rightarrow Cl_2O + H_2O \tag{129}$$

No gas phase HNO_3 ($m/e = 30, 63$) was detected. From previous studies in our laboratory, however, we suspected that any HNO_3 formed would remain bound to the surface. We were able to demonstrate this by warming the ice after reaction and finding nitric acid released from the ice. This is shown in trace B of Figure 27. Notice that if a blank experiment is done, with the ice replaced with a halocarbon wax surface, no HNO_3 is observed upon warming (trace A).

Experiments similar to the above were used to demonstrate that reaction 128 also occurs readily on ice. This led us to confirm that

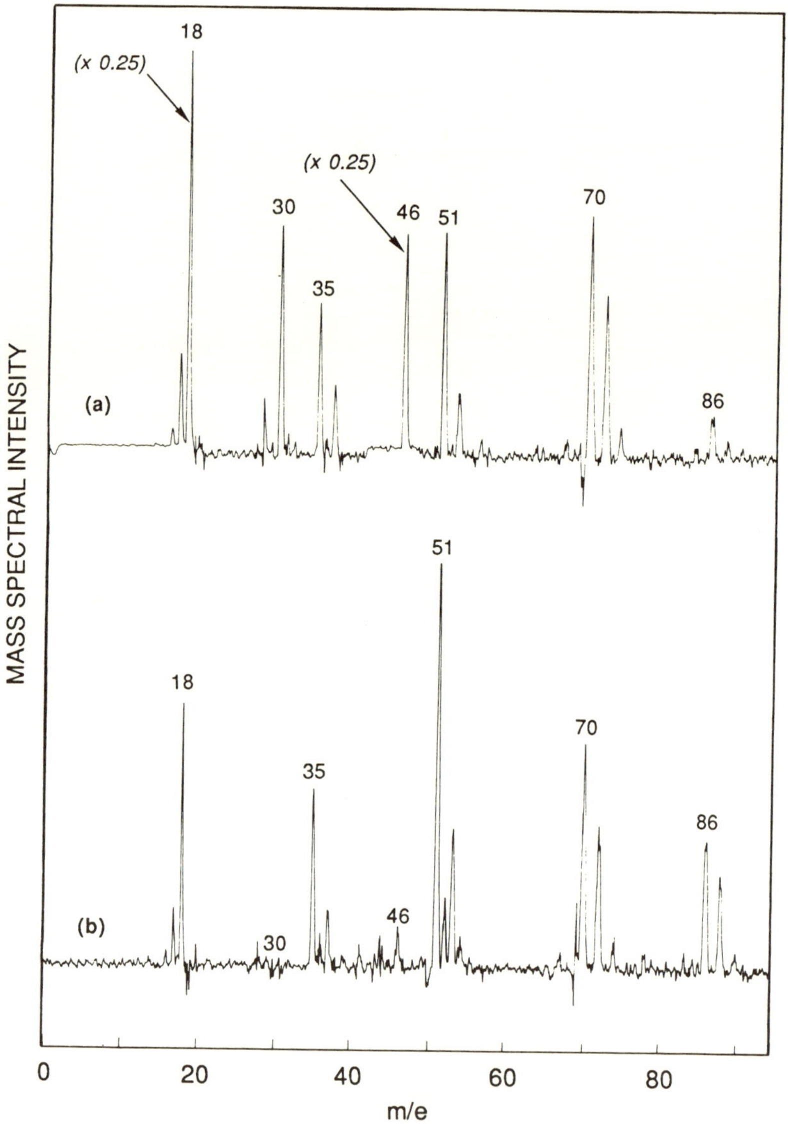

Figure 26. Mass spectrometer scans of the Knudsen cell effluent for (A) $ClONO_2$ before exposure to ice [$P(H_2O)$ = 0.9 mtorr, $P(ClONO_2)$ = 0.5 mtorr] and (B) $ClONO_2$ exposed to an ice surface at 185 K. The two scans are shown on the same intensity scale, except that mass peaks *m/e* 18 and 46 in (A) are shown at one quarter of their actual intensity. The chlorine-atom peaks *m/e* 35 and 37 are due to fragmentation of impurity Cl_2.

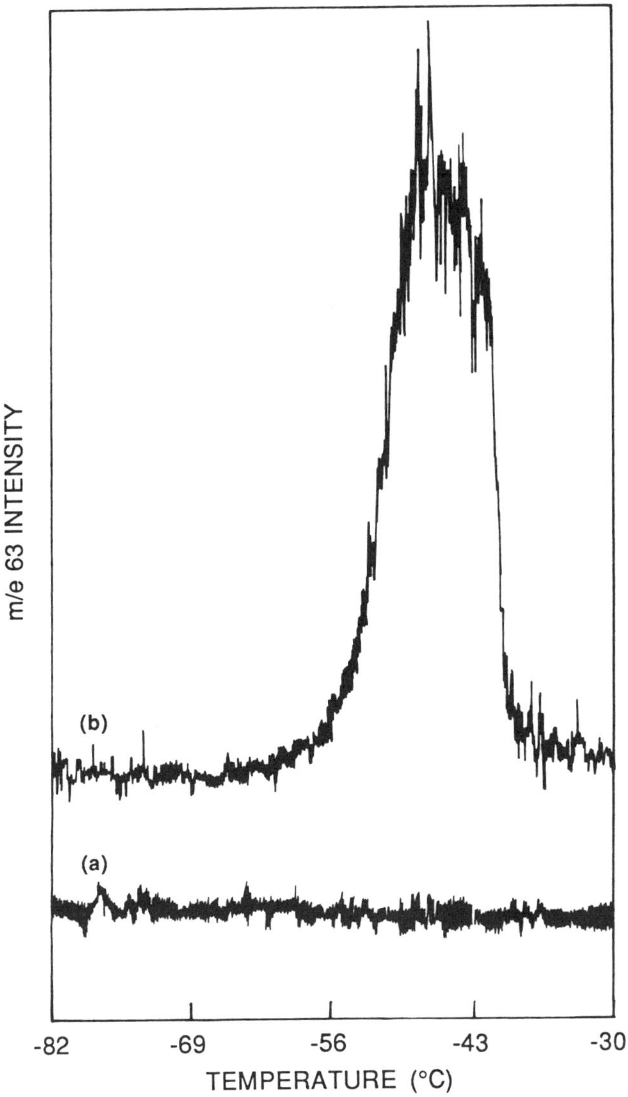

Figure 27. Thermal desorption spectra of HNO_3 monitored by *m/e* 63 after (A) exposure of cold wax-coated copper to $ClONO_2$ for 1 hour and (B) exposure of ice to $ClONO_2$ for 1 hour, shown on the same intensity scale.

(127) and (128) could play important roles in the photochemical mechanism of ozone depletion in the Antarctic spring.[89]

Some unanswered questions remained, however. For example, in order to cause ozone loss, the active chlorine produced in the above reactions must not be scavenged. This requires low levels of NO_x, and thus a sink for these species. Again we returned to study of the conversion of N_2O_5, this time on ice at 185 K:

$$N_2O_5\ (g) + H_2O\ (s) \rightarrow 2\ HNO_3\ (s) \tag{130}$$

$$N_2O_5\ (g) + HCl\ (g) \xrightarrow{\text{ice surface}} ClNO_2\ (g) + HNO_3\ (s) \tag{131}$$

Both of these reactions were shown to occur rapidly on ice.[90] The thermal desorbtion scans of the bound HNO_3 from reaction (131) shown in Figure 28 give information which may help characterize the surface. Two HNO_3 desorbtion peaks (203 K and 230 K) are observed (scan a). Thermal desorbtion scans for water (m/e = 18) with pure ice samples and of ice after reaction (130) are shown in traces c and d. The desorbtion maximum for pure H_2O occurs at 209 K, while, after reaction, H_2O peaks are observed at 209 and 230 K. Since the after reaction surface shows peak maximums at 230 K for both water and HNO_3, it seems likely that these are due to desorbtion of a bound form of these species, perhaps the hydrates $HNO_3{\cdot}H_2O$, $HNO_3{\cdot}2H_2O$, or $HNO_3{\cdot}3H_2O$. The HNO_3 peak at 203 K, with no corresponding water maximum, is more consistent with desorbtion of overlayers of bulk HNO_3. These data suggest that some chemical rearrangement is taking place at the water/HNO_3 interface, although it is not clear if this occurs on initial contact or is a slow process perhaps involving diffusion into the bulk ice.

In any case, both reactions (130) and (131) lead to formation of condensed phase HNO_3, and hence can lead to the required denitrification of the atmosphere. To the extent that HCl is available, reaction (131) could also be a source of $ClNO_2$, which is easily photolyzed in the Antarctic spring to form the active species atomic chlorine.

Polar stratospheric particulate is now thought to be composed largely of ice and nitric acid trihydrate (NAT), with the balance determined by various factors, including temperature and atmospheric composition. In order to better understand and quantify the effects of this partitioning, we have recently examined the uptake coefficients of N_2O_5 on both ice and NAT surfaces at polar temperatures.[91]

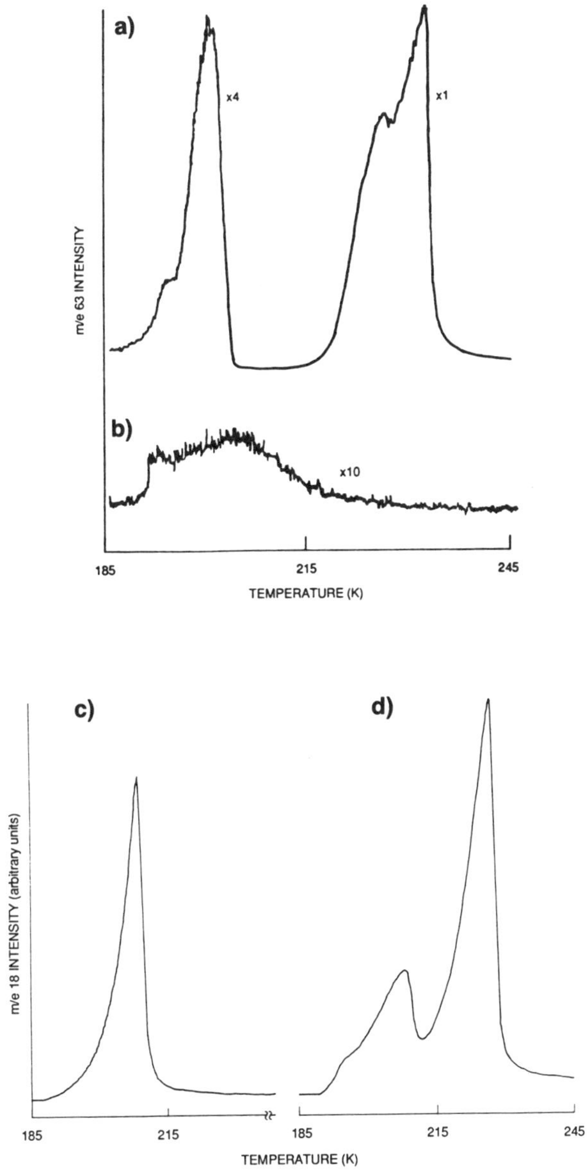

Figure 28. Upper: Thermal desorption scans of HNO_3 monitored by *m/e* 63 after (a) exposure to ice to N_2O_5 for 15 minutes and (b) exposure of cold wax-coated copper to N_2O_5 for 15 minutes, shown on an expanded scale as indicated. Lower: Thermal desorption scans of water monitored by *m/e* 18 after (c) deposition of pure H_2O onto wax-coated copper at 185 K and (d) reaction of N_2O_5 on ice at 185 K.

Current work involves alternative methods of preparing NAT and the use of FTIR to examine the solid phase,[92] with the hope of better surface characterization and further elucidation of products and mechanism. In addition, studies on $H_2SO_4{\cdot}xH_2O$ surfaces are being carried out to elucidate global ozone depletion processes.

Plasma Etching Reactions

Real-Time Experiments. The real-time REMPI detection of transients discussed earlier can also be applied to the measurements of surface processes. In particular a reactive surface can be installed in the reactor and its effect on transient lifetimes determined. We have used such methods to directly probe reactions that are important in the etching of electronic materials.

An example are reactions occurring in fluorocarbon discharges, which are used in the electronics industry to etch Si and SiO_2. The chemical and physical processes in such discharges are very complex, but the general picture is that etching is brought about directly or indirectly via CF_x radicals that arise through electron impact events. In CF_4/O_2 discharges, the CF_x radicals can react with O and O_2 to produce atomic fluorine and the stable products CO, CO_2, and COF_2. This results in a fluorine rich discharge. With CF_4/H_2 discharges, F atoms are depleted by fast reactions to form HF, resulting in a fluorine poor discharge, and leaving CF_x species to dominate the neutral gas and surface processes.

Photolysis- Real-Time REMPI-MBMS. In order to gain a better understanding of some of these events we have recently studied[93] the etching of silicon wafer by CF_3 radicals in the temperature range 560–745 K, using the combined REMPI/MBMS detection techniques.

Figure 29 shows the reactor conformation used for these experiments. The top arm of the reactor holds the high-temperature sample, which is pressed down against a circular knife edge by a spring. Water is used to cool the knife edges and supporting flange so that the sample is the only hot surface exposed to the reactant gases. On the back side of the sample are the heater and thermocouple temperature sensors. The CO_2 laser produces CF_3 radicals from either hexafluoroacetone (HFA) or CF_3I using infrared-multiphoton dissociation (IRMPD). REMPI detection using the excimer pumped dye laser monitors the

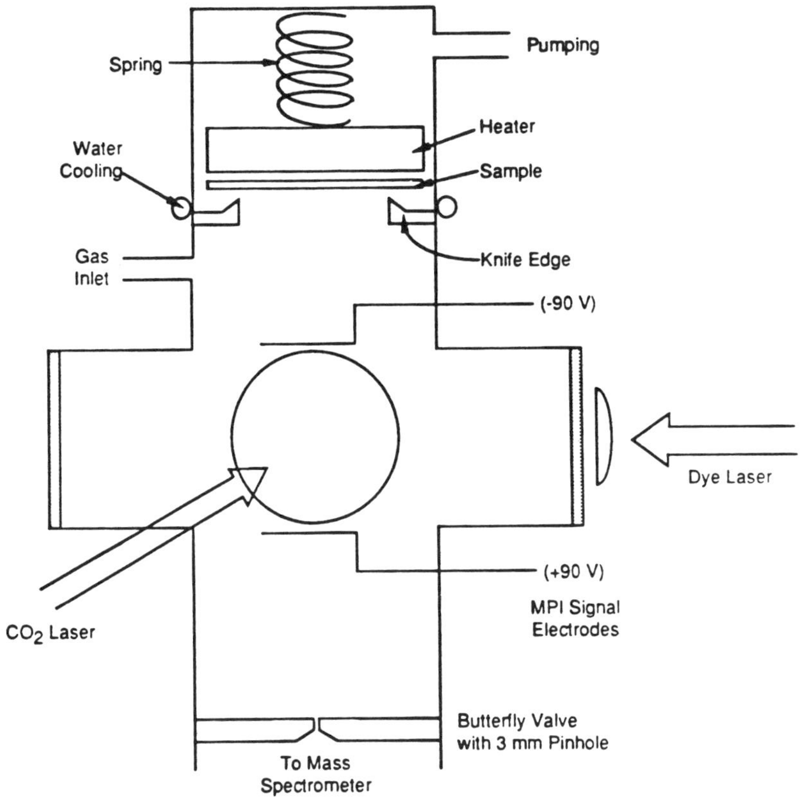

Figure 29. Diagram of the VLPØ reactor used to measure the etching of silicon by CF_3, free radicals. The reactor is a six-way stainless-steel cross. The top and bottom arms have 2 1/2 in. inside diameters (4 1/2 in. con flat), and the four side arms have 1 1/2 in. diameter (2 3/4 inc. conflat). The gas flows out of the reactor through the 3-mm aperture, and the back side of the sample is pumped separately.

density and time history of the radical. Finally, the mass spectrometer is used to measure the extent of dissociation of the precursor gas and the formation of SiF_4, the primary etch product.

One of the advantages of the VLPØ technique is that radicals created with excess energy can be accommodated to the reactor ambient temperature in a relatively short time. This can be seen from the REMPI spectra of CF_3 in Figures 30 and 31 which show the difference in the radical spectra two microseconds after the photolysis laser pulse,

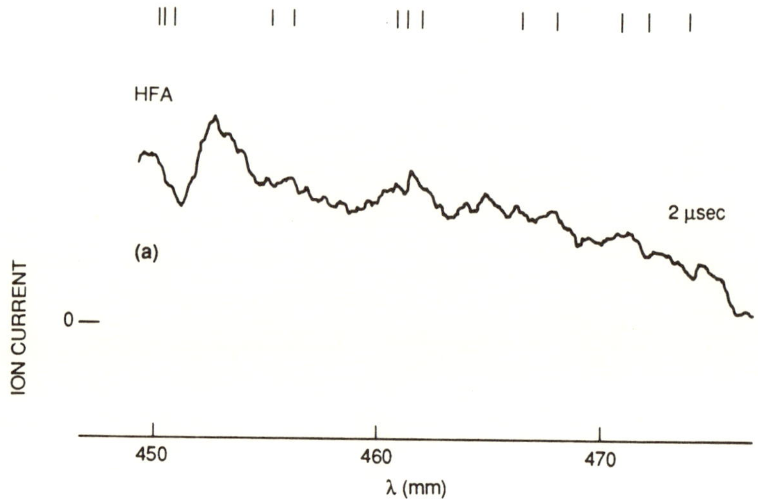

Figure 30. The REMPI spectrum of HFA with the IR laser on and the probe laser delayed by 2 μs. The expected CF_3 REMPI features for CF_3 are marked along the top of the figure.

when the radicals are still internally excited, and after ten milliseconds when they are thermalized due to non-reactive wall collisions.

Etch rates (Å min^{-1} torr of CF_3 precursor) of silicon wafer determined by monitoring SiF_4 production were combined with the REMPI determined CF_3 densities to yield the reaction probability, ε, that is the likelihood that a surface collision will lead to reaction. The reaction probability was found to depend on the surface history. Silicon surfaces etched in situ by F_2 to form a surface free of oxide or carbon initially yield high reaction probabilities. Reactivity of CF_3 towards the surface then drops off, reaching a constant value, as the surface is exposed to the radical bath. If the "aged" surface is then re-exposed to F_2, fluorocarbon species are observed in the effluent. This seems to indicate that inhibition is due to the buildup of a carbon-containing layer on the surface.

These results are summarized in Figure 32. The Xs correspond to CF_3 reaction probability data on "aged" surfaces, while the other data correspond to that observed on "fresh" and partially aged silicon (triangles and boxes represent data with CF_3I and HFA as the CF_3 precursors, respectively). The decrease in etch rate as the surface ages is clearly visible. The data for F and F_2 are from References 94 and 95,

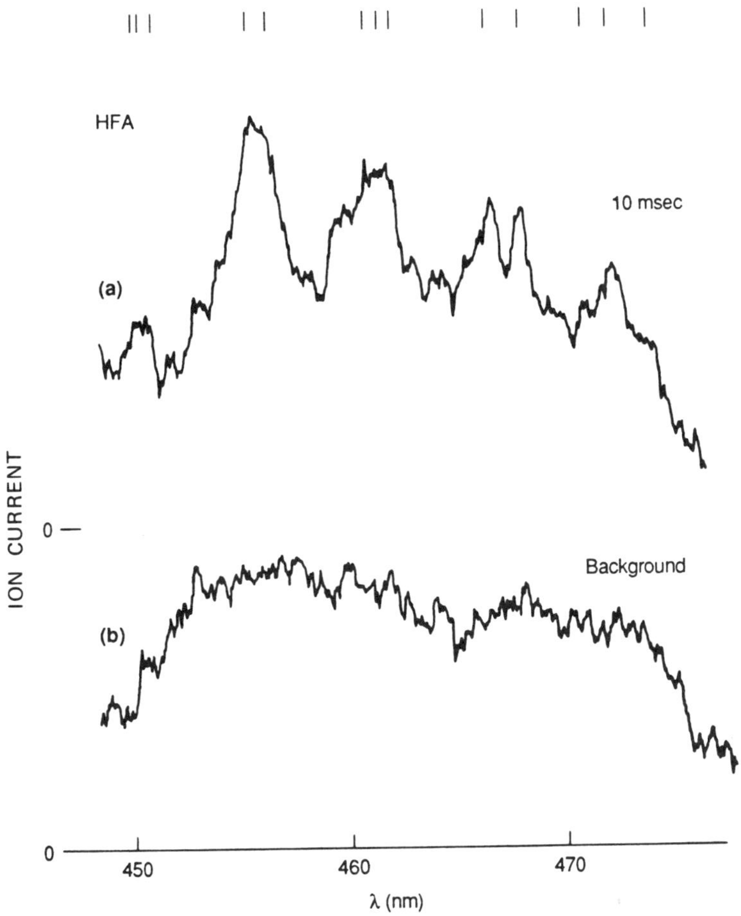

Figure 31. The REMPI spectra for hexafluoroacetone (HFA) with the IR laser off (background) and on (with probe laser delay of 10 ms). The difference between these two spectra is a series of lines that match the expected lines (indicated by the lines across the top of the figure) reported by Duiganan et al.[95] for the REMPI of CF_3. The probe laser wavelength is set at 455 nm for the determination of CF_3 kinetics.

respectively, and correspond to fresh, fluorinated silicon surfaces free of oxide or carbon.

We see that etching by CF_3 (and F_2) is much slower than by F. The activation energy for etching of a fresh surface by F_2 is about 13 kcal/mol,[94] while on an aged surface we find a value of about 8 kcal/mol for

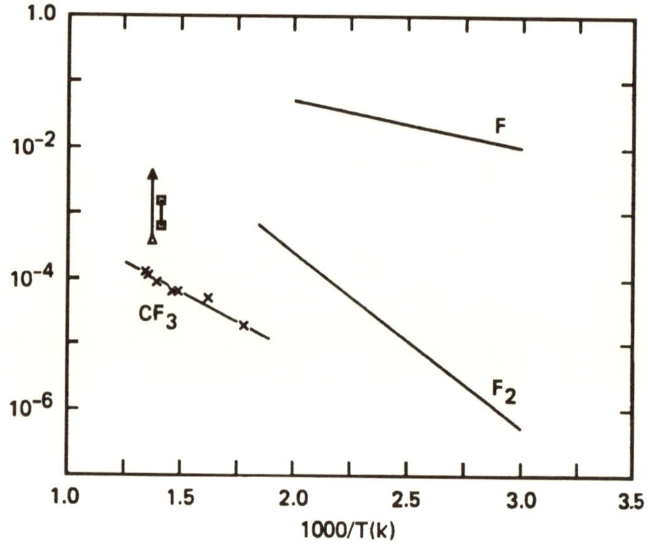

Figure 32. The reaction probability for the etching of silicon, by CF_3, F_2, and F, vs. inverse temperature. The triangles represent $\varepsilon(CF_3)$ data using the precursor CF_3I, and the boxes represent $\varepsilon(CF_3)$ data using HFA. The ×s represent the data, $\varepsilon(CF_3)$, for HFA and a surface that had been exposed for a long period, ε for F (Ref. 94) and F_2 (Ref. 98) are shown for comparison.

CF_3.[93] On similar fresh surfaces, however, $\varepsilon(CF_3)$ and $\varepsilon(F_2)$ are about the same. In separate studies[93] we observed that etching by F_2 is also inhibited as a carbon containing surface layer is formed, so it appears that CF_3 and F_2 have comparable reaction probabilities and probably activation energies, independent of the surface character. Such information can be used in reactive etching models.[96,97]

Pulsed Valve Microwave-Discharge REMPI-MBMS. Two of the requirements that must be met in order to monitor the real time kinetics of transients are (1) that the species of interest must be introduced on a time scale that is short relative to its kinetic decay, and (2) that it must be present in concentrations sufficient for detection by the method of choice. To solve these problems in the case of halogen atoms, a technique developed recently at SRI[99] involves the generation of a constant reservoir of atoms in a microwave discharge external to the reactor, and then the injection of the atoms into the reactor in 300 μs to 300 ms pulses via a solenoid valve. The set up is shown in Figure 33. REMPI detection of the atom and MBMS monitoring of products can be used as discussed above.

One of the features of the microwave discharge is that it produces both ground state Cl ($^2P_{3/2}$) and spin-orbit-excited Cl ($^2P_{1/2}$) in relative

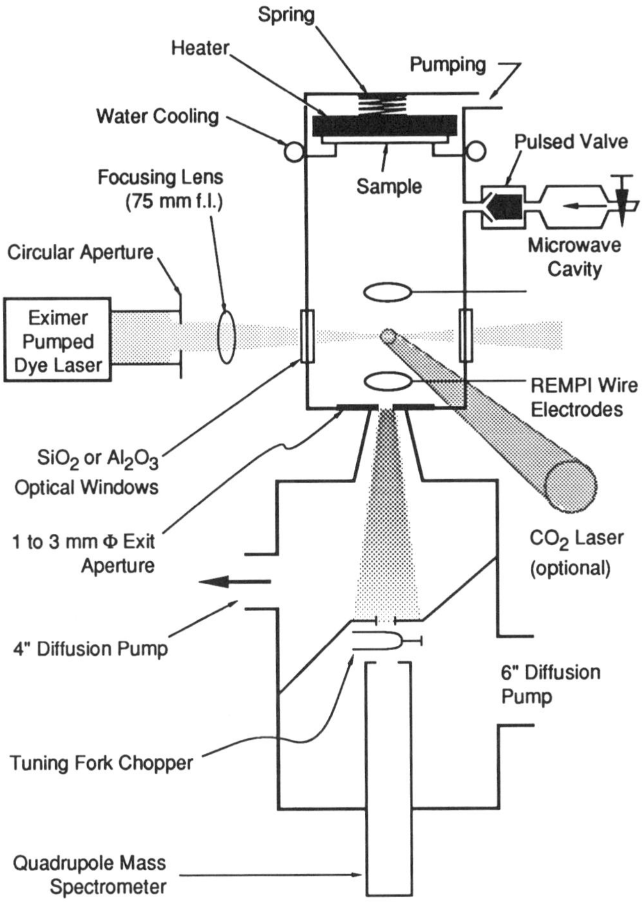

Figure 33. Experimental apparatus used to study surface reaction of halogen atoms in real-time. Atoms are created in the microwave discharge, injected via a pulsed solenoid valve, and monitored with REMPI. The optional CO_2 laser can be used to generate other transients in situ via IRMPD. Products are detected with MBMS.

amounts that, in our present setup, can be varied by a factor of ca. 4 by varying the discharge conditions. These species have distinct REMPI signals, Figure 34, so that it is possible to monitor them separately. Interestingly, in a completely Teflon-coated reactor the REMPI signal decays of both species correspond simply to k_{esc}. That is, the only loss is through the pinhole to the mass spectrometer! Based on the precision of our experiments, we are able to assign maximum values for the uptake coefficients of Cl and Cl* on Teflon of $\gamma \leq 5.0 \times 10^{-6}$.[99] It is perhaps surprising that even spin-orbit excited Cl atoms survive for

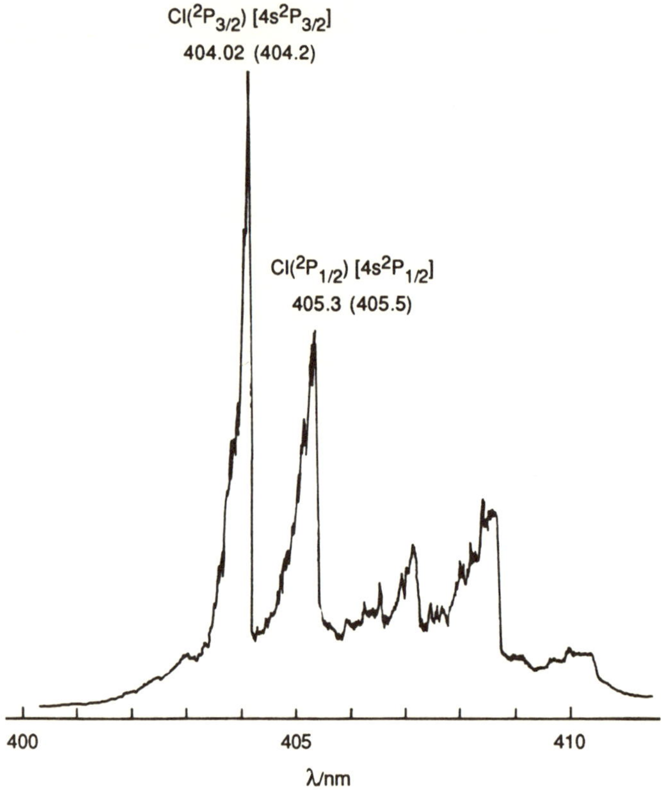

Figure 34. [3 + 2] REMPI spectrum of Cl ($^2P_{3/2}$) and Cl ($^2P_{1/2}$) atoms through the $4s^2P_{3/2}$ and $4s^2P_{1/2}$ resonant states, respectively. Wavelengths in parentheses correspond to the expected, tabulated values for the observed transitions. The probe (dye-) laser power was 8 mJ/pulse, focused through a 70-mm focal length planoconvex lens.

tens of thousands of wall collisions inside our reactor, but this fact gives us the opportunity to study both chlorine species simultaneously.

In work of this kind,[100] chlorine atom REMPI has been used to investigate etching of lightly doped *n*-Si(100) samples from 300 to 600 K. The uptake coefficient of Cl was found to be negligible, with a temperature independent maximum value of $\gamma(\mathrm{Cl}) \leq 5.0 \times 10^{-5}$. Small, but slightly faster values were measured for Cl*: $\gamma(\mathrm{Cl}^*, 304\ \mathrm{K}) = (4.6 \pm 0.6) \times 10^{-4}$, $\gamma(\mathrm{Cl}^*, 597\ \mathrm{K}) = (2.7 \pm 1.7) \times 10^{-4}$. These values correspond to net interaction with the surface (thus physisorbtion, chemisorbtion, surface reaction) and represent so small a reactive flux that we were unable to directly detect any etching product in the

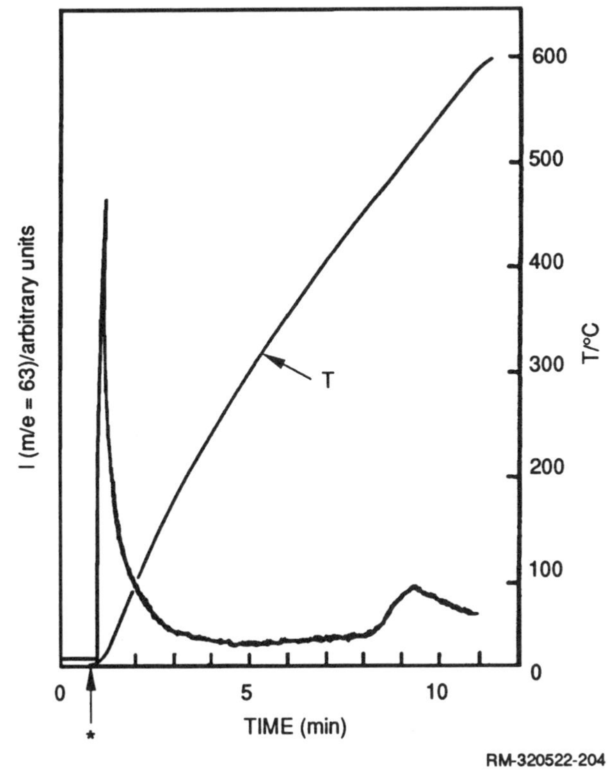

Figure 35. Thermal desorption of the etch product of the Cl/Cl*/*n*-Si(100) reaction, observed at $m/e = 63$ (SiCl^+). The wafer (T = 295 K) was exposed to a steady-state flow of Cl/Cl* for two hours. Then the flow was turned off at $t = 0$. Temperature ramp was started at (*). No product was observed in the absence of Cl atoms (discharge off.)

effluent. Products were therefore determined by first exposing the room temperature wafer to a steady flow of atomic chlorine for two hours and then using thermal desorbtion to detect the etch product. The results monitored at *m/e* = 63 ($SiCl^+$) are shown in Figure 35. Two peaks are observed, one appearing at low temperature upon initial heating and the other occurring with a maximum near 500 °C. This suggests that the low-temperature peak is due to an easily desorbed chlorine-containing multilayer, while that at high temperature corresponds to the Cl-containing adlayer at the Si/Cl interface. A similar but smaller product desorbtion signal could be detected at *m/e* = 98 ($SiCl_2^+$), but no $SiCl_3^+$ signal could be found. We conclude therefore, that the etch product is $SiCl_2$ and not $SiCl_4$.

The very small uptake coefficient for atomic chlorine observed in the above studies is not necessarily typical. In similar studies,[101] we have found that atomic chlorine reacts much faster with, for example, polycrystalline nickel surfaces.

Obviously, the behavior of radicals on surfaces is complex and highly variable. The continuing application of the techniques described herein will be able to provide detailed information on this important class of reactions. It is our hope that the growing database on such reactions will someday allow their classification and description in much the same fashion as homogeneous gas phase chemistry.

V. CONCLUSION

It is hard to actually draw a conclusion from a description of an ongoing research program. We think that we at the Department of Chemical Kinetics of SRI International have demonstrated the usefulness of the application of a sound overview of chemical kinetics to the solution of practical problems of societal interest. This has the double advantage of allowing us to enjoy the fruits of scientifically productive careers at the leading edge of our discipline, while observing the application of the results of that research to the "real world."

ACKNOWLEDGMENTS

We have tried to give an overview of work carried out over a number of years at SRI International. That work was performed, for the most part, by many colleagues whose names appear on the referenced papers. DMG would like to pay particular tribute to his longtime colleagues Don McMillen, Michel Rossi, and Greg Smith, as well as his former colleague John Barker. Recent work related to heterogeneous reactions of possible atmospheric importance has been catapulted into the limelight by the indefatiguable Maggie Tolbert. Sid Benson always looms large in anything done in the Chemical Kinetics Department at SRI, and the memory of Fred Kaufman is cherished.

This work has been funded over the years by AFOSR, ARO, ONR, NASA, NSF, DoE, EPA, GRI, and The Dow Chemical Company.

REFERENCES

1. Nigenda, S. E.; McMillen, D. F.; Golden, D. M. *J. Phys. Chem.*, **1989**, *93*, 1124.
2. Stewart, P. H.; Jeffries, J. B.; Zellweger, J.-M.; McMillen, D. F.; Golden, D. M. *J. Phys. Chem.*, **1989**, *93*, (1989) 3557.
3. Benson, S. W. *Thermochemical Kinetics*; Wiley: New York, 2nd ed., (1976).
4. Lee, R. J.; Rupley, F. M.; Miller, J. A.; "Chemkin II: A Fortran Chemical Kinetics Package for the Analysis of Gas-Phase Chemical Kinetics," Sandia Report SAND89-8009. UC-401, Sandia National Labs, Livermore, CA 94551, 1989.
5. *Selected Values of Properties of Hydrocarbons and Related Compounds: TRC Tables*, Thermodynamics Research Center, Texas A&M University College Station, Texas, 1987.
6. a) Stull, D. R.; Prophet, H.; (Eds.) *JANAF Thermochemical Tables*, 2nd ed., NSRDS-NBS37; National Bureau of Standards: Washington, D.C., 1971; b) ibid., 3rd ed., 1986.
7. Stull, D. R.; Westrum Jr., E. F.; Sinke, G. C. *The Chemical Thermodynamics of Organic Compounds*; Wiley: New York, 1969.
8. Cox, J. D.; Pileher, G. *Thermochemistry of Organic and Organometallic Compounds*; Academic Press: London, 1970.
9. Pedley, J. B.; Naylor, R. D.; Kirby, S. P. *Thermochemical Data of Organic Compounds*, 2nd ed., Chapman & Hall: London, 1986.
10. Orville-Thomas, W. J., Ed. *Internal Rotation in Molecules*; Wiley: New York, 1974.
11. Lister, D. G.; MacDonald, J. N.; Owen, N. L. *Internal Rotation and Inversion*; Academic Press: London, 1978.
12. Lewis, G. N.; Randall, M. *Thermodynamics*; 2nd ed., revised by Pitzer, K. S., and Brewer, L.; McGraw-Hill: New York, 1961.
13. Tsang, W. *J. Phys. Chem.*, **1984**, *88*, 2812.
14. Brouard, M.; Lightfoot, P. D.; Pilling, M. J. *J. Phys. Chem.*, **1986**, *90*, 445.
15. Chen, Y.; Rauk, A.; Tschuikow-Roux, E. *J. Phys. Chem.*, **1990**, *94*, 2775.
16. Jacox, M. E. *J. Phys. Chem. Ref. Data.* **1984**, *13*, 945.
17. Ibid., **1988**, *17*, 269.
18. Ibid., **1990**, *19*, 1387.

19. McQuarrie, D. A. *Statistical Mechanics*; Harper and Row: New York, 1976; pp. 104–108.
20. Herzberg, G. *Molecular Spectra and Molecular Structure, II. Infrared and Raman Spectra of Polyatomic Molecules*; Van Nostrand Company: New York, 1956; p. 221.
21. Laane, J. *Appl. Spectrosc.* **1970**, *24*, 73.
22. Lloyd, R. V. In *Chemical Kinetics of Small Organic Radicals*, Alfassi, Z. B., Ed.; CRC Press: Boca Raton, 1988; Vol. 1, Chap. 1.
23. Kafafi, S.; Hudgens, J. W. *J. Phys. Chem.*, **1989**, *93*, 3474.
24. Luke, B. T.; Loew, G. H.; McLean, A. D. *J. Am. Chem. Soc.* **1987**, *109*, 1307.
25. Pacansky, J.; Coufel, H. *J. Chem. Phys.* **1980**, *72*, 5285.
26. Pacansky, J.; Dupuis, M. *J. Am. Chem. Soc.* **1982**, *104*, 415.
27. Chettur, G.; Snelson, A. *J. Phys. Chem.*, **1987**, *91*, 3483.
28. Pacansky, J.; Schraedar, B. S. *J. Chem. Phys.* **1983**, *78*, 1033.
29. Luo, Y-R.; Benson, S. W. *J. Phys. Chem.*, **1989**, *93*, 3304.
30. Tschuikow-Roux, E.; Chen, Y. *J. Amer. Chem. Soc.* **1989**, *111*, 9030.
31. Müller-Markgraf, W.; Rossi, M. J.; Golden, D. M. *J. Amer. Chem. Soc.* **1989**, *111*, 957.
32. Tsang, W. *J. Phys. Chem.*, **1984**, *88*, 2812.
33. McMillen, D. F.; Golden, D. M. *Annu. Rev. Phys. Chem.* **1982**, *33*, 493.
34. *CRC Handbook of Chemistry and Physics*; CRS Press: Boca Raton, 1989.
35. Colussi, A. J. *Chemical Kinetics of Small Organic Radicals*, Alfassi, Z. B. Ed., CRC Press: Boca Raton, 1988; Vol. 1, Chap. 2.
36. Cohen, N.; Westberg, K. R. *Int. J. Chem. Kinet.* **1986**, *18*, 99.
37. Cohen, N.; Benson, S. W. *J. Phys. Chem.*, **1987**, *91*, 162.
38. Stewart, P. H.; Rothem, T.; Golden, D. M. *Twenty-Second Symposium (International) on Combustion*; The Combustion Institute: 1988; p. 943.
39. Wallington, T. J.; Neuman, D. M.; Kurylo, M. J. *Int. J. Chem. Kinet.* **1987**, *19*, 725.
40. Warnatz, J. *Combustion Chemistry*; Gardiner, W. C., Ed.; Springer-Verlag: New York, 1984; p. 243.
41. Michael, J. V.; Keil, D. G.; Klemm, R. B. *Nineteenth Symposium (International) on Combustion*; The Combustion Institute: 1983; p. 39.
42. Tully, F. P.; Drogege, A. T.; Koszykowski, M. L. O.; Melius, C. F. *J. Phys. Chem.*, **1986**, *90*, 691.
43. Tully, F. P.; Koszykowski, M. L.; Binkley, S. J. *Twentieth Symposium (International) on Combustion*, The Combustion Institute: 1985; p. 715.
44. Kerr, J. A.; Parsonage, M. J. *Evaluated Kinetic Data on Gas-Phase Hydrogen Transfer Reactions of Methyl Radicals*; Butterworths: London, 1976.
45. Halstead, M. P.; Konar, R. S.; Leathard, D. A.; Marshall, R. M.; Purnell, J. H. *Proc. Roy. Soc. (London) Ser. A* **1969**, *310*, 525.
46. Pacey, P. D. *Can. J. Chem.* **1973**, *51*, 2415.
47. Tsang, W.; Hampson, R. F. *J. Phys. Chem. Ref. Data* **1986**, *15*, 1087.
48. Troe, J. *J. Chem. Phys.* **1977**, *66*, 4758.
49. Troe, J. *J. Phys. Chem.*, **1979**, *83*, 114.

50. Troe, J. *Ber. Bunsenges. Phys. Chem.* **1983**, *87*, 161.
51. Troe, J. *Ber. Bunsenges. Phys. Chem.* **1983**, *87*, 169.
52. Larson, C. W.; Patrick, R.; Golden, D. M. *Comb. and Flame* **1984**, *58*, 229.
53. Gilbert, R. G.; Smith, S. C. *Theory of Unimolecular and Recombination Reactions;* Blackwell Scientific: Oxford, 1990.
54. Smith, G. P.; Golden, D. M. *Int. J. Chem. Kinet.* **1978**, *10*, 489.
55. Stewart, P. N.; Smith, G. P.; Golden, D. M. *Int. J. Chem. Kinet.* **1989**, *21* 923.
56. Chen, C-J; Back, M. H.; Back, R. A. *Can. J. Chem.*, **1975**, *53*, 3580.
57. Brouard, M.; MacPherson, M. T.; Pilling, M. J. *J. Phys. Chem.*, **1989**, *93*, 4047.
58. Tabagashi, K.; Bauer, S. H. *Comb. Flame* **1979**, *34*, 63.
59. Patrick, R.; Barker, J. R.; Golden, D. M. *J. Phys. Chem.*, **1984**, *88*, 128.
60. Smith, G. P.; McEwen, A. B.; Golden, D. M.; Vaghjianai, G. L.; Ravishankara, A. R.; Tully, F. P. Unpublished manuscript.
61. Baulch, D. L.; Drysdale, D. D.; Duxbury, J.; Grant, S. *Evaluated Kinetic Data for High Temperature Reactions*, Butterworths: London, 1976, Vol. 3.
62. Paraskevopoulos, G.; Irwin, R. S. *J. Phys. Chem.* **1984**, *80*, 259.
63. Stewart, P. H.; Larson, C. W.; Golden, D. M. *Comb. and Flame* **1989**, *75*, 25.
64. Slagle, I. R.; Gutman, D.; Davies, J. W.; Pilling, M. J. *J. Phys. Chem.* **1988**, *92*, 2455.
65. Wagner, A. F.; Wardlaw, D. M. *J. Phys. Chem.* **1988**, *92*, 2462.
66. Rossi, M. J.; McMillen, D. F.; Golden, D. M. J. Phys. Chem., **1984**, *88*, 5031.
67. Rossi, M. J.; Golden, D. M. *Int. J. Chem. Kinet.* **1979**, *11*, 969.
68. Rossi, M. J.; King, K. D.; Golden, D. M. *J. Am. Chem. Soc.* **1979**, *101*, 1223.
69. Rossi, M. J.; Golden, D. M. *Int. J. Chem. Kinet.* **1979**, *11*, 775.
70. Rossi, M. J.; Golden, D. M. *Int. J. Chem. Kinet.* **1983**, *15*, 1283.
71. Russell, J. J.; Seetula, J. A.; Timonen, R. S.; Gutman, D.; Nava, D. F. *J. Amer. Chem. Soc.* **1988**, *110*, 3084.
72. Selamoglu, N.; Rossi, M. J.; Golden, D. M. *Chem. Phys. Lett.* **1986**, *124*, 68.
73. Golden, D. M.; Rossi, M. J.; Baldwin, A. C.; Barker, J. R. *Accts of Chem. Res.* **1981**, *14*, 56.
74. Robertson, R. M.; Golden, D. M.; Rossi, M. J. *J. Phys. Chem.*, **1988**, *92*, 5338.
75. Tsang, W. *Shock Waves in Chemistry;* Lifshitz, A., Ed.; Marcel Dekker: New York, 1981, p. 60.
76. Lewis, K. E.; Golden, D. M.; Smith, G. P. *J. Amer. Chem. Soc.* **1984**, *106*, 3905.
77. Connor, J. A. *Curr. Top. Chem.* **1977**, *71*, 71.
78. Engelking, P. C.; Lineberger, W. C. *J. Amer. Chem. Soc.* **1979**, *101*, 5569.
79. Stevens, A. E.; Feigerle, C. S.; Lineberger, W. C. *J. Amer. Chem. Soc.* **1982**, *104*, 5026.
80. Day, J. P.; Basolo, F.; Pearson, R. G. *J. Amer. Chem. Soc.* **1968**, *90*, 6927.
81. Stewart, P. H.; Jeffries, J. B.; Zellweger, J.-M.; McMillen, D. F.; Golden, D. M. *J. Phys. Chem.* **1989**, *93*, 3557.

82. Smith, G. P.; Fairchild, P. W.; Jeffries, J. B.; Crosley, D. R. *J. Phys. Chem.* **1985**, *89*, 1269.
83. Zellner, R.; Lorenz, K. *J. Phys. Chem.* **1984**, *88*, 984.
84. Ohta, T. *J. Phys. Chem.* **1983**, *87*, 1209.
85. Atkinson, R.; Pitts, J. N. *J. Chem. Phys.* **1975**, *63*, 3591.
86. Smith, G. P. *Int. J. Chem. Kinet.* **1987**, *19*, 269.
87. Reihs, C. M.; Golden, D. M.; Tolbert, M. A. *J. Geophys. Res.* **1990**, *95-D10*, 16545.
88. Brouwer, L.; Rossi, M. J.; Golden, D. M. *J. Phys. Chem.* **1986**, *90*, 4599.
89. Tolbert, M. A.; Rossi, M. J.; Malhotra, R.; Golden, D. M. *Science* **1987**, *238*, 1258.
90. Tolbert, M. A.; Rossi, M. J.; Golden, D. M. *Science* **1988**, *240*, 1018.
91. Quinlan, M. A.; Reihs, C. M.; Golden, D. M.; Tolbert, M. A. *J. Phys. Chem.* **1990**, *94*, 3255.
92. Tolbert, M. A.; Middlebrook, A. M. *J. Geophys. Res.* **1990**, *95-D13*, 22423.
93. Robertson, R. M.; Golden, D. M.; Rossi, M. J. *J. Vac. Sci. Technol.* **1988**, *B6*, 1632.
94. Vasile, M. J.; Stevie, F. A. *J. Appl. Phys.* **1982**, *53*, 3799.
95. Duignan, M. T.; Huyddgens, J. W.; Wyatt, J. R. *J. Phys. Chem.* **1982**, *86*, 4156.
96. Flamm, D. L.; Donnelly, V. M.; Ibbotson, D. E. *VLSI Electron Microstructure Sci.* **1984**, *8*, 190.
97. Oehrlein, G. S.; Williams, H. L. *J. Appl. Phys.* **1987**, *62*, 662.
98. Selamoglu, N.; Mucha, J. A.; Flamm, D. L.; Ibbotson, D. E. *J. Appl. Phys.* **1987**, *62*, 1049.
99. Müller-Markgraf, W.; Rossi, M. J. *Rev. Sci. Instrum.* **1990**, *61*, 1217.
100. Müller-Markgraf, W.; Rossi, M. J. *J. Vac. Sci. Technol. A* **1991**, *5*, 217.
101. Müller-Markgraf, W.; Rossi, M. J. *J. Phys. Chem.* **1991**, *95*, 825.

BIOGRAPHICAL SKETCHES OF THE EDITOR AND CONTRIBUTORS

John R. Barker was born in Tennessee and raised in Orlando, Florida. He received his B.S. degree in 1965 from Hampden-Sydney College and his M.S. and Ph.D. degrees in 1969 from Carnegie-Mellon University, where he worked with Joe V. Michael. It was there that he first heard the siren song of chemical kinetics experiments and theory. During postdoctoral research with B. S. Rabinovitch at the University of Washington, he carried out experiments on inter- and intramolecular vibrational energy transfer and on unimolecular reactions. In additional postdoctoral research with Ralph E. Weston, Jr., and William Spindell at Brookhaven National Laboratory, he investigated electronic to vibrational energy transfer and laser isotope separation. In 1974, he joined Sidney W. Benson's group at Stanford Research Institute (now known as SRI International), where he carried out research on atmospheric reactions, collisional energy transfer, and other topics, prior to becoming a Professor of Atmospheric Science and of Chemistry at the University of Michigan in 1985. He is also associated with the Space Physics Research Laboratory at Michigan. His present research interests center on the mechanisms of collisional energy transfer, atmospheric chemistry, and interpretation of the molecular infrared emissions observed from interstellar space.

David M. Golden is Vice-President for the Physical Sciences Division at SRI International. He has previously been Laboratory Director of the Chemistry Laboratory and Department Director of the Chemical Kinetics Department. He is the Editor of the International Journal of Chemical Kinetics. He received the A.B. degree from Cornell University in 1956 and the Ph.D. from the University of Minnesota in 1960.

His research interests are in chemical kinetics, atmospheric chemistry, combustion chemistry, and codification and extrapolation of rate data for use in modeling complex chemical systems.

Yong-Eun Koo received her B.S. degree in Chemistry (1982) and M.S. degree (1984) in Physical Chemistry from Seoul National University, and her Ph.D. in Chemistry (1991) from the University of Michigan. She has worked as a Research Fellow at the University of Michigan since 1991. Her current research interests are in diffusion-controlled reaction kinetics in small, restricted systems. Koo's main research contribution is in carrying out the first chemical reaction kinetics experiment showing the segregation of reactants.

Raoul Kopelman received his B.S.E. degree (1955), Chemical Engineering Diploma (1956), and M.S. degree (1957) in Physical Chemistry from the Israel Institute of Technology, and his Ph.D. in Chemistry (1960) from Columbia University. After working as a Research Associate at Harvard and a Senior Research Fellow at Cal-Tech, he joined the faculty of the University of Michigan in 1966. He currently holds the position of Professor of Chemistry and of Physics. He also is a member of the Applied Physics and Biophysics programs. Kopelman is the author of about 300 scientific papers in solid state spectroscopy and chemical reaction kinetics. Current research interests are in nonclassical reaction kinetics, scanning photon and exciton tunneling microspectroscopy, polymer morphology, and ultrasmall optochemical sensors. Kopelman's research contributions include the discovery of pseudolocalized phonons; the measurement of exciton exchange and superexchange energies in molecular crystals; the demonstration of quantum percolation via excitation trapping and annihilation; the initiation of nonclassical reaction kinetics experiments, simulations, and theory; the concept and practice of active subwavelength light sources and their application to near-field scanning optical microscopy and spectroscopy; and the construction and application of subwavelength optical sensors for chemical and biochemical analysis.

Jeffrey A. Manion received his Bachelor of Science degree from the University of Delaware, where he conducted undergraduate research on organic reaction mechanisms under the direction of Dr. Harold Kwart. His graduate studies with Professor Dr. Robert Louw at Leiden University, The Netherlands, involved the gas-phase hydrogenolysis of organic chlorine-containing molecules. He received his doctorate in 1989 and is currently a researcher at SRI International with interests in the application of thermochemical kinetics to environmental and industrial problems, atmospheric chemistry, coal conversion, and the modeling of complex systems.

J. V. Michael received his B.A. in Chemistry from Wabash College and his Ph.D. in Chemistry from the University of Rochester. He has been concerned with the absolute rates of chemical reactions in thermally reacting systems throughout his career. Most of the work has been in support of practical disciplines such as atmospheric chemistry or the chemistry of combustion. He has used a variety of techniques including discharge flow tube and flash or laser photolysis methods with atomic resonance absorption and/or fluorescence spectrometric detection. When appropriate, laser-induced fluorescence has also been used to detect reacting species. In recent years, these methods have been applied to high-temperature chemistry, and absolute values of thermal rate constants can now be directly determined by using the shock tube technique. A review of some of the data obtained by these methods is the subject of the chapter in this volume. Dr. Michael is currently a Senior Scientist at the Chemistry Division of Argonne National Laboratory.

Hanna Reisler was born in Tel-Aviv, Israel, and obtained her B.Sc. and M.Sc. degrees from the Hebrew University of Jerusalem in 1964 and 1966, respectively. After her military service, she obtained a Ph.D. in Physical Chemistry from the Weizmann Institute, Rehovot, Israel, in 1972. She was awarded an Atomic Energy Agency Fellowship to pursue postdoctoral studies at Johns Hopkins University during 1972–1974. During 1974–1977 she was a group leader of the Chemical Laser Project at the Soreq Nuclear Research Center, Yavne, Israel. She came to USC in 1977 and held research positions in the Departments of Electrical Engineering, Physics, and Chemistry. She was appointed Associate Professor of Chemistry in 1987 and Professor in 1991. Her research interests include (1) photodissociation dynamics and unimolecular reactions of small gas-phase molecules; (2) the influence of reagent translational and vibrational energy on reaction dynamics; and (3) photoinitiated reactions and collision-induced dissociation at surfaces.

James J. Valentini is Professor of Chemistry at Columbia University. His research interests are in dynamics of chemical reactions, photochemistry, and applications of coherent and resonance Raman spectroscopy. Professor Valentini received his B.S. degree summa cum laude at the University of Pittsburgh in 1972. His graduate work was done as an NSF Predoctoral Fellow with Yuan T. Lee, first at the University of Chicago, where he received the M.S. degree, and subsequently at the University of California, Berkeley, from which he obtained his Ph.D. in 1976. He was a Weizmann Fellow with Dudley Herschbach at Harvard and an Oppenheimer Fellow at Los Alamos, where he stayed as a staff member for several years. Before assuming his present position at Columbia he had been Professor of Chemistry at the University of California, Irvine.

Curt Wittig was born in Chicago, Illinois, on September 14, 1943. He was an undergraduate at the University of Illinois at Chicago and received the Ph.D. degree in Electrical Engineering from the University of Illinois at Urbana in 1970. He carried out postdoctoral research in Chemistry at the University of Cambridge and the University of California at Berkeley and joined the USC faculty in 1973, where he was appointed Professor in 1979. Since then he received several awards for research, directed the Center for the Study of Fast Transient Processes, served on the Editorial Boards of the Journal of Physical Chemistry, Chemical Physics Letters, and the Journal of Chemical Physics, and participated in many administrative and curriculum development activities. Sabbatical leaves at Harvard (1981), Oxford (1981), MIT (1988), and the Hebrew University of Jerusalem (1988) improved the quality of his research and helped preserve his sanity. His research interests are (1) photoinitiated reactions in weakly bonded complexes, (2) femtosecond studies of reaction dynamics including the use of weakly bonded complexes, (3) collision-induced dissociation in hyperthermal molecule-surface interactions and photoinitiated reactions on surfaces, (4) reactive and inelastic scattering studies using the high-resolution Rydberg time-of-flight method, and (5) high-resolution laser spectroscopy of floppy molecules. Visitors to his laboratories are often struck by the sign on the door: *Idle people are dead people that you can't bury.*

Transfer Chemistry of Monoamine Oxidase, *Richard B. Silverman, Northwestern University.* **Photolyase. DNA Repair by Photoinduced Electron Transfer,** *Aziz Sancar, University of North Carolina, Chapel Hill.*

Volume 3, In preparation, Summer 1993
ISBN 1-55938-320-8 Approx. $90.25

CONTENTS: Preface, *Patrick S. Mariano.* **Recent Advances in Light-Induced Election Transfer Involving Inorganic Systems,** *Nick Serpone and Rita Terzian, Concordia University, Jean-Marie Hermann, Ecole Central Lyon, France.* **Photoinduced Electron Transfer in Dye-Polymer Conjugates,** *Guilford Jones, Boston University.* **Sequential Electron Transfer in Oxidation Reactions Catalyzed by Cytochrome P-450 Enzymes,** *Peter Guengerich, Vanderbilt University and Timonty L. Macdonald, University of Virginia.* **Inner Shell Relaxation Effects on Electron Transfer Reactions of Amino Centered Systems,** *Stephen F. Nelsen, University of Wisconsin.*

Volume 4, In preparation, Summer 1994
ISBN 1-55938-506-5 Approx. $90.25

CONTENTS: Intramolecular Photochemical Electron Transfer (PET) Induced Bond Cleavage Reactions in Some Sulfonium Salt Derivatives, *Franklin D. Saeva, Corporate Research Laboratories, Rochester, NY.* **Aspects of Electron Transfer Promoted Stilbene Photocycloaddition Reactions,** *Kevin Peters, University of Colorado, Boulder.* **Mechanistic and Synthetic Aspects of Amine-Eone SET-Photochemistry,** *Ung Chan Yoon, Pusan National University and Richard S. Givens, University of Kansas.*

JAI PRESS INC.
55 Old Post Road - No. 2 P.O. Box 1678
Greenwich, Connecticut 06836-1678
Tel: (203) 661-7602 Fax: (203)661-0792